SLURRY FLOW

BUTTERWORTH-HEINEMANN SERIES IN CHEMICAL ENGINEERING

SERIES TITLES

Bubble Wake Dynamics in Liquids and Liquid-Solid Suspensions *Liang-Shih Fan* and *Katsumi Tsuchiya*
Chemical Process Equipment: Selection and Design *Stanley M. Walas*
Chemical Process Structures and Information Flows *Richard S.H. Mah*
Computational Methods for Process Simulations *W. Fred Ramirez*
Constitutive Equations for Polymer Melts and Solutions *Ronald G. Larson*
Fluidization Engineering, Second Edition *Daizo Kunii* and *Octave Levenspiel*
Fundamental Process Control *David M. Prett* and *Carlos E. Garcia*
Gas-Liquid-Solid Fluidization Engineering *Liang-Shih Fan*
Gas Separation by Adsorption Processes *Ralph T. Yang*
Granular Filtration of Aerosols and Hydrosols *Chi Tien*
Heterogeneous Reactor Design *Hong H. Lee*
Interfacial Transport Processes and Rheology *David A. Edwards, Howard Brenner,* and *Darsh T. Wasan*
Introductory Systems Analysis for Process Engineers *E. Bruce Nauman*
Microhydrodynamics: Principles and Selected Applications *Sangtae Kim* and *Seppo J. Karrila*
Modelling with Differential Equations *Stanley M. Walas*
Molecular Thermodynamics of Nonideal Fluids *Lloyd L. Lee*
Phase Equilibria in Chemical Engineering *Stanley M. Walas*
Physicochemical Hydrodynamics: An Introduction *Ronald F. Probstein*
Slurry Flow: Principles and Practice *Clifton A. Shook* and *Michael C. Roco*
Transport Processes in Chemically Reacting Flow Systems *Daniel E. Rosner*
Viscous FLows: The Practical Use of Theory *Stuart W. Churchill*

REPRINT TITLES

Advanced Process Control *W. Harmon Ray*
Applied Statistical Mechanics *Thomas M. Reed* and *Keith E. Gubbins*
Elementary Chemical Reactor Analysis *Rutherford Aris*
Kinetics of Chemical Processes *Michel Boudart*
Reaction Kinetics for Chemical Engineers *Stanley M. Walas*

SLURRY FLOW

Principles and Practice

C.A. Shook
University of Saskatchewan

M.C. Roco
University of Kentucky and
National Science Foundation

Butterworth–Heinemann
Boston London Oxford Singapore Sydney Toronto Wellington

We dedicate this book to our wives:
Kate Shook and Cathy Roco

Recognizing the importance of preserving what has been written, it is the policy of Butterworth–Heinemann to have the books it publishes printed on acid-free paper, and we exert our best efforts to that end.

Library of Congress Cataloging-in-Publication Data

Shook, C.A.
Slurry flow: principles and practice / C.A. Shook, M.C. Roco.
p. cm. — (Butterworth-Heinemann series in chemical engineering)
Includes bibliographical references and index.
ISBN 0–7506–9110–7 (alk. paper)
1. Hydraulic conveying. 2. Slurry. 3. Two-phase flow.
I. Roco, M.C. II. Title. III. Series
TJ898.S53 1991
621.8′67—dc20 91–7514
CIP

British Library Cataloguing in Publication Data

Shook, C.A. (Clifton A.)
Slurry Flow: Principles and practice.
1. Slurry. Flow.
I. Title II. Roco, M.C. (M.C.)
532.51

ISBN 0–7506–9110–7

Butterworth–Heinemann
80 Montvale Avenue
Stoneham, MA 02180

10 9 8 7 6 5 4 3 2 1

Printed in the United States of America

Contents

Preface

This book has been written for the designer and the plant engineer. Our purpose has been to summarize the current state of knowledge in areas of interest to those designing and operating slurry pipelines. In so doing we have assumed a background in fluid mechanics and mathematics, which would be provided by introductory undergraduate courses.

The range of topics is extensive, reflecting the variety of problems associated with slurry flows. In selecting these topics we have drawn upon our experience in testing industrial slurries for purposes of plant design, and in teaching short courses to plant engineers and designers. We realize that many readers, interested in selecting equipment, will turn directly to Chapters 4 to 6 and 8 to 11. Since some of the concepts of fluid, particle, and slurry behavior may be unfamiliar, we hope that Chapters 1 to 3 will provide useful clarifications or direct the reader to the appropriate literature. No description of slurry rheology is complete without an introduction to the role played by surface phenomena and we have tried to do this for readers with a limited background knowledge of chemistry.

There can be little doubt that as research proceeds, mechanistic flow models will provide the framework for generalizing experimental measurements and we have tried to discuss the basis for these models. The modeling task remains as a challenge to the researcher, however, and Chapter 7 indicates the scope of this task.

We have tried to make the book a self-contained source of information for the engineer with a limited technical library. Since our treatment of many concepts must be cursory, we have tried to indicate the literature to which one can turn for further amplification and discussion. The bibliography includes references whose stature would justify their inclusion in a small technical library.

Chapter 1

Basic Concepts for Single-Phase Fluids and Particles

1.1 STEADY PIPE FLOW

The basic physical principles which govern all flows are conservation of mass, conservation of momentum, and conservation of energy.

In dealing with these physical laws, we choose either a microscopic or a macroscopic control volume. A microscopic control volume is shown in Appendix 1 and the appropriate equations are given there. A macroscopic control volume, which contains the familiar items of equipment found in fluid transport systems, is illustrated in Figure 1–1. The equations expressing the physical laws for a macroscopic control volume are simpler as long as the flow is steady and one dimensional, i.e., one velocity component is of dominant importance.

For steady flow, the mass conservation relationship is

$$\text{Mass flow rate } w = (\text{density } \rho \times \text{velocity } V \times \text{area } A) = \text{constant} \tag{1–1}$$

In terms of the mean values of these quantities at the inlet (1) and outlet (2) of the control volume this means

$$(\rho AV)_1 = (\rho AV)_2$$

The averaging process, by which these mean values are related to the point or local values for incompressible fluid–particle mixtures, is discussed in Appendix 2.

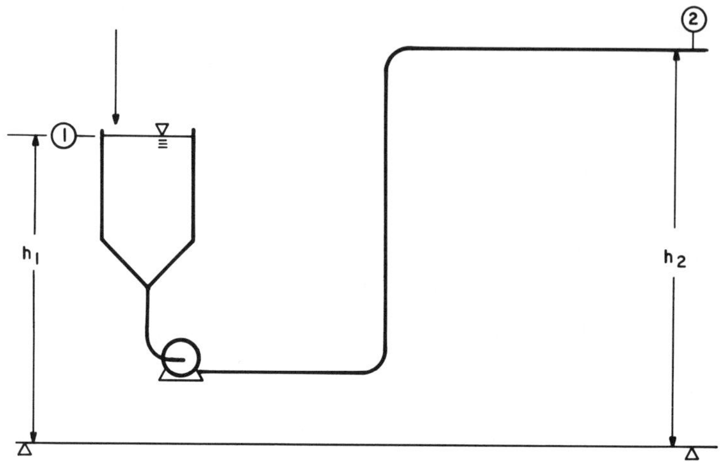

Figure 1–1. A macroscopic control volume.

In a single-phase system the averaging is simpler than it is in a slurry because the density of a single-phase fluid is constant over the cross section. However, the velocity varies with position so that the mean or bulk velocity V is computed by integration of the local velocity v_x in the mean flow (x) direction:

$$V = \left(\frac{1}{A}\right)\int_A v_x \, dA = \left(\frac{2\pi}{A}\right)\int_0^{D/2} r v_x \, dr = \frac{Q}{A} \qquad (1\text{–}2)$$

Equation 1–2 shows that V is the volumetric flow rate per unit area. ρAV is the mass flow rate that is used in material balance calculations.

The linear momentum equation for the quantities averaged over a pipe cross section may be derived from the differential linear momentum equation for the x-direction, Equation A1–6 in Appendix 1, by integrating over the cross section of the pipe (e.g., Longwell, 1966) to give

$$\rho\left[\frac{\partial V}{\partial t} + V\frac{\partial V}{\partial x} + g\frac{\partial h}{\partial x}\right] + \frac{\partial P}{\partial x} + 4\frac{\tau_w}{D} = 0 \qquad (1\text{–}3)$$

where t denotes time, P is the average pressure, and τ_w is the resisting wall shear stress. The sign convention for surfaces and stresses is given in Appendix 1. A positive surface of a control volume is defined as one whose normal, in a positive coordinate direction, points *into* the volume. A positive stress acts in a positive coordinate direction on a positive surface or in a negative coordinate direction on a negative surface.

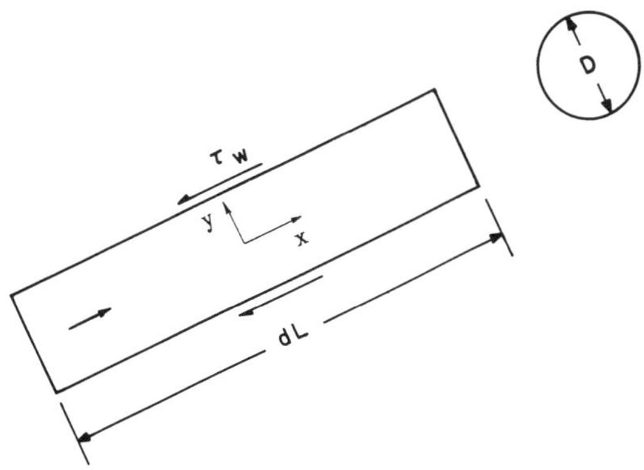

Figure 1–2. An elemental control volume for a pipeline.

The extended form of Bernouilli's equation is obtained by integrating the steady flow form of Equation 1–3 in the axial (x) direction for the macroscopic control volume. The most useful form of the resultant energy equation is given in Equation 1–4 for a simple pipeline consisting of sections of length L_i and diameter D_i. A differential fluid element is shown in Figure 1–2. The curved surface of the element is a negative surface.

$$\int_1^2 \frac{dP}{\rho} + \left[\left(\frac{V^2}{2\alpha}\right)_2 - \left(\frac{V^2}{2\alpha}\right)_1\right] + g(h_2 - h_1) - gH + \sum \frac{4\tau_w L}{\rho D} = 0 \quad (1\text{–}4)$$

The terms in this equation are all expressed in J/kg of fluid. They represent, from left to right, pressure change, kinetic energy change, potential energy change, pump energy input, and frictional dissipation effects. H is the "head" generated by the pump (H is negative for a turbine) and τ_w is the shear stress that opposes flow at the wall of each pipe section.

α is a correction factor that allows for the fact that all the fluid particles passing through a plane in a given time do not have the same kinetic energy per unit mass. Its value depends on the distribution of the axial velocity $v_x(r)$, where r is the radial coordinate. α ranges between 0.5 (parabolic distribution) and 1.0 (flat).

The wall shear stress τ_w opposes fluid motion and its sense is therefore opposite to that of V. τ_w is expressed in terms of the Fanning friction factor f as

$$\tau_w = 0.5\, f\, V|V|\rho \quad (1\text{–}5)$$

The absolute value restriction in Equation 1–5 is usually ignored, except in pipe network calculations when the direction of flow is unknown. For Newtonian fluids (defined in Appendix 1), f can be calculated from the Reynolds number $\mathrm{Re} = D|V|\rho/\mu$ and the equivalent sand roughness of the wall k with Churchill's equation (Churchill, 1977), which is equivalent to the well-known Moody diagram:

$$f = 2\left[\left(\frac{8}{\mathrm{Re}}\right)^{12} + (A + B)^{-1.5}\right]^{1/12} \tag{1–6}$$

where

$$A = \left\{-2.457 \ln\left[\left(\frac{7}{\mathrm{Re}}\right)^{0.9} + \frac{0.27k}{D}\right]\right\}^{16}$$

and

$$B = \left(\frac{37530}{\mathrm{Re}}\right)^{16}$$

1.2 TURBULENT PIPE FLOW

With its rapid mixing of neighboring fluid elements, turbulent flow consumes a great deal more energy than the simple gliding of fluid layers in laminar flow. This increased energy consumption accompanies an increase in the wall shear stress τ_w and in the other stresses within the flow. There are numerous turbulence models and new theories and computational techniques are still being developed (Lumley, 1990). In this section we summarize a few simple formulas that are used in subsequent sections of this book.

For the simplest flow situation, where the x-wise velocity varies only with y, the stress τ_{yx} in turbulent flow is expressed in terms of the velocity fluctuations associated with the turbulence, v_x' and v_y', by Equation 1–7.

$$\tau_{yx} = \rho\overline{v_x' v_y'} - \mu \frac{dv_x}{dy} \tag{1–7}$$

where y is measured from the axis of symmetry of the flow. The two terms on the right-hand side of Equation 1–7 represent the inertial or turbulent Reynolds stress and the viscous shear contributions to the total stress. The bar superscript denotes a time average and the sign convention for stresses is the same as that used by Bird et al. (1960).

Strictly speaking, v_x in Equation 1–7 should have a bar superscript too, but this was omitted in Equation 1–2. The term $\rho\overline{v_x' v_y'}$ is dominant except in the thin viscous sublayer at the wall.

The fact that a momentum equation can be written either for instantaneous values or time-averaged values of velocities, with alterations in the meaning of the stresses, is potentially confusing. It is usually necessary to inspect a derivation carefully to see which approach is being used. We shall see that the same considerations apply to fluid–particle systems.

In principle, one integrates Equation 1–7 in the direction normal to the wall to find V in terms of ΔP if one can relate $\rho\overline{v_x'v_y'}$ or, for pipes, $\rho\overline{v_x'v_r'}$, to the gradient of v_x. The integration gives v_x as a function of r. This velocity distribution should reproduce the experimentally determined Law of the Wall. For smooth walls, the experimental data for homogeneous fluids fit the equation

$$\frac{v_x}{u_*} = \left(\frac{1}{\kappa}\right) \ln\left(\frac{u_*(0.5D - r)\rho}{\mu}\right) + A \qquad (1\text{–}8)$$

where u_* is the friction velocity $(\tau_w/\rho)^{0.5}$. Approximate values of κ and A are 0.4 and 5.75. An expression for both smooth and rough pipes (Roco, 1980) is

$$\frac{v_x}{u_*} = 8.48 - 5.75 \log_{10}\left(\frac{k}{0.5D - r} + \frac{3.3\mu}{\rho u_*(0.5D - r)}\right) \qquad (1\text{–}9)$$

An eddy kinematic viscosity ν_t is used in the intermediate steps of the prediction of v_x as a function of position. In its simplest form it is defined by Equation 1–10

$$\nu_t\frac{dv_x}{dy} = -\overline{v_x'v_y'} \qquad (1\text{–}10)$$

ν_t is a function of position as well as u_*. A useful empirical expression for ν_t in fully turbulent pipe flow (Roco and Frasineanu, 1977) is

$$\nu_t = 0.146\, u_*\left(\frac{D}{2}\right)\left[1 - \left(\frac{2r}{D}\right)^2\right]\left[\left(\frac{2r}{D}\right)^2 + 0.54\right] \qquad (1\text{–}11)$$

1.3 PARTICLE SIZE DISTRIBUTIONS

The size, density, shape, and surface texture of particles affect their behavior in a fluid–particle mixture. The size can be defined in a variety of ways, which depend upon the method used to make the measurements. These include:

1. an equivalent diameter, determined from particle volume, V_p, or surface, S_p. There are at least two possibilities:

$$d_V = \left(\frac{6V_p}{\pi}\right)^{1/3} \qquad (1\text{–}12a)$$

$$d_S = \left(\frac{S_p}{\pi}\right)^{1/2} \qquad (1\text{–}12b)$$

2. an equivalent diameter, determined from the settling velocity of a particle and a drag force relationship known to apply to spheres. This method is used frequently for very fine particles.
3. the sieve size, determined by the width of the minimum square aperture through which the particle will pass. This is the method that is used most often for coarse particles.

Since a mixture of sizes is invariably present, a size distribution curve should always be used to specify a mixture. Although log probability coordinates are sometimes used to detect bimodal distributions, the simple cumulative distributions shown in Figure 1–3 are usually sufficient to display measurements. If it is desirable to describe the size distribution quantitatively, several possible distributions are available (Allen, 1981). The simplest of these distributions are summarized in Table 1–1 as values of the cumulative fraction P_s finer than screen size d. The parameters are d_{50} (median size) and m (dimensionless parameter) or the standard deviation σ. Frequently, more than one distribution appears to fit a given set of measurements fairly well.

Figure 1–3 shows some typical size distributions (solid lines) for industrial slurries and (dashed lines) Rosin–Rammler distributions, plotted as percentages coarser than size d. A Rosin–Rammler distribution with $m < 1$ can be considered broad, whereas one with $m > 2$ is comparatively narrow.

The particle shape and size is described more completely using image analysis (Beddow et al., 1984). This approach requires a high-resolution optical system and a computer.

Table 1–1. Two-Parameter Particle Size Distributions (Data from Allen, 1981)

Distribution	$P_s = f(d)$
Rosin–Rammler	$1 - \exp\left[-\ln 2\left(\frac{d}{d_{50}}\right)^m\right]$
Gates–Gaudin–Schumann	$0.5\left(\frac{d}{d_{50}}\right)^m$
Gaudin–Meloy	$1 - \left[1 - \left(\frac{d}{d_{50}}\right)\left(1 - \sqrt[m]{0.5}\right)\right]^m$
Log normal	$\left(\frac{1}{\sqrt{2\pi}}\right)\int_{-\infty}^{x} \exp\left(-\frac{t^2}{2}\right)dt$
	where $x = \frac{\ln(d/d_{50})}{\ln \sigma}$

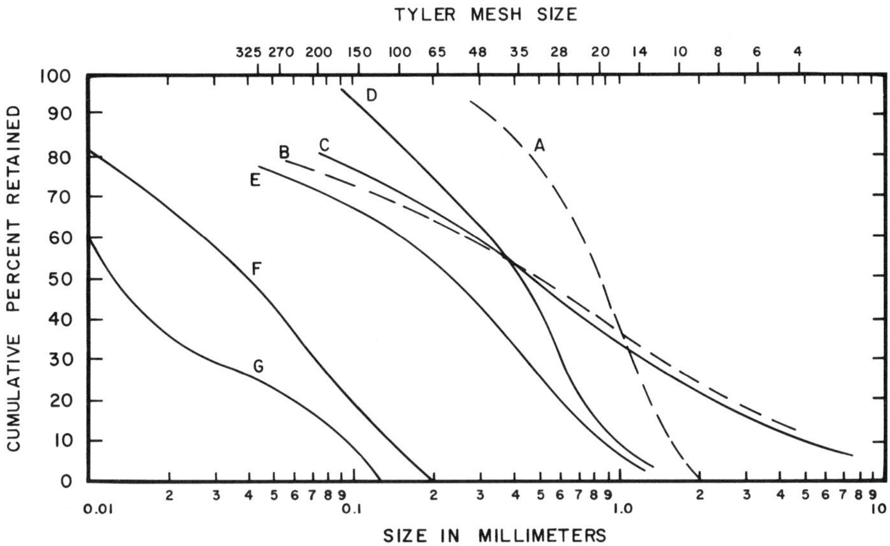

Figure 1–3. Some typical particle size distributions: A: Rosin–Rammler, $m = 2$; B: Rosin–Rammler, $m = 0.5$; C: -50 mm washed coal; D: potash (KCl) mill slurry; E: "long distance pipeline" coal slurry; F: boiler feed coal slurry; G: copper ore flotation cell underflow.

1.4 PACKING OF SOLID PARTICLES IN CONTAINERS

For uniform spheres, six regular arrangements are possible. The simplest of these are the *stacked* and the *close-packed* patterns. These are illustrated schematically in Figure 1–4 which shows the location of the particle centers in successive layers. The volume fractions of solids in the basic repeating units of these patterns are 0.5236 (stacked) and 0.7406 (close-packed rhombohedral). Other packing arrangements give concentrations between these limits. Random packing of large (negligible surface forces) spheres gives concentrations ranging between 0.59 (loose) and 0.64 (close) in the absence of wall effects. Smooth isometric particles, such as sand grains, give similar concentrations if the particle size distribution is narrow (Rosin–Rammler $m > 2$ or thereabouts).

Wall effects are important since the center of a basic packing unit cannot lie within $d/2$ of a wall and the volume fraction of solids at the tip of a sphere in contact with a wall approaches zero. This produces a lower volume fraction of the solid phase (higher local voidage) in the nearest layers as shown in Figure 1–5. As the ratio of particle diameter to container diameter increases, this effect causes the mean solid concentrations for randomly packed spheres to fall below the values quoted above.

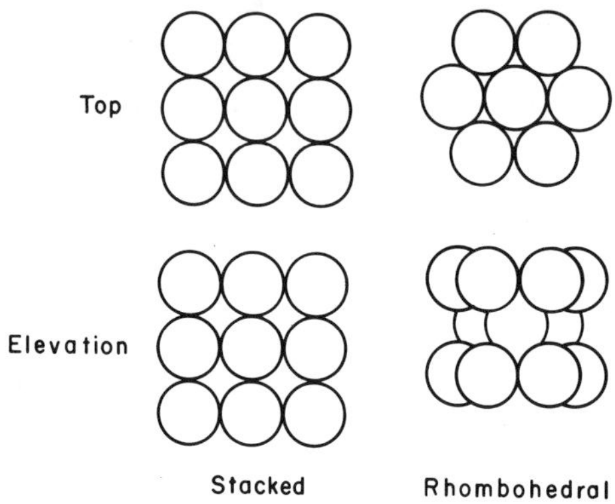

Figure 1–4. Two particle packing arrangements.

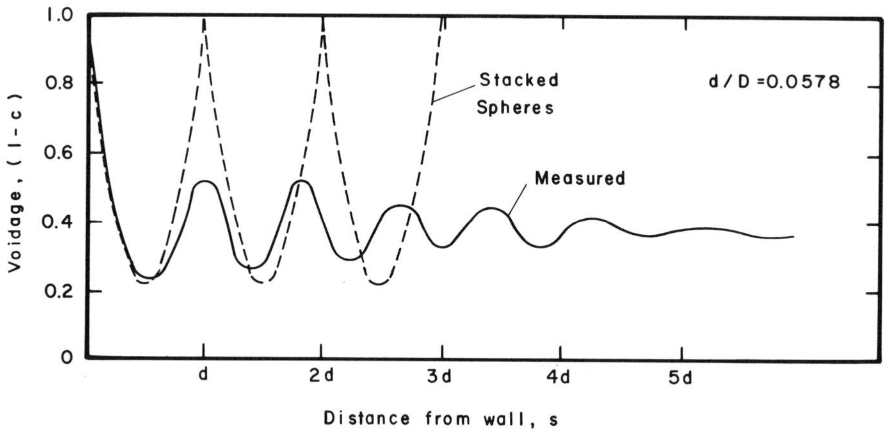

Figure 1–5. Voidage $(1 - c)$ as a function of distance from the pipe wall. (Data from Ridgeway and Tarbuck, 1968.)

The *sphericity* of a particle is defined as the surface area of a sphere divided by the surface area of a particle of the same volume. As the sphericity decreases below its limiting value of unity, the concentration of packed beds of these particles also decreases. However, the concentrations differ less between the loose-packed and the close-packed conditions. Factors that increase the forces acting between particles, such as increased surface roughness or the presence of moisture to form bridges, favor the formation of voids of low concentration and cause the mean concentration of a packed bed to fall below the limits quoted earlier.

If a significant range of particle sizes is present, the concentration in a randomly packed bed can increase considerably. For example, a binary mixture of particles can contain more than 80% solids by volume if there is a very large difference in particle sizes. This is seen easily by considering a mixture of total volume equal to unity, in which the volume occupied by the larger particles is c_2.

Since there is a great difference in sizes, the smaller particles can be assumed to be uniformly distributed in the spaces between the large ones so that the volume occupied by the smaller particles is $(1 - c_2)c_1$, where c_1 is volume fraction of a fine particle in the space it occupies. The total volume of solids in this limiting case is then $c_2 + (1 - c_2)c_1$. With $c_2 = c_1 = 0.6$, this gives a total volumetric concentration of 0.84. High volumetric concentrations of mixtures with broad size distributions were considered by Asszonyi et al. (1972).

With sets of particles and diameters chosen to fill the interstitial spaces as fully as possible, even higher concentrations will be achieved. With spheres in an ordered arrangement, one obtains $c = 0.961$ using the so-called Horsfield packing.

1.5 FORCES ACTING ON A SINGLE PARTICLE IN A DILUTE SUSPENSION

A single particle moves in a fluid under the influence of the particle inertial force $\mathbf{F}_i$, the body force $\mathbf{F}_b$, and the net surface force $\mathbf{F}_f$.

$$\mathbf{F}_i + \mathbf{F}_b + \mathbf{F}_f = 0 \tag{1–13}$$

$\mathbf{F}_b$ is the resultant of a number of effects including fluid drag $\mathbf{F}_D$, fluid lift force $\mathbf{F}_L$, fluid inertia (added mass $\mathbf{F}_{\text{am}}$ and Basset history force $\mathbf{F}_{\text{Bh}}$), and the Brownian diffusion effect. The body force is caused by an external field (gravitational, electrostatic, or magnetic). The surface and inertial forces are relatively well understood for single particles but, as we shall see in subsequent chapters, at higher concentrations it is often necessary to estimate their magnitudes from the results of idealized calculations or experiments.

The motion of a single small particle in a turbulent fluid is described by the Basset–Boussinesq–Oseen equation (BBO equation) as explained, for example, by Wallis (1969) or Hinze (1975). In this context, "small" implies particle dimensions much smaller than the characteristic dimension of a turbulent eddy. However, to solve the equation, a description of the turbulent fluid motion is required and this is still a subject of research.

We shall see in subsequent chapters that the expressions for forces acting on particles in dilute systems are particularly useful in constructing dimensionless groups with which to characterize slurries and their flows.

1.6 DRAG FORCE ON IMMERSED OBJECTS

If v is the velocity of a fluid relative to a particle of projected cross-sectional area A_p, the drag force in the direction of v_r is calculated from an expression which defines the drag coefficient C_D.

$$\mathbf{F}_D = 0.5 C_D A_p \rho_L \mathbf{v}_r |v_r| \tag{1–14}$$

Equations 1–14 and 1–5 show the resemblance of C_D to the friction factor f. C_D depends on the shape and the Reynolds number $d|v_r|\rho_L/\mu$, where d is a characteristic dimension of the particle and ρ_L is the liquid density. C_D also depends on the particle surface roughness, the degree of turbulence in the fluid, and the acceleration of the fluid relative to the particle. For steady flow past spheres, the relationship is given by Equation 1–15.

$$C_D = \frac{24}{\text{Re}_p} \qquad \text{for } \text{Re}_p = d|v_r|\rho_L/\mu < 0.2$$

$$C_D = \frac{24}{\text{Re}_p}(1 + 0.15\ \text{Re}_p^{0.687}) \qquad \text{for } 0.2 < \text{Re}_p < 1000 \tag{1–15}$$

$$C_D = 0.44 \qquad \text{for } 1000 < \text{Re}_p < 3 \times 10^5$$

A single equation alternative expression for C_D is

$$C_D = \frac{24}{\text{Re}_p} + \frac{3.5}{\text{Re}_p^{0.3}} + 0.23\ k\ \log_{10}\left(\frac{\text{Re}_p}{1500}\right) \tag{1–16}$$

for $\text{Re}_p < 7 \times 10^4$, k is 0 for $\text{Re}_p < 1500$ and 1 otherwise.

One of the simplest methods for determining C_D is to measure the terminal falling velocity of a single particle in a large container. When the particle density is greater than that of the fluid, the immersed weight of the particle acts downward and V is an upward velocity numerically equal to the terminal velocity because the fluid velocity is negligible. Since the forces must balance, we have

$$(\rho_s - \rho_L)g\left(\frac{\pi d^3}{6}\right) = 0.5 C_D\left(\frac{\pi d^2}{4}\right)\rho_L\ v_r|v_r|$$

Using the symbol V_∞ for the value of v_r in this situation (infinite dilution), we have the important expression

$$V_\infty = \left(\frac{4gd(S_s - 1)}{3C_D}\right)^{0.5} \tag{1–17}$$

When settling takes place in a tube, the container boundary increases the drag force by retarding the fluid in its upward motion. If V_s is the measured settling velocity in a container of diameter D, the infinite dilution velocity V_∞ is given approximately by Francis's correction ($\text{Re}_p < 1$):

$$V_\infty = V_s\left[\left(1 - \frac{0.475d}{D}\right) \Big/ \left(1 - \frac{d}{D}\right)\right]^4 \tag{1–18}$$

Corrections for other conditions are given by Clift, Grace, and Weber (1978). These corrections are important when C_D is determined from measurements of V_∞.

For irregular particles, settling is much more complicated because a measurement will determine the drag force for a preferred orientation reflecting the balance of moments of the forces exerted by the fluid on the surface of the particle. The terminal velocity is no longer a function of a single characteristic dimension of the particles but depends on shape and orientation.

For flowing slurries, settling is important because it influences the concentration distribution and because settling measurements are convenient techniques for particle characterization.

Heywood's volumetric shape factor (Clift et al., 1978) probably provides the most useful correlation for the effect of shape on the terminal velocity. The shape factor k is defined in terms of the particle volume V_p and the projected area diameter d_A:

$$k = \frac{V_p}{d_A^3} \tag{1–19}$$

where

$$d_A = \left(\frac{4A_p}{\pi}\right)^{0.5}$$

The particle shape can be characterized by other shape coefficients, such as the sphericity Ψ. If S_p is the surface area of the particle, in terms of the diameters d_V and d_S defined in Equations 1–12a and 1–12b,

$$\Psi = \pi\left(\frac{6V_p}{\pi}\right)^{2/3} \Big/ S_p = \frac{d_V^2}{d_S^2} \tag{1–20}$$

Frequently, the particle shape is included in the equivalent particle diameter. For instance, the Stokes diameter is defined as

$$d_{\text{Stokes}} = \left(\frac{18\mu V_\infty}{(\rho_s - \rho_L)g}\right)^{0.5} \tag{1–21}$$

for $\text{Re}_p < 0.2$.

Of course, these methods of describing particle shapes must be related. For example, Pettyjohn and Christiansen (1948) found that, for isometric particles settling at very low Reynolds numbers,

$$\frac{d_{\text{Stokes}}^2}{d_V^2} = 0.842 \log_{10}\left(\frac{\Psi}{0.065}\right) \tag{1–22}$$

Although, in principle, the shape factor can be used as a correlating parameter to determine particle drag coefficients, the simplest way to determine particle settling velocities is to measure them.

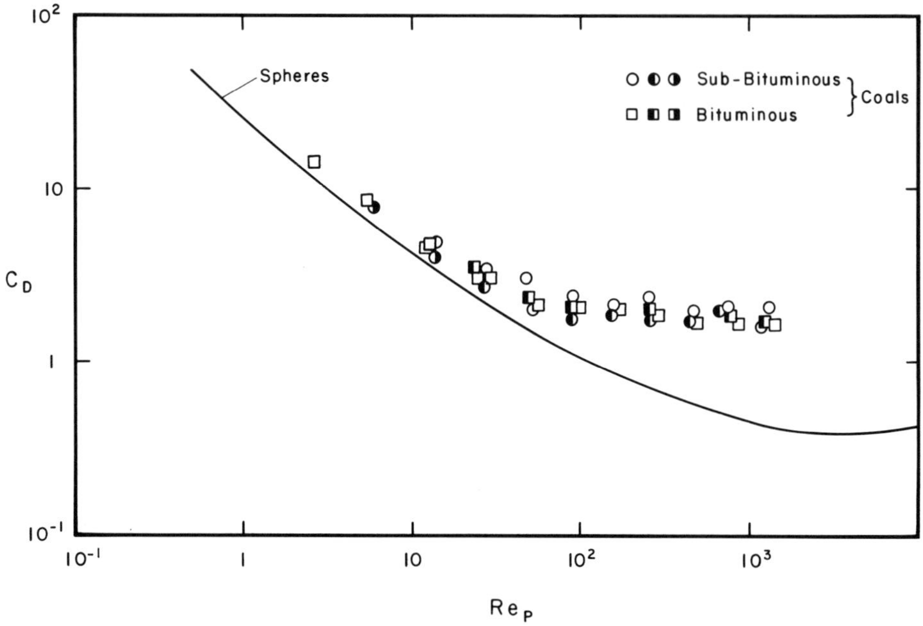

Figure 1–6. Particle drag coefficients for coals.

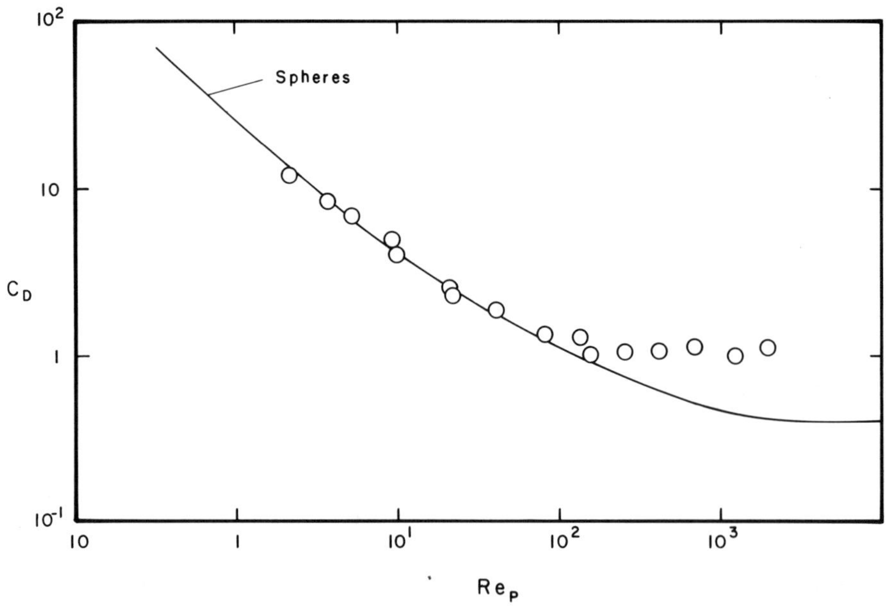

Figure 1–7. Particle drag coefficients for sands and gravels.

Table 1–2. Particle Drag Coefficients

Particle	*Range*	a_1	b_1
Coal	Ar < 24	576	−1
	24 < Ar < 4660	128	−0.482
	4660 < Ar	2.89	−0.0334
Sand	Ar < 24	576	−1
	24 < Ar < 2760	80.9	−0.475
	2760 < Ar < 46100	8.61	−0.193
	46100 < Ar	1.09	0

For high Reynolds numbers, ($10^3 < \mathrm{Re}_p < 10^5$), C_D becomes independent of fluid viscosity so that empirical correlations for settling velocity can relate V_∞ to particle size and the density ratio $S_s = \rho_s/\rho_L$. An empirical formula for large natural sand and gravel particles is

$$V_\infty = \left(\frac{8.925}{d}\right)\left\{\left[1 + 95(S_s - 1)\, d^3\right]^{0.5} - 1\right\} \tag{1–23}$$

where V_∞ is expressed in mm/s and d in mm.

In practice, V_∞ should be measured if reliable values are required. Figures 1–6 and 1–7 show typical experimental results for coal, and sand and gravel particles. C_D is correlated conveniently with the Archimedes number $C_D\mathrm{Re}_p^2 = 4d^3g(S_s - 1)\rho_L^2/3\mu_L^2 = \mathrm{Ar}$ that removes the necessity for iteration in calculating particle settling velocities.

Table 1–2 gives the coefficients of the piecewise fits of these results to equations of the form

$$C_D = a_1\left(C_D\mathrm{Re}_p^2\right)^{b_1} \tag{1–24}$$

These coefficients provide an indication of the importance of particle shape in settling measurements. The coal particles were fractured by mining and cleaning processes and had sharp edges and corners. The sand particles were rounded to some extent and had lower drag coefficients than the coals. V_∞ values for the coarsest of the sands and gravels of Table 1–2 were about 20% higher than those predicted by Equation 1–23 and this difference reflects shape differences between particles from different sources.

Pairs and other clusters settle more quickly than single particles because the effective immersed weight of the aggregate (including any entrapped fluid) increases more rapidly with its volume than does the effective surface area. Such clusters form spontaneously in fluid–particle mixtures of low and moderate concentration. This complicates the tendency toward a reduced settling velocity which normally occurs as the concentration of the dispersed phase increases, and which is discussed in Chapter 2.

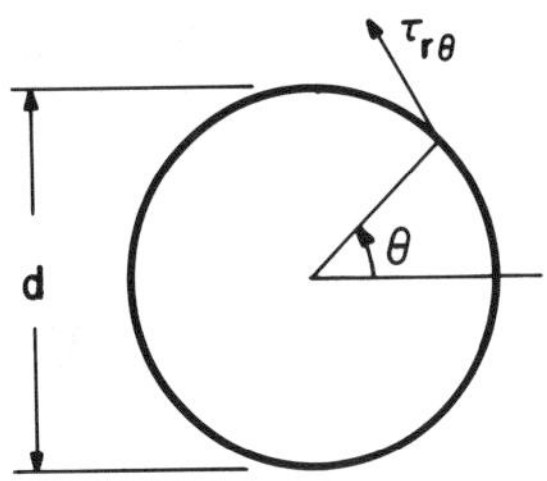

Figure 1–8. Effect of a fluid stress in supporting a particle.

For settling to occur in a fluid which has a yield stress τ_y, the resisting stress in sheared fluid must be at least equal to τ_y. (A definition of the yield stress can be found in Appendix 1.) If we assume the particle surface stress is everywhere equal to τ_y, we can calculate the minimum particle diameter for which settling will occur. The vertical component of the shear force exerted on the surface is given in terms of the yield stress by the integral (see Figure 1–8)

$$0.5\pi d^2 \tau_y \int_{-\pi/2}^{\pi/2} \cos^2\theta \, d\theta$$

Since this force could balance the immersed weight of the sphere, we obtain Equation 1–25 for the diameter of the smallest particle which will settle.

$$d_{\min} = \frac{1.5\pi\tau_y}{(\rho_s - \rho_L)g} \tag{1–25}$$

Experiments appear to confirm that Equation 1–25 applies for particles immersed in stationary fluids with yield stresses. However, if motion of the particle relative to the fluid occurs, the critical size can be considerably different (Atapattu et al., 1988). The discrepancy is probably due to the fact that the surface shear stress need not be τ_y in the region of sheared fluid which forms at the surface of the particle (Beris et al., 1985).

When the fluid is non-Newtonian the drag coefficient will depend on all the rheological parameters of the fluid. For Bingham fluids, the drag coefficient is usually expressed in terms of the Reynolds number $\mathrm{Re}_p = dV_\infty \rho_L/\mu_p$ and the Bingham number $\mathrm{Bi} = \tau_y d/V_\infty \mu_p$. These two groups arise naturally if one defines a Reynolds number as the ratio of inertial forces, proportional to ρV_∞^2, to viscous forces that are proportional to $(\tau_y + K'\mu V_\infty/d)$, where K' is a constant. Ansley and Smith (1967) evaluated the constant K' to obtain the group $\mathrm{Re}_p/(1 + 7\pi\mathrm{Bi}/24)$ as an appropriate correlating parameter for drag coefficients. Alternatively, Dedegil (1986) suggested, in effect, that the denominator in the parameter could be taken as $(1 + \mathrm{Bi})$.

Hanks and Sen (1983) generalized the Ansley and Smith criterion to obtain the settling parameter q for yield–power law fluids. The definitions of τ_y, n, and K are given in Appendix 1.

$$q = \frac{2^{6-2n} d^{2n-2} V_\infty^{4-2n} K^{(2/n)-2} \rho [n/(3n + 1)]^{2n}}{(7\pi/24)\tau_y^{(2/n)-1} + 2^{3-n} d^{n-2} V_\infty^{2-n} K^{(2/n)-1} [n/(3n + 1)]^n} \tag{1–26}$$

For power law fluids, this simplifies to

$$q = \frac{2^{(3-n)} d^n V_\infty^{2-n} \rho [n/(3n + 1)]^n}{K} \tag{1–27}$$

1.7 RELAXATION TIME

If a solid particle of velocity v_s is suddenly introduced into a fluid with velocity v_L, the main forces acting on the particle are drag and inertia. The acceleration process is described by the equation

$$\left(\frac{\pi d^3}{6}\right) \rho_s \frac{dv_r}{dt} = -0.5 C_D \left(\frac{\pi d^2}{4}\right) \rho_L v_r |v_r|$$

where $v_r = v_L - v_s$. In terms of t_r, the "relaxation time" for momentum transfer between fluid and particle, this can be written

$$\frac{dv_r}{dt} = \frac{-v_r}{t_r}$$

where

$$t_r = \left(\frac{4}{3}\right)\left(\frac{S_s}{C_D}\right) d \Big/ |v_L - v_s| \tag{1–28}$$

t_r is the interval in which the velocity difference is reduced from $v_L - v_s$ to $(v_L - v_s)/e$. For small particles in the Stokes law range,

$$t_r = \frac{d^2 \rho_s}{18 \mu_L} \tag{1–29}$$

If the relaxation time is small compared to the time scale of other motions, it is often assumed that liquid–solid mixtures can be considered to act as homogeneous fluids.

1.8 LIFT FORCE ON A ROTATING PARTICLE (MAGNUS FORCE)

When a relative velocity exists, a rotating sphere experiences a lift force. The force is normal to both the relative velocity (slip velocity) $\mathbf{v}_r = \mathbf{v}_L - \mathbf{v}_s$ and the particle rotation vector $\boldsymbol{\omega}_s$. The magnitude of the force is expressed in terms of the lift coefficient:

$$C_L = \frac{F_L}{0.5\rho_L v_r^2(\pi d^2/4)} \tag{1-30}$$

where $\pi d^2/4$ is the projected area of the sphere.

A lift force will exist even in an inviscid fluid. The *circulation* Γ is defined as the integral of the fluid velocity component parallel to the path for a closed path S around the particle.

$$\Gamma = \oint \mathbf{v}_L \cdot dS$$

If the circulation is finite and if a relative velocity v_r exists between the particle and the fluid, a lift force is produced by the pressure variation around the path. Such a path could be drawn around a cylinder of radius R rotating with angular velocity ω. The general expression for the lift force per unit length is

$$F_L = \rho_L v_r \Gamma$$

For the rotating cylinder (see Figure 1–9) the coefficient of lift is then

$$C_L = \frac{2\pi R\omega}{v_r} \tag{1-31}$$

In real (viscous) fluids, the lift coefficient is affected by the boundary layer on the immersed object. In a viscous fluid, C_L depends upon the ratio of the equatorial velocity $(d\omega/2)$ to the relative velocity v_r. At very low Reynolds numbers, C_L for a sphere is equal to $(d\omega/2v_r)$, according to Brenner (1966). At very high Reynolds numbers, C_L is small and *negative* (Goldstein, 1938) for $d\omega/2v_r < 0.5$, i.e., the force acts in the opposite direction. However, C_L increases rapidly thereafter and levels off for $d\omega/2v_r > 1.5$. The drag coefficient is increased by particle rotation in the high Reynolds number region.

In addition to the lift force produced by simultaneous slip and externally induced spin, a lift force also exists when slip and shear occur together. If $\dot{\gamma}$ is the shear rate, the lift force on a sphere as calculated by Saffman (1965) is (Figure 1–10)

$$F_L = 1.615d^2(\mu_L\rho_L\dot{\gamma})^{0.5}(v_s - v_L) \tag{1-32}$$

The force on a particle denser than the fluid acts toward the pipe axis in upward flow. Here the particle rotation is induced by fluid shear.

When a sphere moves through a stationary fluid parallel to a plane wall, lubrication theory can be used to estimate the normal force F_r repelling the particle from the wall. This is, of course, a form of lift force, which depends upon the

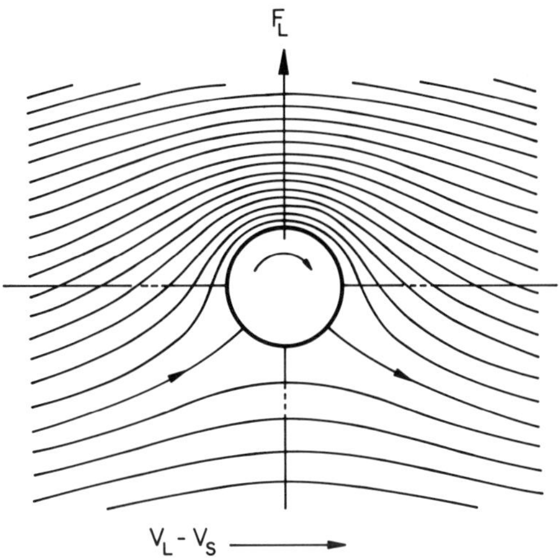

Figure 1–9. Circulation and lift force on a rotating cylinder.

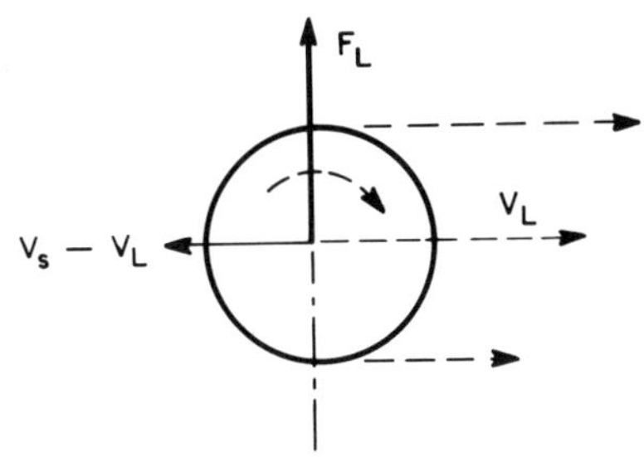

Figure 1–10. The Saffman (shear–slip) force.

sphere diameter, the fluid viscosity, and the clearance h_0 between the sphere and the wall (Cameron, 1982).

$$F_r = \frac{1.9 v_r \mu_L d^{1.5}}{h_0^{0.5}} \tag{1–33}$$

Experimental measurements of lift forces on spheres held in a flow close to a plane wall are reported by Willets and Naddel (1976).

1.9 FLUID INERTIA EFFECT

The *added mass* effect results from the pressure gradient in the fluid when there is a relative acceleration of the particle with respect to the fluid. It can be estimated from potential flow theory in which viscous effects are neglected. The force depends upon the shape and orientation of the particle:

$$\mathbf{F}_{am} = C_s\left(\frac{\pi d^3}{6}\right)\rho_L\left(\frac{d\mathbf{V}}{dt}\right) \tag{1–34}$$

where C_s is a dimensionless shape factor ($= 0.5$ for spheres). The added mass $C_s(\pi d^3/6)\rho_L$ is the mass of fluid which would cause the same inertial effect if it were accelerated with the solid particle.

A particle moving in a fluid is also subject to a transient viscous drag force during the development of the fluid velocity field around the particle. The so-called "Basset force" is a function of the history of the relative velocity in the interval $t_0 < t' < t$:

$$\mathbf{F}_{\mathrm{Bd}} = 1.5d^2\sqrt{\pi\rho_L\mu_L}\int_{t_0}^{t}\left(\frac{d\mathbf{V}}{dt'}\right)(t - t')^{-0.5}\,dt' \tag{1–35}$$

These two forces act in combination during transient flows. Fluid turbulence is also known to affect the drag force exerted on an immersed particle (Torobin and Gauvin, 1959; Lee, 1987) so that the drag coefficient may deviate significantly from the steady flow values because of these effects.

1.10 BROWNIAN DIFFUSION

Very fine particles diffuse in a fluid as a result of their collisions with the fluid molecules. This diffusion flux component in the x-direction can be described by Fick's first law:

$$J_x = -D\frac{\partial c}{\partial x} \tag{1–36}$$

where c is the particle concentration and D is the diffusion coefficient (m^2/s). To predict D, one employs the fact that the particle kinetic energy is determined by the thermal energy of the fluid molecules (which, in turn, is proportional to the absolute temperature). The particle motion is assumed to be resisted by a drag force that obeys Stokes's law so that the expression for D is

$$D = \frac{kT}{3\pi\mu d} \tag{1–37}$$

where k is Boltzmann's constant.

1.11 ELECTROMAGNETIC BODY FORCES

A single particle of charge q (Coulombs) and velocity $\mathbf{V}_s$, moving in an electric field of intensity $\mathbf{E}$ (V/m) and one of magnetic induction $\mathbf{B}$ (Teslas), experiences a force $\mathbf{F}$ (Newtons)

$$\mathbf{F} = q(\mathbf{E} + \mathbf{V}_s \times \mathbf{B}) \tag{1–38}$$

The current carried by the particle is $q\mathbf{V}_s$ (amp).

1.12 HEAT AND MASS TRANSFER TO OR FROM SPHERES

The flux of heat $\dot{q}$(W/m^2) from the surface of a particle is expressed in terms of the difference in temperature between the surface T_s and the bulk of the surrounding fluid T_L, and a heat transfer coefficient h:

$$\dot{q} = h(T_s - T_L) \tag{1-39}$$

h depends on particle properties, fluid properties, and the Reynolds number of the relative motion between the phases. With no relative motion, the limiting value of h is determined by thermal conduction through the fluid (thermal conductivity k_L) as a limiting value of the Nusselt number:

$$\text{Nu} = \frac{hd}{k_L} = 2$$

With relative motion between fluid and particles, the Nusselt number depends on the Reynolds number $(d|v_L - v_s|\rho_L/\mu_L)$ and the fluid Prandtl number $(\mu_L C_p/k_L)$.

The molar flux $\dot{m}_A$ from the surface (kg moles/m^2 s) of a transferred species A in a binary mixture (A and B) is related to the mole fraction of A at the surface, x_0, and that in the bulk of the fluid, x_∞. If the rate of mass transfer is low, the fluid velocity distribution in the vicinity of the particle is not altered and the flux is expressed in terms of the mass transfer coefficient k_x:

$$\dot{m}_A - x_0(\dot{m}_A + \dot{m}_B) = k_x(x_0 - x_\infty) \tag{1-40}$$

For low mass transfer rates, the diffusion processes for heat and mass in the fluid are analogous. The dimensionless group for mass transfer that corresponds to the Nusselt number is the Sherwood number $\text{Sh} = k_x d/C_t D_{AB}$, where C_t is the total molar density and D_{AB} is the diffusivity. The Schmidt number $\text{Sc} = \mu_L/\rho_L D_{AB}$ for mass transfer corresponds to the Prandtl number for heat transfer. Because the heat transfer and mass transfer processes are analogous, the equations for the transfer coefficients are the same when written in terms of these dimensionless groups. For low mass transfer rates the coefficients are given by the Froessling equation:

$$\text{Nu} = 2 + 0.6\text{Re}_p^{1/2}\text{Pr}^{1/3}$$

or

$$\text{Sh} = 2 + 0.6\text{Re}_p^{1/2}\text{Sc}^{1/3} \tag{1-41}$$

At high mass transfer rates, the velocity field in the vicinity of the particle is distorted and the coefficients depend upon the magnitude and direction of the mass transfer. Theoretical studies for flat plates show that h, k_x, and the drag coefficient increase when rapid mass transfer takes place toward the surface and decrease when the flux has the opposite sense.

For very dilute solutions (essentially all species B), the mass flux of A is calculated from the approximation

$$\dot{m}_A M_A = k_x(\rho_{A_0} - \rho_{A_\infty}) \tag{1–42}$$

M_A is the molecular weight of A, so that the left-hand side of Equation 1–42 is the mass flux of A (kg/m^2 s). ρ_A is the mass concentration of A in the solution (kg/m^3). Further explanation of these concepts is given by Bird et al. (1960).

1.13 SURFACE FORCES BETWEEN DISPERSED PARTICLES

To understand these forces, we must first consider conditions at the surface of an individual particle. As the diameter of a particle decreases and its surface area per unit volume increases, surface forces grow in importance compared to other forces to which the particle may be subjected. Of course, it is impossible to give an exact limit, but for particle diameters below about 10 μm, surface effects can be expected to begin to become significant. These effects result from the fact that conditions at the surface of the particle are different from those in the bulk of the fluid phase. Surfaces are usually charged, as a result of ionization, isomorphous substitution of ions, or adsorption of species from the fluid.

The charged surface alters the properties of the fluid in its immediate vicinity so that a region of predominantly oppositely charged species forms. This region is conveniently visualized as a *double layer,* the inner layer (Stern layer) being more tightly bound to the particle than the outer one (diffuse layer). Figure 1–11 shows these layers.

The space charge density ρ is related to the potential ψ in the diffuse layer by Poisson's equation. Since the diffuse layer is often thin compared to the diameter of the particle and since other approximations must be made, the curvature of the surface is often neglected and Poisson's equation is written as

$$\frac{d^2\psi}{dx^2} + \frac{\rho}{\epsilon} = 0 \tag{1–43}$$

where ϵ is the permittivity (C^2/N m^2) of the medium and x is the distance measured perpendicular to the surface. The permittivity of the medium is the product of the (dimensionless) dielectric constant and the permittivity of free space. ρ depends upon the potential because the charge density is the algebraic sum of the charges per unit volume for all the ionic species which are present:

$$\rho = \sum z_i n_i e$$

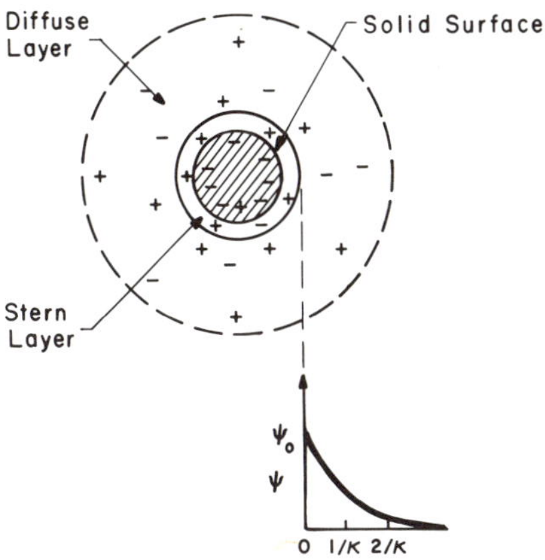

Figure 1–11. The charge distribution in the vicinity of a particle.

Since the particle has a charge (usually negative), ions of the opposite charge ("counter ions") predominate around the particle so that the summation will be finite although the solution is electrically neutral as a whole. e is the charge of an electron and n_i is the concentration of species i. n_i depends upon the concentration in the bulk of the solution, n_{i_∞}, the charge $z_i e$, and the potential ψ, in accordance with Boltzmann's distribution,

$$n_i = n_{i_\infty} \exp\left(\frac{-z_i e \psi}{kT}\right)$$

where k is Boltzmann's constant.

Substituting these expressions, one can solve Equation 1–43 using the conditions $\psi = \psi_0$ at $x = 0$ and $\psi = 0$ as $x \to \infty$. As long as ψ_0 is not too high, the solution may be approximated as

$$\psi = \psi_0 \exp(-\kappa x) \tag{1–44}$$

where $1/\kappa$ is a measure of the thickness of the double layer

$$1/\kappa^2 = \left(\frac{\epsilon k T}{e^2 \sum n_{i_\infty} z_i^2}\right) \tag{1–45}$$

Equation 1–45 shows that the thickness of the double layer decreases considerably if electrolyte concentration is high, and especially so for counter ions of high charge. The double layer is effectively a capacitor so that the surface charge and the potential ψ_0 are proportional and have the same sign. The capacitance is proportional to the product of ϵ and κ so that if the double-layer thickness decreases when the surface charge remains constant, the surface potential decreases.

A measure of the surface charge or potential can be obtained by determining the particle velocity (mobility) when it is driven through a fluid by an applied electric field. The force applied to the particle is determined by the product of the field strength and the charge on the particle in accordance with Equation 1–38. The resisting force is due to fluid drag.

The resisting force in a mobility measurement is determined by the relative velocity, fluid viscosity, and the shape and orientation of the particle. Shear is presumed to occur at the junction between the two layers so that the surface potential determined from the balance of forces is the potential at this junction. The viscosity and permittivity of the bulk of the fluid are assumed to apply to the diffuse layer. The potential calculated from mobility measurements is called the *zeta potential* ζ and is a reasonable approximation to ψ_0. It is the parameter with which the properties of colloidal dispersions are correlated most frequently. For the reasons described earlier, counter ions of high concentration and/or high charge reduce the zeta potential.

We now consider the effect of these forces on the properties of a mixture of fluid and particles. The property of these dispersions which has been studied most thoroughly is the tendency of the particles to aggregate. This tendency is explained in terms of the total potential energy of an agglomerate. The total potential energy is the sum of contributions resulting from attractive (London) forces between surfaces and repulsive (electrostatic) forces between diffuse layers of neighboring particles. The attractive force is independent of the surface potential ψ_0 but the repulsive force increases strongly with it. Both forces decrease as the separation s between the surfaces increases.

For separations which are small compared to the radii of the particles, the attractive potential energy depends on the Hamaker constant A of the surfaces and the separation s between them:

$$\text{Attraction P.E.} = -\frac{Ad}{24s} \tag{1–46}$$

For separations which are large compared to the "thickness" of the double layer, the repulsion potential energy is, very roughly,

$$\text{Repulsion P.E.} = \left(\frac{\epsilon\psi_0^2 d}{4}\right) \ln[1 + \exp(-\kappa s)] \tag{1–47}$$

Figure 1–12 shows a distribution of the total potential energy as a function of separation s. The latter is expressed as multiples of $1/\kappa$, for the low ψ_0 case which can be represented by the approximations of Equations 1–46 and 1–47. Since the favored state is that of minimum potential energy, we see that only particles with very small separations are likely to form strong aggregates. The higher the (net repulsion) peak, in the energy diagram, the greater the amount of energy which must be provided (say from Brownian collisions) for an agglomerate to form. Equation 1–47 shows that high ψ_0 values, or high zeta potentials, are associated with a low tendency to form agglomerates (high stability). The following table gives an indication of the relationship with zeta potential (Riddick, 1968).

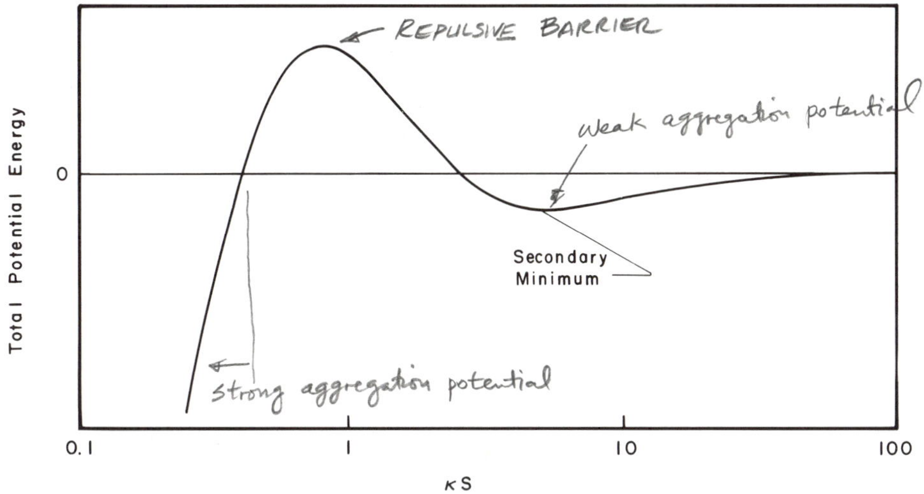

Figure 1–12. Agglomerate potential energy as a function of separation between surfaces.

Stability characteristics	*Zeta potential (mV)*
Maximum agglomeration	0 to +3
Strong agglomeration	+5 to −5
Threshold of agglomeration	−10 to −15
Threshold of dispersion	−16 to −30
Moderate stability	−31 to −40
Fairly good stability	−41 to −60
Very good stability	−61 to −80
Excellent stability	−81 to −100

Addition of electrolyte can affect the properties of fine-particle dispersions profoundly. These changes are usually explained in terms of the repulsive forces since electrolyte is not considered to affect the attractive forces. Small quantities of electrolyte may alter surface potentials by adsorption, whereas higher concentrations reduce the thickness of the double layer and the surface charge as described previously.

The effect of the charge of the counter ions is summarized by the Schulze–Hardy rule which states that the electrolyte concentrations required to produce aggregation by monovalent, divalent, and trivalent ions are in the ratio

$$1 : (1/2)^6 : (1/3)^6 \quad \text{or} \quad 100 : 1.6 : 0.13$$

Since most particles are negatively charged, Fe^{+++}, Al^{+++}, and Ca^{++} are useful coagulants.

In addition to agglomeration that occurs in the primary minimum of the potential energy diagram, a weaker association (flocculation) can also form in the secondary minimum region if the particle repulsion energy is high. This case is also shown in Figure 1–12.

Nonionic long-chain molecules can also promote aggregation if they are adsorbed on the particle surfaces and form linkages between particles. Conversely, an adsorbed species can reduce the tendency toward aggregation through mutual (steric) repulsion of adsorbed layers on adjacent particles.

The state of aggregation in a dilute fluid–particle mixture can be inferred from the settling velocity of the particles, since fully dispersed particles will settle more slowly than aggregates. However, at high concentrations, interference between aggregates becomes likely and their settling rates can be very low. The effect of aggregation on slurry rheology is profound and this is discussed in Chapter 3. Further discussion of these concepts can be found in Kruyt's (1952) classical textbook.

1.14 PARTICLE ROTATION

Particle rotation about an axis passing through the particle itself, or particle spin, can play an important role in momentum and energy transfer in two-phase flow, particularly close to containing walls. The rotation reflects the mean deformation of the flow field, particle–particle, and particle–wall interactions. Forces arising

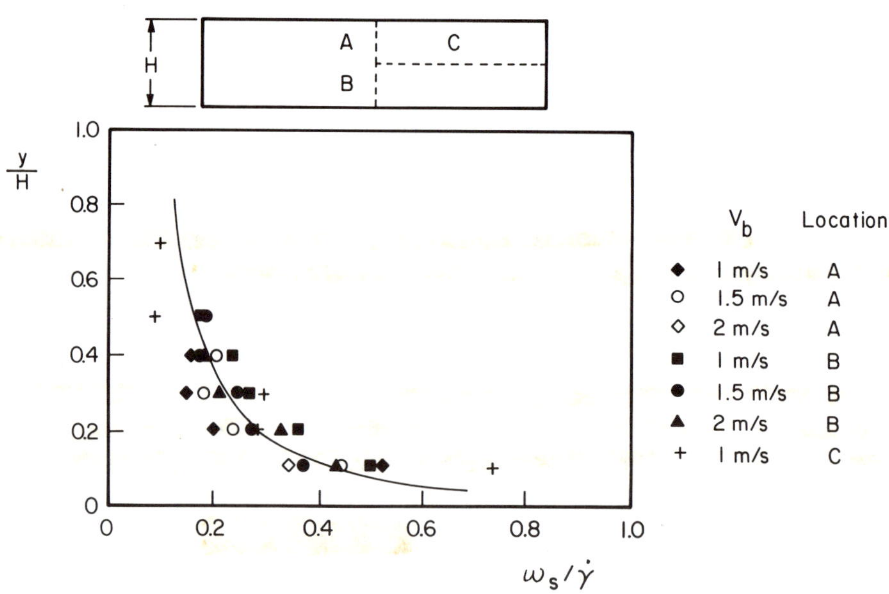

Figure 1–13. Particle angular velocity close to a wall. (Adapted from Ye et al., 1990.)

from rotation will modify particle trajectories and should be included in force balance models describing concentration distributions (discussed in Chapter 7). Ye et al. (1990) used neutrally buoyant particles ($d = 6.35$ mm) in a turbulent Couette flow apparatus to study particle rotation. A cross-section of their apparatus and results are shown in Figure 1–13: the lower boundary is moving at velocity V_b. A law of the wall was obtained for particle rotation in dilute suspensions

$$|\omega_s/\dot{\gamma}| = 0.11(y/H)^{-0.6} \tag{1–48}$$

where ω_s is the angular velocity of the particle, $\dot{\gamma}$ is the mean shear rate in the cross section of the Couette flow, $y \leq H/2$ is the distance from the wall, and $H = 50$ mm is the gap of the Couette section. This result is a limiting case for negligible particle–particle interactions.

Chapter 2
Fluid–Particle Mixtures

2.1 DEFINITIONS FOR SLURRY FLOWS

As in the case of single-phase flows, many slurry engineering problems can be simplified because one velocity component is dominant. Pipe flow certainly falls into this category and we begin by considering the equations for a finite control volume.

The mean or bulk velocity V for a multiphase mixture is defined in the same way as it was for a pure fluid: the (total) volumetric flow rate divided by the area $A = \pi D^2/4$. In terms of the flow rates (m^3/s), Q_s and Q_L of solids and liquid, we have a two-phase version of equation 1–2:

$$V = \frac{Q_s + Q_L}{A} \tag{2–1}$$

The delivered mixture contains a fraction C_v (or C_{vd}) of the solids defined by

$$C_v = \frac{Q_s}{Q_s + Q_L} \tag{2–2}$$

In terms of the time-averaged local concentration and the velocities of the phases at a point in the pipe cross section (see Figure 2–1), the flow rates are

$$Q_s = \int_A v_s c \, dA \tag{2–3}$$

$$Q_L = \int_A v_L (1 - c) \, dA \tag{2–4}$$

where c is the point in situ or spatial concentration by volume. The density of a mixture of solids concentration c is often of interest:

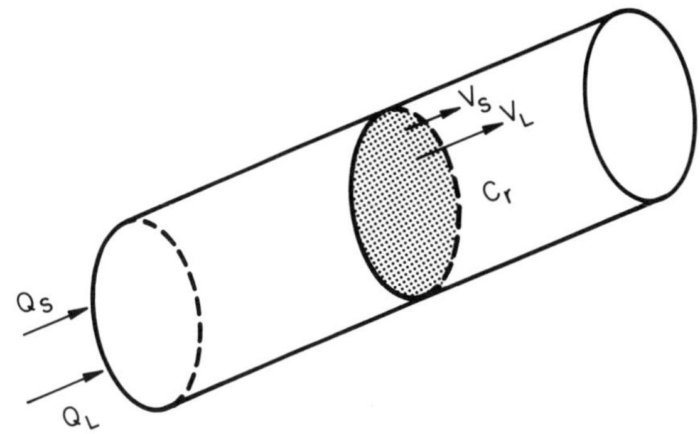

Figure 2–1. Volumetric flow rates and the mean in situ concentration C_r.

$$\rho_m = \rho_s c + \rho_L(1 - c) = \rho_L[1 + c(S_s - 1)] \tag{2–5}$$

where S_s is the ratio of solids density to that of the fluid.

The mass (or "weight") fraction solids in the delivered mixture C_w is frequently a specified quantity in design situations:

$$C_w = \frac{\rho_s Q_s}{\rho_s Q_s + \rho_L Q_L} \tag{2–6}$$

It is sometimes convenient to regard Q_s as the product of the pipe area A, a (measurable) area-averaged and time-averaged in situ or spatial concentration C_r, and a mean solids velocity V_s. This introduces the definitions:

$$C_r = \left(\frac{1}{A}\right)\int_A c\, dA \tag{2–7}$$

$$V_s = \left(\frac{1}{AC_r}\right)\int_A c v_s\, dA = \frac{Q_s}{AC_r} \tag{2–8}$$

so that

$$\frac{C_v}{C_r} = \frac{V_s}{V} \tag{2–9}$$

Similarly, for the liquid we have a defined mean velocity

$$V_L = \frac{Q_L}{A(1 - C_r)} \tag{2–10}$$

Although v_s and v_L may have very similar values at all points in a flow, the variation of c, v_L, and v_s over the cross section frequently causes V_s and V_L to be significantly different from V in horizontal pipes. As an illustration, Figure 2–2 shows c as a function of position for a sand slurry ($d = 0.25$ mm) flowing in a

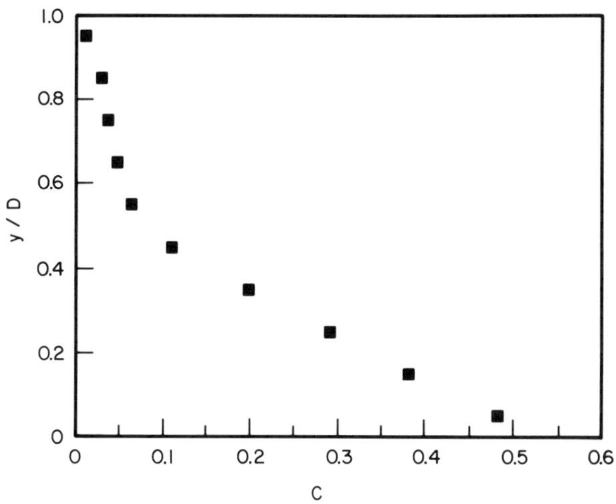

Figure 2–2. Concentration as a function of elevation for a sand slurry.

315-mm pipeline at $V = 3.69$ m/s. Although the difference between v_s and v_L was negligible at all points, measurements showed $C_v = 0.116$ and $C_r = 0.151$. From Equations 2–8 and 2–10, $V_s = 2.83$ m/s and $V_L = 3.84$ m/s.

2.2 CONSERVATION EQUATIONS FOR ONE-DIMENSIONAL FLOWS

The conservation equations can be expressed in terms of the time-averaged concentration c and the velocities of the individual phases. Equation A2–7 and the corresponding expression for the fluid yield ($v_s = v_{sx}$, $v_L = v_{Lx}$)

$$\frac{\partial(c\rho_s)}{\partial t} + \frac{\partial(c\rho_s v_s)}{\partial x} = 0$$

$$\frac{\partial[(1 - c)\rho_L]}{\partial t} + \frac{\partial[(1 - c)\rho_L v_L]}{\partial x} = 0$$

The densities of liquids and solids are essentially constant unless we are considering pressure wave phenomena so that the equations we will use are

$$\frac{\partial c}{\partial t} + \frac{\partial(cv_s)}{\partial x} = 0 \tag{2–11}$$

$$\frac{\partial(1 - c)}{\partial t} + \frac{\partial[(1 - c)v_L]}{\partial x} = 0 \tag{2–12}$$

The momentum equations for the particles and the fluid are given in Appendix 2. It should be noted that several different forms of the momentum equation for multiphase flows have been used; the differences arise in modeling the phase interaction, solids stress, and pressure terms. We consider one-dimensional forms of these equations that are obtained from Equations A2–8b and A2–11. Using Equations 2–11 and 2–12, the momentum equations are, when gravity is the only body force,

$$\left(\frac{1}{c}\right)\rho_s\left(\frac{\partial(cv_s)}{\partial t} + \frac{v_s\partial(cv_s)}{\partial x}\right) = -\left(\frac{1}{c}\right)\frac{\partial(cP)}{\partial x} - \rho_s g\frac{\partial h}{\partial x} + f_{sL} + f_{ss} + f_{sw} \tag{2–13}$$

$$\left(\frac{1}{1-c}\right)\rho_L\left(\frac{\partial(1-c)v_L}{\partial t} + v_L\frac{\partial(1-c)v_L}{\partial x}\right) = -\left(\frac{1}{1-c}\right)\frac{\partial(1-c)P}{\partial x} - \rho_L g\frac{\partial h}{\partial x} + f_{Ls} + f_{LL} + f_{Lw} \tag{2–14}$$

For one-dimensional flows, Wallis wrote the momentum equations in the form

$$\rho_k\frac{Dv_k}{Dt} = -\nabla P + \rho_k b + f_k \tag{2–15}$$

where k denotes a particular phase and f_k is the interaction force. For liquid–solid flows with small concentration gradients, the difference between Equation 2–15 and 2–13 or 2–14 is small.

The terms f_{ss} and f_{sw} in Equation 2–13 are related because they arise from forces acting on different portions of the boundary surface of a control volume. The same is true of f_{LL} and f_{Lw}. In many one-dimensional slurry flow problems we may employ an area-averaged or integral formulation and use Wallis's equation in the form

$$\rho_s\frac{DV_s}{Dt} = -\frac{\partial P}{\partial x} - \rho_s g\frac{\partial h}{\partial x} + \langle f_{sL}\rangle + \langle f_{sw}\rangle \tag{2–16}$$

where h is the elevation above a datum. Similarly, for the liquid phase

$$\rho_L\frac{DV_L}{Dt} = -\frac{\partial P}{\partial x} - \rho_L g\frac{\partial h}{\partial x} + \langle f_{Ls}\rangle + \langle f_{Lw}\rangle \tag{2–17}$$

The braces $\langle\ \rangle$ denote an area-averaged quantity and are often omitted. The reciprocal drag condition (Equation A2–12) relates f_{sL} and f_{Ls}.

$$(1-c)f_{Ls} + cf_{sL} = 0 \tag{2–18}$$

As we observed in the case of single-phase flows, the interaction forces $\langle f_{sw} \rangle$ and $\langle f_{Lw} \rangle$ include inertial contributions because the derivatives D/Dt are applied to time-averaged fluid and particle velocities.

If we consider a truly homogeneous flow, one in which the particles are distributed uniformly throughout the mixture, we can obtain a simple momentum equation. We assume $V_s = V_L = V$ and that the concentration c is independent of position. Combining Equations 2–16 and 2–17, using Equation 2–18, and integrating over the cross section of the pipe, we obtain the familiar momentum equation (Equation 1–3) in terms of the mean density ρ_m and the boundary shear stress τ_w:

$$\rho_m\left(\frac{\partial V}{\partial t} + V\frac{\partial V}{\partial x} + g\frac{\partial h}{\partial x}\right) + \frac{\partial P}{\partial x} + \frac{4\tau_w}{D} = 0 \tag{2–19}$$

where ρ_m is defined in Equation 2–5 and

$$\frac{4\tau_w}{D} = -\frac{1}{A}\int_A [c f_{sw} + (1 - c) f_{Lw}]\, dA \tag{2–20}$$

We observe that τ_w is a positive quantity because the forces f_{sw} and f_{Lw} are negative, opposing motion of the phases past the boundary.

2.3 MULTIPARTICLE DRAG RELATIONSHIPS

In dilute suspensions, homogeneity is difficult to achieve and clusters of particles form. These clusters behave rather like large particles and settle as much as two or three times as rapidly as the individual particles. As the concentration of a slurry increases beyond 5% to 10% by volume, the crowding effect of surrounding particles upon the drag force becomes progressively stronger. This requires us to modify the drag force equation for a particle, Equation 1–14. A reasonable estimate of the magnitude of the concentration effect can be made with theoretical models which visualize a particle embedded in a cage or a porous medium. However, the experimental results of Richardson and Zaki (1954) as generalized by Rowe (see Wallis, 1969) are most useful. For single-size spheres in a steady flow, the drag force is

$$\mathbf{F}_D = \frac{C_{Ds}\rho_L(\mathbf{v}_L - \mathbf{v}_s)|v_L - v_s|A_p}{2(1 - c)^{1.7}}$$

The corresponding force per unit volume for one-dimensional flows is

$$f_{sL} = \frac{3C_{Ds}\rho_L(v_L - v_s)|v_L - v_s|}{4d(1 - c)^{1.7}} \tag{2–21}$$

The drag coefficient is given by

$$C_{Ds} = \frac{24}{\text{Re}_s} \qquad \text{for } \text{Re}_s = \frac{d(1-c)|v_L - v_s|\rho_L}{\mu_L} < 0.2$$

$$C_{Ds} = \frac{24}{\text{Re}_s(1 + 0.15\,\text{Re}_s^{0.687})} \qquad \text{for } 0.2 < \text{Re}_s < 1000 \tag{2–22}$$

$$C_{Ds} = 0.44 \qquad \text{for } 1000 < \text{Re}_s$$

Equation 2–22 reduces to Equation 1–15 at infinite dilution. It can be used for all concentrations lower than the close-packed case. At high concentrations, an alternative expression is provided by the Ergun (1952) equation. Although it was originally intended for packed bed flows where $v_s = 0$, we generalize the original expression by reasoning that it is the velocity of the fluid relative to the particles which determines the drag force. Our expression for the force of the fluid on the particles is then expressed in terms of an interfacial friction factor C_{Lb}:

$$f_{sL} = \frac{C_{Lb}\rho_L(v_L - v_s)|v_L - v_s|}{d} \tag{2–23}$$

where

$$C_{Lb} = 1.75 + \frac{150\mu_L c}{d\rho_L|v_L - v_s|(1-c)} \tag{2–24}$$

If an effective mean particle diameter d cannot be defined for a multiphase mixture, it is necessary to use Darcy's law, generalized with the same reasoning:

$$f_{Ls} = -\frac{\mu_L(v_L - v_s)(1-c)}{k} \tag{2–25}$$

Equation 2–25 defines the permeability k of the medium (m^2). The negative sign results from the fact that the equation has been written with the same velocity difference as the expressions for f_{sL}.

2.4 FORCES IN TRANSIENT FLOWS

The fluid inertia effects shown in the previous chapter are altered in mixtures. Zuber (1964) examined the effect of concentration on the forces in the laminar regime. His expressions for the incremental contributions to f_{Ls} from added mass and Basset force effects are, for axes fixed with respect to a particle,

$$\text{added mass: } C_s \rho_L \left(\frac{1 + 2c}{1 - c}\right) \frac{d(v_L - v_s)}{dt} \tag{2–26}$$

$$\text{Basset force: } 1.5\, d^2 \left(\frac{\pi \rho_L \mu_L}{(1 - c)^{2.5}}\right)^{0.5} \int_0^t \left(\frac{d(v_L - v_s)}{dt^{\cdot}}\right)(t - t^{\cdot})^{-0.5}\, dt^{\cdot} \tag{2–27}$$

2.5 SETTLING OF MONODISPERSE SUSPENSIONS

Here we consider the ideal case of particles which are of a single size and do not aggregate. Although the situation is idealized, the concepts involved have been shown to apply to multispecies dispersions containing particles of various sizes and densities. Figure 2–3 shows a vertical cylinder in which the particles are settling with velocity v_s.

We assume that the concentration in the settling region is independent of time and distance. Equations 2–11 and 2–12 then show that v_s and v_L must also be constant in this region so that the inertial forces in the momentum equations can be neglected. Comparing terms in the momentum equations, we see that unless the concentration c is very low or the particles are very large, the interfacial drag forces are likely to be very much greater than the wall-produced forces. We, therefore, neglect f_{sw} and f_{Lw}.

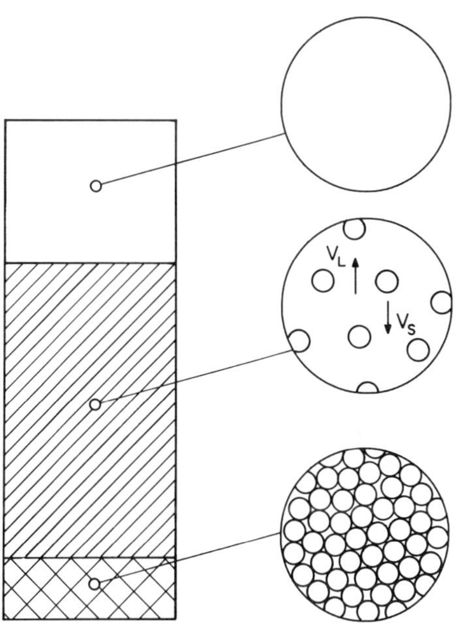

Figure 2–3. Settling of a slurry.

The momentum equations then reduce to a balance of gravity, pressure, and drag forces. Equations 2–13, 2–14, and 2–18 can be combined to give a general expression for steady one-dimensional flows with no wall-produced forces (f_{ss}, f_{sw}, f_{LL}, f_{Lw} all zero). If the upward direction corresponds to increasing values of x, then

$$\frac{f_{sL}}{1 - c} = (\rho_s - \rho_L)g \tag{2–28}$$

To apply this to a settling slurry, we use the fact that the net volumetric flow through any plane must be zero:

$$cv_s + (1 - c)v_L = 0 \tag{2–29}$$

Using Equations 2–21, 2–22, and 2–29, we can substitute for f_{sL} in terms of v_s. If Re_s is less than 0.2, we obtain

$$-\frac{v_s}{V_\infty} = (1 - c)^{4.7} \tag{2–30}$$

whereas if Re_s is greater than 1000, we find

$$-\frac{v_s}{V_\infty} = (1 - c)^{2.35} \tag{2–31}$$

V_∞ is the terminal velocity of a single particle, calculated from Equation 1–17. v_s is negative because the x-direction was arbitrarily chosen to be positive upward.

We note that the effect of buoyancy arose naturally when we obtained Equation 2–28 by eliminating the pressure gradient between the momentum equations. If the pressure gradient in the settling suspension is calculated from the above equations, we find that the density of the mixture determines the pressure gradient because the particle weight is balanced by the drag force:

$$-\frac{dP}{dx} = [\rho_s c + \rho_L(1 - c)]g \tag{2–32}$$

The high exponent of the $(1 - c)$ term in Equation 2–30 reflects the fact that the crowding of a particle by its neighbors (increasing the value of $v_L - v_s$ at a given falling velocity) increases both surface shear stresses and pressure drag on a particle.

The exponent on the $(1 - c)$ term may be different for nonspherical particles than for spheres. In the Stokes' law region, the value of 4.7 in Equation 2–30 has been found to be slightly higher (about 5.3) for sand particles. With coarse and reasonably isometric particles, the exponent in Equation 2–31 is similar to that for spheres. In recognition of this uncertainty, we regard the exponent 1.7 in Equation 2–21 as subject to some uncertainty and represent it by the symbol n.

Equations 2–30 and 2–31 clearly fail when the concentration reaches that of a packed or fully settled bed. In a settled deposit, the wall-produced interparticle force f_{sw} is important so that one of the assumptions used in deriving Equations 2–30 and 2–31 is invalid. The condition where f_{sw} is significant may not arise abruptly as c increases toward its limiting value so that Equations 2–30 and 2–31 must be used with caution at high concentrations.

2.6 FLOCCULATED SLURRIES

The tendency of particles to form agglomerates increases as particle size decreases. We note that a distinction between coagulation and flocculation is often not made so that the aggregates are usually called flocs. An idealized clay floc is illustrated in Figure 2–4. Since the agglomerates include a substantial quantity of fluid, this causes the maximum concentration in a sediment to decrease. For very fine particles, the settling velocities in dilute slurries can be correlated with concentration by

$$-\frac{v_s}{V_{\text{floc}}} = (1 - Kc)^{4.7} \tag{2–33}$$

Kc can be considered to be the effective concentration of the floc. V_{floc} can be interpreted with Stokes' law in terms of a floc diameter, where the floc density is $\rho_L(K + S_s - 1)/K$. K depends upon particle size, shape, and the surface and fluid compositions.

As the concentration increases, flocs interfere with each other during settling and Equation 2–33 eventually fails. Typical values of Kc at which this occurs are greater than 0.35 (Scott, 1984). At high concentrations, channels must develop in an aggregated mass of flocs for fluid to escape upward and permit settling. Gentle stirring can destroy these channels or form new ones by breaking solid–solid linkages. Thus, the settling rate can either increase or decrease with stirring, in contrast with dilute suspensions. For the latter, settling is unaffected by gentle agitation.

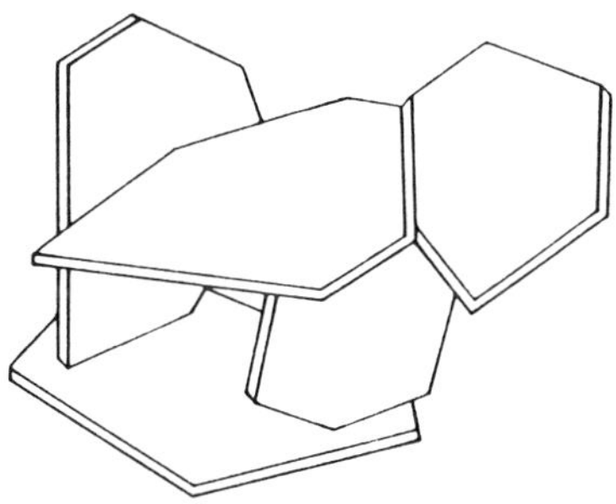

Figure 2–4. An idealized clay floc.

To correlate settling velocity–concentration measurements, one can use Scott's (1984) equation (2–34):

$$-\frac{v_s}{V_{\text{floc}}} = \left[(1 - Kc)^{4.7} + \left(\frac{E}{Kc}\right)\{1 - \exp[0.5(-\lambda + |\lambda|)]\}\right] \quad (2\text{–}34)$$

where

$$\lambda = BK(c - c_d)$$

c_d is the concentration at which deviation from Equation 2–33 begins. E is the maximum deviation (at $Kc \rightarrow 1$) and B is an empirical parameter.

In addition to their importance for particles in slurries left in pipelines at shutdown, settling velocity measurements complement rheological measurements for slurries because they provide quantitative evidence of the extent of flocculation.

2.7 FLUIDIZATION OF MONODISPERSE MIXTURES

By examining the forces acting on the particles, we can calculate the liquid flow rate which is necessary to fluidize a stationary packed bed of particles or to initiate flow in a vertical pipe containing a settled bed. The situation is shown schematically in Figure 2–5. Assuming that flow changes are made gradually, we can use the momentum equations 2–13 and 2–14 (or 2–16 and 2–17) with the inertial terms omitted. As in the previous case, h is measured upward.

When the fluid is stationary, the axial forces f_{Ls} and f_{Lw} are zero. Since both phases are stationary, f_{sL} is also zero. This means that the pressure gradient in the fluid in the packed bed depends only on the fluid density, in contrast with the situation in a settling suspension where the immersed weight of the particles is borne by drag forces.

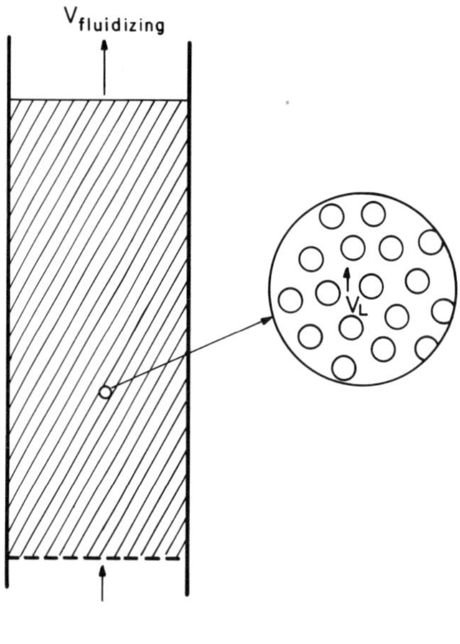

Figure 2–5. Fluidization of a stationary deposit.

Eliminating the pressure gradient, we have, because f_{Lw} is very small,

$$f_{sw} = (\rho_s - \rho_L)g - \frac{f_{sL}}{1 - c} \tag{2–35}$$

The force is positive, decreasing as the fluid velocity and f_{sL} increase. Eventually, a point is reached at which f_{sw} is zero and the condition of Equation 2–28 is established. Further increases in fluid velocity cause the bed to expand, decreasing c and maintaining the condition of Equation 2–28. As long as net particle motion does not occur, V is given by $V = v_L(1 - c)$ with c equal to c_*, the concentration in the settled bed.

At Re_s values less than 0.2, we find the fluidizing velocity is

$$\frac{V_{\text{fluidizing}}}{v_\infty} = (1 - c_*)^{4.7} \tag{2–36}$$

and at $\text{Re}_s > 1000$

$$\frac{V_{\text{fluidizing}}}{v_\infty} = (1 - c_*)^{2.35} \tag{2–37}$$

The pressure gradient in the fluidized bed is given by Equation 2–32. As the bed expands, the pressure gradient falls but the total pressure drop remains nearly constant, balancing the immersed weight of the particles. Since the drag force balances gravity for both fluidization and sedimentation, we have the same drag force and the same *relative velocity*, $v_r = v_L - v_s$ in the two cases, where

$$v_r = \left(\frac{4gd(S_s - 1)}{3C_{Ds}}\right)^{0.5} (1 - c)^{0.5(n+1)} \tag{2–38}$$

and n is the exponent which was 1.7 in Equation 2–21.

At high velocities the bed expansion is not uniform and large variations of concentration occur. A model considering this regime has been proposed by Ding and Gidaspow (1990).

2.8 MULTISPECIES SYSTEMS

The mean density of a mixture ρ_m is calculated from the densities and the volume fractions of c_i of the individual species as

$$\rho_m = \rho_L(1 - C) + \sum \rho_{si} c_i \tag{2–39}$$

where the C is the total solids concentration

$$C = \sum c_i \tag{2–40}$$

In the multispecies mixture, we assume that Equations 2–21 and 2–22 still apply, with the total concentration C replacing c [following the suggestion of Richardson and Meikle (1961)]. This means that for spheres the interfacial drag force per unit volume is

$$f_{sL} = \frac{18\mu_L(v_L - v_s)(1 + 0.15\,\mathrm{Re}_s^{0.687})}{d^2F(1 - C)} \tag{2–41}$$

where $F(1 - C) = (1 - C)^{2.7}$.

When inertia and wall forces are negligible, as in sedimentation or fluidization, the interfacial drag forces balance the pressure and gravitational forces on the components of the mixture. Using the Wallis forms of the mass conservation and momentum equations, Masliyah (1979) obtained a generalized expression for slip velocity of species i in these flows:

$$v_{si} - v_L = -\left(\frac{(gd_i^2/18\mu_L)(\rho_{si} - \rho_m)F(1 - C)}{1 + 0.15\,\mathrm{Re}_{si}^{0.687}}\right) \tag{2–42}$$

where ρ_m is the local mixture density and $F(1 - C)$ is defined above. Equation 2–42 can be used to show that fine particles denser than the fluid can move *upward* in a multispecies mixture undergoing sedimentation. Equation 2–42 clarifies the confusion which has existed concerning the appropriate density to use in calculating the settling velocity of a particle in a mixture. One can use Equation 2–42 and the condition that no net volumetric flow occurs:

$$v_L(1 - C) + \sum c_i v_{si} = 0 \tag{2–43}$$

or one can use the full set of momentum equations for the particles and the fluid. The following example illustrates how these concepts can be used.

Example

A homogeneous mixture of three discrete particle sizes ($d_1 = 16$ mm, $d_2 = 4$ mm, $d_3 = 1$ mm) settles in water at 20°C. The particle densities are all 2650 kg/m^3 and settling tests with the individual particles show that $C_D = 1.0$ for all three sizes. The concentrations c_i are all 10% by volume so that $C = 0.30$. We must find the particle velocities v_{s1}, v_{s2}, and v_{s3} and the liquid velocity v_L.

We can combine Equations 2–16 and 2–17 written for particle species 1, using the simplification of no inertial forces and no wall forces. This yields, if the positive direction is measured upward,

$$f_{sL1} = (\rho_{s1} - \rho_L)g + f_{Ls}$$

Similar equations are obtained for the other two species. Since the net interfacial force is zero,

$$c_1 f_{sL1} + c_2 f_{sL2} + c_3 f_{sL3} + f_{Ls}(1 - C) = 0$$

where C is defined by Equation 2–40. Using the individual expressions for f_{sLi} in this equation provides an equation for f_{Ls}. We note that in this particular case ρ_{si} is a constant, so that

$$f_{Ls} = -(\rho_s - \rho_L)g\sum c_i$$

With this expression for f_{Ls}, our expression for f_{sL1} is

$$f_{sL1} = (\rho_s - \rho_L)g(1 - C)$$

and using Equation 2–21 we obtain a particular form of Equation 2–38 or 2–42:

$$(v_L - v_{s1})|v_L - v_{s1}| = \left(\frac{4gd_1(\rho_s - \rho_L)}{3C_{Ds1}}\right)(1 - C)^{2.7}$$

Analogous equations apply for the other two species. Solving these simultaneously with Equation 2–43, using $C_{Di} = 1.0$ for all three species, we find

$$v_L = 0.0636 \text{ m/s}, \quad v_{s1} = -0.299 \text{ m/s},$$

$$v_{s2} = -0.118 \text{ m/s}, \quad v_{s3} = -0.0272 \text{ m/s}$$

These results are qualitatively reasonable: the fluid moves up and the particles move down. However, the Re_s values should be calculated and used with Figure 1–7 to verify the drag coefficients. Considering the smallest particles, species 3, $\text{Re}_s = (0.001)(999)(0.0636 + 0.0272)(1 - 0.3)/(0.001) = 63.6$.

This is somewhat lower than the value which would occur in single particle settling ($\text{Re}_p = 147$). To obtain better predictions, the drag coefficient should be obtained as a function of Re_s for these particles.

The three particle velocities at infinite dilution are -0.587, -0.294, and -0.147 m/s. Comparing the results for the slurry to these values, we observe that the finest particles are affected most strongly by the upward fluid flow. If these particles were much finer, they would actually move upward in the high concentration slurry.

The preceding calculation has illustrated the role of the conservation of mass relationship using Wallis's approach to evaluate the drag forces. An alternative simple method to predict relative velocities in multispecies mixtures has been given by Patwardhan and Tien (1985).

This example demonstrates what can be expected to occur in a pipeline containing a slurry if the flow is stopped. It also illustrates a problem which can arise when slurries are prepared by adding solids with a wide range of particle sizes to a fluid. Fine particles can be carried upward by the fluid while the coarser ones settle. Agitation of a mixing vessel to disperse these fines is often essential.

With two particle species of low settling velocity, settling is considerably more complicated than the preceding sample calculation would suggest. An initially homogeneous mixture develops fingers or streams in which one species moves through the other (Weiland and McPherson, 1979; Nasr-El-Din et al., 1988; Law et al., 1987).

2.9 PARTICLE–PARTICLE FORCES IN THE MOMENTUM EQUATION

The momentum equation for a single-phase fluid can be written as

$$\rho_L \frac{Dv_L}{Dt} = -\frac{\partial P}{\partial x} - \rho_L g \frac{\partial h}{\partial x} + f_{Lw}$$

If v_L is the instantaneous velocity, f_{Lw} is the viscous force arising because of drag at the wall and is transmitted through the fluid. It is expressed per unit volume of fluid and could also be denoted as f_{LL}. For the x-direction, this force is

$$-(\nabla \cdot \mathbf{T})_x = -\left(\frac{\partial \tau_{xx}}{\partial x} + \frac{\partial \tau_{yx}}{\partial y} + \frac{\partial \tau_{zx}}{\partial z}\right)$$

In the two-phase mixture we have forces on both phases which arise because of the presence of the wall. As we noted in connection with Equations 2–13, 2–14 and 2–16, 2–17, these forces could be split into those transmitted by direct contact with the wall and those transmitted by liquid–liquid (in f_{LL}) and solid–solid (in f_{ss}) interaction. Since we will be considering forces at the pipe wall, we are interested in f_{sw} and f_{Lw}.

We now exclude any inertial contributions to these forces so that we can examine the other effects. We adopt the interpenetrating continuum model for both phases and average the stresses over the whole surface of the elemental volume (N/m^2 system). In this way we compute the forces in terms of the x-components of the effective stresses $\mathbf{T}_s$ and $\mathbf{T}_L$. Then

$$f_{swx} = -\left(\frac{1}{c}\right)\left(\frac{\partial \tau_{sxx}}{\partial x} + \frac{\partial \tau_{syx}}{\partial y} + \frac{\partial \tau_{szx}}{\partial z}\right) \tag{2–44}$$

$$f_{Lwx} = -\left(\frac{1}{1-c}\right)\left(\frac{\partial \tau_{Lxx}}{\partial x} + \frac{\partial \tau_{Lyx}}{\partial y} + \frac{\partial \tau_{Lzx}}{\partial z}\right) \tag{2–45}$$

We observe that the units of f_{sw} are N/m^3 solids (the same as those of the other terms in Equation 2–16) since c has units m^3 solids/m^3 system. The stresses and

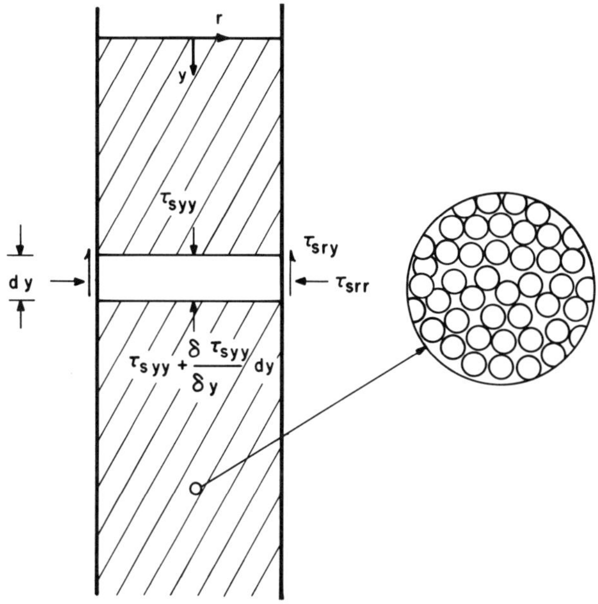

Figure 2–6. A vertical pipe containing a settled slurry.

derivatives are defined in terms of distances measured in the mixture. To proceed further, we require methods for evaluating the stress components in Equations 2–44 and 2–45. These have to be inferred from the results of simple experiments.

Consider a vertical pipe containing stationary liquid and solids as shown in Figure 2–6. For convenience, we will consider the positive y-direction to be downward so that $\partial h/\partial y$ is -1. We recall that a positive stress component acts in a positive coordinate direction on a positive surface or a negative direction on a negative surface, with other combinations giving negative stresses. We also recall that a positive surface is one whose normal in a positive coordinate direction points *into* the phase.

Using the definitions of f_{sw} and f_{Lw} from Equations 2–44 and 2–45, the momentum equations 2–16 and 2–17 in this situation are, for the vertical direction,

$$0 = -\frac{dP}{dy} + \rho_s g + f_{sL} - \left(\frac{1}{c}\right)\left[\frac{\partial \tau_{syy}}{\partial y} + \left(\frac{1}{r}\right)\frac{\partial (r\tau_{sry})}{\partial r}\right]$$

$$0 = -\frac{dP}{dy} + \rho_L g + f_{Ls} - \left(\frac{1}{1-c}\right)\left[\frac{\partial \tau_{Lyy}}{\partial y} + \left(\frac{1}{r}\right)\frac{\partial (r\tau_{Lry})}{\partial r}\right]$$

We have used the condition that both phases are stationary to eliminate inertial forces. f_{sL} and f_{Ls} are also zero, as are all the components of the fluid stress.

Eliminating the pressure gradient, we obtain

$$\frac{\partial \tau_{syy}}{\partial y} + \left(\frac{1}{r}\right)\frac{\partial (r\tau_{sry})}{\partial r} = (\rho_s - \rho_L)gc \tag{2–46}$$

Averaging the stress terms over the cross section,

$$2\pi\int_{r=0}^{D/2} r\left(\frac{\partial \tau_{syy}}{\partial y}\right)dr = \left(\frac{\pi D^2}{4}\right)\frac{\partial \langle \tau_{syy}\rangle}{\partial y}$$

$$2\pi\int_{r=0}^{D/2}\left[\partial\frac{(r\tau_{sry})}{\partial r}\right]dr = 2\pi\int_{r=0}^{D/2} d(r\tau_{sry}) = \pi D(\tau_{sry})_{\text{wall}}$$

The braces on $\langle \tau_{syy}\rangle$ denote an area-averaged stress. At the wall, τ_{sry} and τ_{srr} are related by Coulomb's law through the coefficient of friction η_s.

$$(\tau_{sry})_{\text{wall}} = (\tau_{srr})_{\text{wall}}\,\eta_s \tag{2–47}$$

With these substitutions, Equation 2–46 may be written as an ordinary differential equation:

$$\frac{d\langle \tau_{syy}\rangle}{dy} + \left(\frac{4\eta_s}{D}\right)(\tau_{srr})_{\text{wall}} = (\rho_s - \rho_L)gc$$

This equation can be solved after assuming that the mean axial stress and the wall stress are proportional, viz.,

$$(\tau_{srr})_{\text{wall}} = K\langle \tau_{syy}\rangle \tag{2–48}$$

Solving Equation 2–46 with the condition that $\langle \tau_{syy}\rangle = 0$ at $y = 0$ gives Janssen's equation for axial stress as a function of depth.

$$\langle \tau_{syy}\rangle = \left(\frac{(\rho_s - \rho_L)gcD}{4K\eta_s}\right)\left[1 - \exp\left(-\frac{4K\eta_s y}{D}\right)\right] \tag{2–49}$$

Equation 2–49 is shown in dimensionless form in Figure 2–7. We see that the axial stress (and the radial stress at the wall) approach limiting values asymptotically. With dry particles, the method of filling the container will also affect the stresses.

Equations 2–48 and 2–49 show how the axial and wall stresses in a container of slurry vary with depth when settling is complete. The fully fluidized pressure gradient is given by Equation 2–32. Equation 2–49 shows that the vertical stress increases linearly with depth for shallow depths.

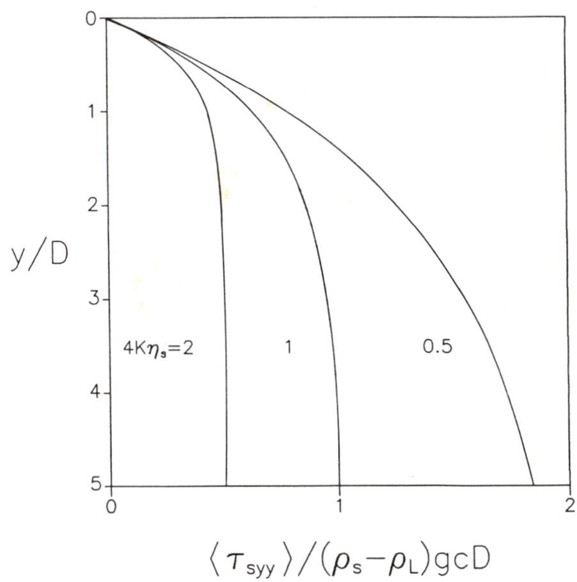

Figure 2–7. Dimensionless mean axial stress as a function of depth in a settled slurry.

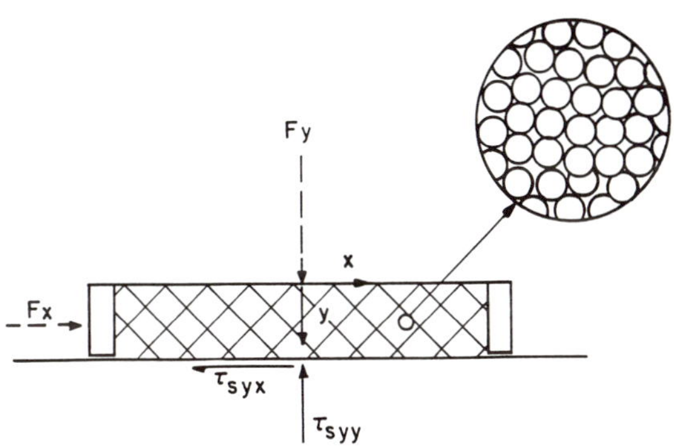

Figure 2–8. A simple direct shear experiment.

2.10 STRESSES IN FLOWING GRANULAR SOLIDS

Equation 2–47 (Coulomb's law) has also been shown to apply to situations in which motion occurs. The simple shear cell of Figure 2–8 is an apparatus which can be used to determine the coefficient of friction

$$\eta_s = \frac{\tau_{syx}}{\tau_{syy}}.$$

This coefficient depends upon the nature of the two surfaces (the particles and the substrate).

If the substrate is replaced by a layer of particles, the ratio of the stresses in the mixture (analogous to K in Equation 2–48, but for incipient motion) is a property of the granular mass. The coefficient of proportionality is denoted as tan ϕ, where ϕ is the angle of internal friction.

$$\tan \phi = \frac{\tau_{syx}}{\tau_{syy}} \tag{2–50}$$

ϕ is the angle at which the granular material will slide over itself when forming a free surface. It is approximately the angle of repose. This can be seen from the force balance for the thin layer of particles on the inclined surface of Figure 2–9. In practice, the fact that surface particles are loosely packed produces some difference between the angle of internal friction and the angle of repose.

When shear takes place within a granular mass of particles, as in the shear cells of Figures 2–8 and 2–10, the relative motion consists of sliding and rolling of particles in contact as well as rotation of aggregates and abrupt slippage between interparticle contacts. If motion is produced by movement of the boundaries, shear often does not take place uniformly within a granular mass, as it does for a liquid. Instead, shear is confined to regions a few particles in width, with the remaining solids moving *en bloc*. When a liquid is present in the pore space the contacts are lubricated, which could alter the stress ratio.

Although flowing slurries usually have solids concentrations considerably lower than those in stationary beds, the direct shear apparatus is of interest because it represents a limiting case in which a coefficient of friction can be measured. If there is no lubrication effect and if the contact angles between particles do not change, the value of η_s between the particles and a wall does not depend on the velocity of displacement. Model calculations which show the effects of changing interparticle contact angles and lubrication are discussed by Roco et al. (1989), Castillo and Williams (1979), and Ma and Roco (1990).

We are also interested in the stress components in mixtures flowing at high shear strain rates. Although there have been innumerable viscometric studies of sheared fluid–particle mixtures, there have been few attempts to measure the normal stresses since Bagnold (1954) discovered their existence. However, considerable theoretical progress has been made. Two regimes of shear occur, distinguished by the value of the Bagnold number B which represents the ratio of inertial to shear forces in the mixture.

$$B = \frac{\rho_s \lambda^{1/2} d^2 \dot{\gamma}_s}{\mu_L} \tag{2–51}$$

$\dot{\gamma}_s$ is the shear rate in the mixture. λ is a linear concentration, the ratio of the particle diameter to the distance between surfaces (the ratio d/b in Figure 2–11).

$$\lambda = \left[\left(\frac{c_{max}}{c} \right)^{1/3} - 1 \right]^{-1} \tag{2–52}$$

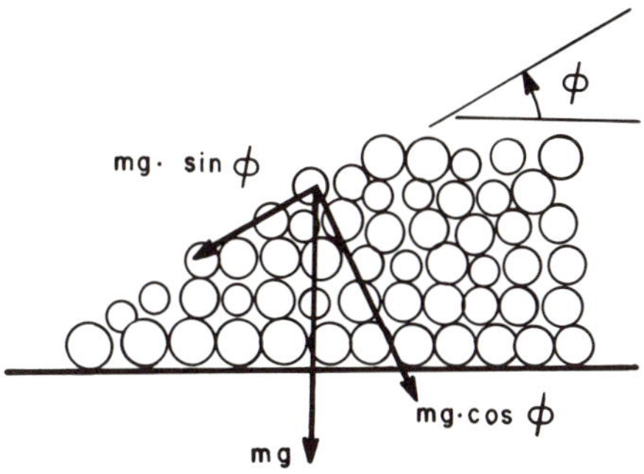

Figure 2–9. The angle of repose of a granular material.

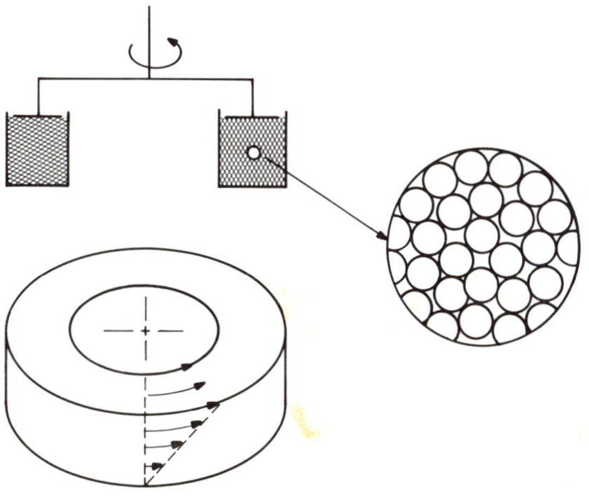

Figure 2–10. A ring shear cell.

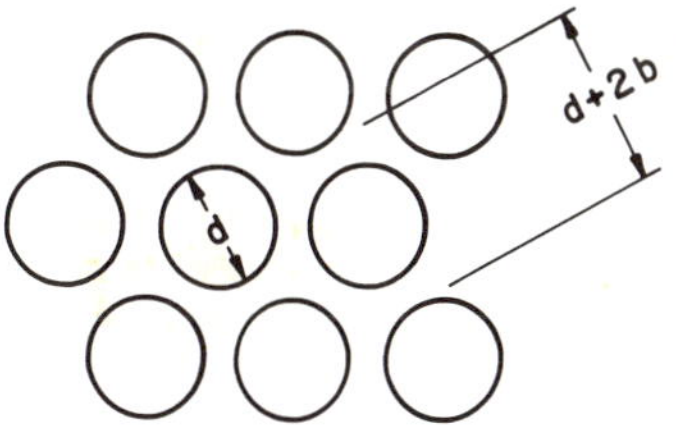

Figure 2–11. Idealized array of particles.

At low values of B (experiments suggest $B < 40$) the shearing process is said to be "macroviscous" and the shear stress varies with the first power of the shear strain rate:

$$\tau_{syx} = \mu_L f(\lambda) \dot{\gamma}_{syx} \tag{2–53}$$

When a slurry is sheared in a viscometer under laminar conditions, the particle stress in Equation 2–53 is a component of the total measured stress. At high B

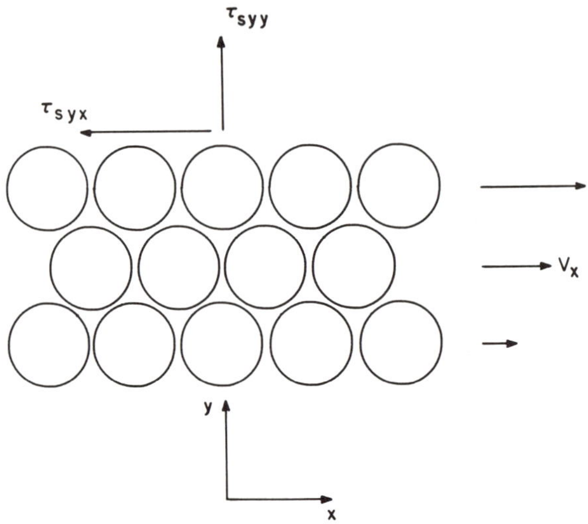

Figure 2–12. Idealized sheared array of particles.

values (above 450), the shear is said to be "inertial" and the relationship of τ to $\dot{\gamma}$ changes:

$$\tau_{syx} = K_1 \rho_s (\lambda d)^2 \dot{\gamma}_{syx} |\dot{\gamma}_{syx}| \tag{2–54}$$

The "inertial" stress arises as a result of collisions between particles moving in layers at different velocities; this stress does not include the effect of particle exchange between layers. The situation is shown schematically in Figure 2–12. The collision process has been examined theoretically and experimentally in a number of subsequent investigations. For inertial shear, the factor K_1 is evidently a function of the coefficient of restitution of the particles and of the fluctuating component of the velocity. The latter may be affected by turbulent fluid motion in pipe flow. Normal stresses occur during shear, and Bagnold suggested that the ratio

$$\left| \frac{\tau_{syx}}{\tau_{syy}} \right| = \tan \alpha \tag{2–55}$$

was constant in each of the two (viscous and inertial shear) regions. α is the angle of internal friction for the sheared particle mass. More recent work (Savage and McKeown, 1983; Jenkins and Savage, 1983; Hanes and Inman, 1985) suggests that in the inertial region, tan α depends on the properties of the particles, the velocity fluctuations, and $\dot{\gamma}_s$. Numerical simulations for dry particles lead to the same conclusions (Walton and Brown, 1986; Campbell and Brennen, 1985) and the model has been extended to dilute suspensions by Richmann (1988).

The dynamics of collisions between elastic spheres have been worked out with the goal of predicting the transport properties (viscosity, thermal conductivity, and diffusivity) in the kinetic theory of fluids (Chapman and Cowling, 1952). In a fluid, the distribution of the sphere (molecule) velocities which cause collisions is related to the temperature. A similar approach is employed in theoretical treatments of granular media flow, although in the latter case the contacts between the spheres are inelastic. As a result, the energy balance equation for particles contains an additional term which represents the energy dissipated by the inelastic collisions.

When the particles are flowing, the total energy per unit mass is the sum of the internal energy U, the translational and rotational kinetic energy of the mean flow, E_{mf}, and the kinetic energy associated with particle velocity fluctuations.

$$E = U + E_{mf} + E_p \tag{2–56}$$

The fluctuation kinetic energy E_p is related to the *pseudotemperature* Θ_p of the aggregate in a manner which is analogous to the relationship between molecular velocities and the thermodynamic temperature.

$$\Theta_p = \frac{\overline{v'^2}}{3} \tag{2–57}$$

where v' is the velocity fluctuation. The velocity fluctuations are produced by collisions between particles. Unlike the collisions which occur between molecules in dense gases at rest, the distribution of collisions is not uniform for the whole surface of a particle: In the arrangement of Figure 2–12, the upstream quadrant of a particle is most likely to receive a collision from a faster moving particle above it.

For a simple flow in the x-direction, Jenkins and Savage (1983) derive for Bagnold's inertial region

$$\begin{aligned} \tau_{syx} &= -0.2\kappa_p(2 + \alpha)\dot{\gamma}_{syx} \\ \tau_{syy} &= \frac{\kappa_p(\pi\Theta_p)^{1/2}}{d} \end{aligned} \tag{2–58}$$

where α is a dimensionless parameter in the collision distribution function and κ_p is the pseudothermal conductivity. This, in turn, depends on the coefficient of elasticity e of the particles.

$$\kappa_p = \frac{\rho_s d(1 + e)(\Theta_p/\pi)^{1/2}c^2(2 - c)}{(1 - c)^3} \tag{2–59}$$

The agreement between Equation 2–59 and the experimental observations is encouraging. Since v' is a function of d and $\dot{\gamma}_{syx}$, Equation 2–58 shows the factors with which tan α in Equation 2–55 is likely to vary in the inertial region. Further discussion of these effects can be found in the work of Savage (1979), Savage and McKeown (1983), and Johnson and Jackson (1987).

2.11 LIQUEFACTION AND COMPACTION

As a result of earthquakes, explosions, or other shock waves imparted to sediments whose pores are completely filled with liquid, the sediments behave like liquids for a short period of time. The momentum equations used in predicting the dynamic response differ only slightly from those we have seen previously. For the liquid phase, assuming no "boundary" forces

$$\rho_L \frac{D[(1 - c)v_{Lx}]}{Dt} = - \frac{\partial[P(1 - c)]}{\partial x} - \rho_L(1 - c)g\frac{\partial h}{\partial x} + (1 - c)f_{Ls} \tag{2–14}$$

and Equation 2–25 may be used to evaluate f_{Ls}. Because the concentration is high, the boundary drag effect is negligible compared to the interfacial drag force.

For the solids, the boundary effect is significant and in this case the momentum equation (Equation 2–13, see also Appendix 2) is

$$\rho_s \frac{D(cv_{sx})}{Dt} = - \frac{\partial(cP)}{\partial x} - c\rho_s g\frac{\partial h}{\partial x} + cf_{sL} - \nabla \cdot (c^a \mathbf{T}_s)_x$$

The tensor $\mathbf{T}_s$ is the effective stress in the saturated granular medium. Stress–strain relationships are required to permit the response to be predicted.

The behavior of a sediment during slow compaction can be analyzed by neglecting inertial forces. For a one-dimensional compaction, neglecting $P(\partial c/\partial x)$, Equations 2–13 and 2–14 yield

$$0 = - \frac{\partial P}{\partial x} - \rho_L g\frac{\partial h}{\partial x} + f_{Ls}$$

$$0 = - \frac{\partial P}{\partial x} - \rho_s g\frac{\partial h}{\partial x} + f_{sL} - \left(\frac{1}{c}\right)\left(\frac{\partial c}{\partial x}\right)\Big/\left(\frac{\partial c}{\partial \tau_{sxx}}\right)$$

$(\partial c/\partial \tau_{sxx})$ is a material property which can be determined from a compression test. Wallis (1969) shows how these relationships can be used to describe the compaction process. A simplified version of this description is the diffusion equation

$$\frac{\partial c}{\partial t} = K\frac{\partial^2 c}{\partial x^2} \tag{2–60}$$

2.12 PRESSURE WAVE PROPAGATION

The velocity of propagation of a pressure wave in a mixture depends on the properties and the concentrations of the phases. The wave velocity is of interest because it affects the performance of measuring devices and the response of a pipeline to imposed flow changes. To illustrate the most important effects, we can rework the relationships which apply to single-phase flows.

For a homogeneous mixture, the velocity of a pressure wave can be derived from the conservation of mass and momentum equations. We first assume that the interfacial drag force f_{sL} is very high so that the relaxation time is very short and the slip fraction $(v_s - v_L)/v_L$ is negligible. This means that no variation in solids concentration occurs. In addition, we assume that the flow velocity V is much less than the velocity of propagation of the disturbance so that $V(\partial/\partial x) \ll \partial/\partial t$ in all cases.

With these simplifications, the area-averaged velocities V_s and V_L are identical to V. Retaining the pipe cross-sectional area as a dependent variable, the mass conservation equations then become (c is identical to C_v)

$$\left(\frac{c}{\rho_s}\right)\frac{\partial \rho_s}{\partial t} + c\frac{\partial V}{\partial x} + \left(\frac{c}{A}\right)\frac{\partial A}{\partial t} = 0$$

$$\left(\frac{1 - c}{\rho_f}\right)\frac{\partial \rho_L}{\partial t} + (1 - c)\frac{\partial V}{\partial x} + \left(\frac{1 - c}{A}\right)\frac{\partial A}{\partial t} = 0$$

These may be combined into a single equation as long as the pressure disturbance is isentropic. The isentropic compressibility B of a phase is the physical property that governs the pressure wave propagation.

$$\left(\frac{1}{\rho}\right)\left(\frac{\Delta\rho}{\Delta P}\right)_s = B \tag{2–61}$$

where the subscript s denotes constant entropy. Using the compressibilities of the individual phases, we can obtain a combined mass conservation equation.

$$\left[CB_s + (1 - c)B_L + \left(\frac{1}{A}\right)\left(\frac{dA}{dP}\right)\right]\frac{\partial P}{\partial t} + \frac{\partial V}{\partial x} = 0 \tag{2–62}$$

The momentum equation for this homogeneous flow is Equation 2–19 which contains the wall friction term which produces damping of the pressure wave. Since wall friction does not affect the velocity of propagation, if we are only interested in the wave velocity we can use the undamped form. Neglecting wall friction and gravity and using the assumption that pressure changes take place rapidly, i.e., $\partial/\partial t \gg V(\partial/\partial x)$, the momentum equation for the mixture is

$$\rho_m\frac{\partial V}{\partial t} + \frac{\partial P}{\partial x} = 0 \tag{2–63}$$

Treating the mean density ρ_m and the bracketed quantity in Equation 2–62 as constants we can eliminate the velocity between these two equations to obtain the one-dimensional wave equation which describes the undamped pressure wave:

$$\frac{\partial^2 P}{\partial t^2} = a^2 \frac{\partial^2 P}{\partial x^2} \tag{2–64}$$

a is the wave velocity, defined by

$$a^2 = \left\{ \rho_m \left[cB_s + (1 - c)B_L + \left(\frac{1}{A}\right)\left(\frac{dA}{dP}\right) \right] \right\}^{-1} \tag{2–65}$$

For gas–liquid mixtures the wave velocity a in the mixture can be much lower than that in either phase when it flows alone. Figure 2–13 compares wave velocities for air (1 atmosphere)–water and limestone–water dispersions in a pipe for which the wall expansion (dA/dP) contribution is negligible.

Since Equation 2–65 has employed the homogeneous fluid approximation, the model can be improved by considering the velocities of the individual phases and the concentration changes which accompany variations in slip velocity. The fact that the pressure change occurs so rapidly is a serious complication because the interfacial drag will include added mass and Basset force effects. Furthermore, if the particles are large enough to give substantial slip velocities, they will also tend to settle in the pipeline so that the flow will not be axisymmetric.

Figure 2–13 shows that a small quantity of gas in a slurry can affect the wave velocity significantly. It is difficult to prepare slurries that are entirely free from dispersed gases and measurements of wave velocity can be used to determine the gas content (Shook and Liebe, 1976). Since the volume of a gas will vary with pressure, its effect upon slurry rheology can be substantial.

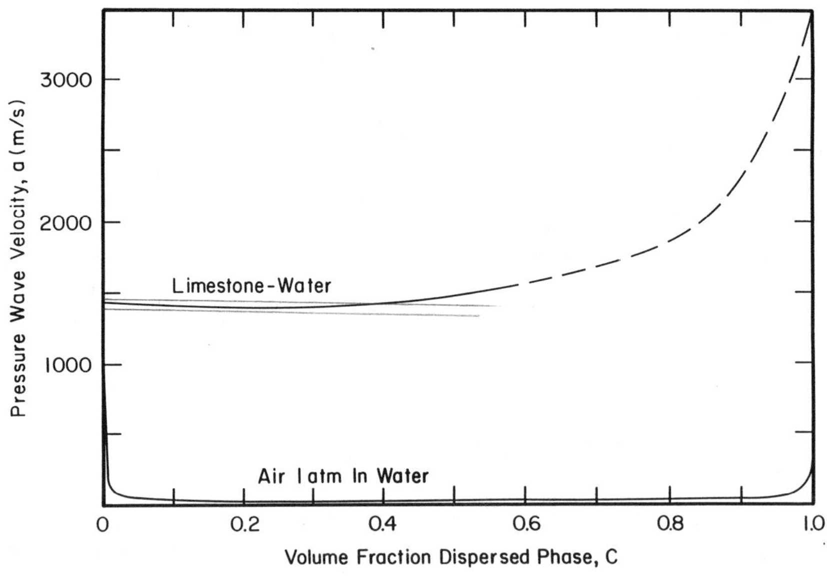

Figure 2–13. Homogeneous model predictions of pressure wave velocities in pipelines.

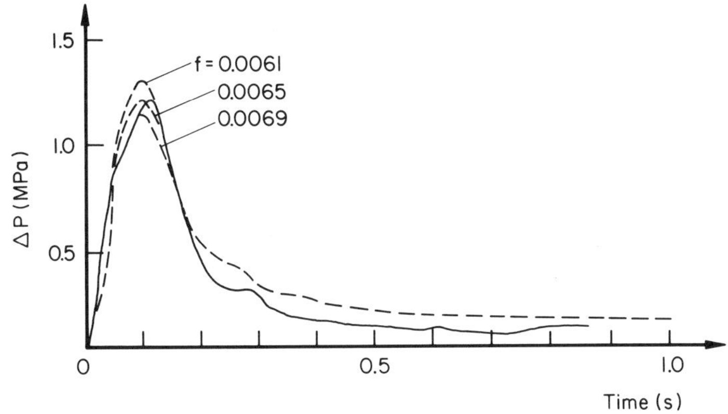

Figure 2–14. Predicted and measured pressure variations for a homogeneous slurry subjected to a flow disturbance. (Data of Hubbard, 1972.)

For a homogeneous mixture in which a velocity decrease ΔV occurs instantaneously, the pressure surge ΔP in the absence of damping or reflection complications is

$$\Delta P/\rho_m = a\Delta V \tag{2–66}$$

If the wall friction and pipe distortion terms in the equations are retained, the methods used to predict pressure changes and flow velocities for single-phase flows can be used for slurries. Figure 2–14 compares predicted and measured pressure variations for flow changes induced in a 50-mm i.d. pipeline of equivalent length approximately 100 m by a valve at the outlet. The pipeline was gravity driven and the velocity was changed from 1.64 to 0.5 m/s in 0.1 s. In these simulations, the full momentum equation (Equation 2–19) was used instead of Equation 2–63. A slurry friction factor was used to calculate the wall shear stress τ_w. The peak pressure predictions are fairly strongly dependent on the damping effect produced by wall friction.

More sophisticated expressions for the velocity of propagation of the pressure wave have been derived (Liou, 1984), including the added-mass contribution to the drag between the phases. For most slurries the difference between the wave velocity in the clear fluid and in the mixture is predicted to be less than 5%, provided the gas volume is low. This fact, and the homogeneous mixture expression for the pressure surge, Equation 2–66, illustrates that of all the properties of the phases (particle diameter, densities, viscosity, etc.) it is the slurry density that exerts the major influence on ΔP (Kao and Wood, 1978). Slip between the phases tends to reduce the pressure surge.

For transient pipe flows which remain laminar, such as those encountered in startup of long pipelines containing non-Newtonian fluids, the momentum and mass conservation equations (A1–1 and A1–6) may be simplified by assuming v_x is the only significant velocity component. Neglecting any slip between the fluid and solids, the equations to be solved are

$$\rho_m \frac{\partial v_x}{\partial t} = -\frac{\partial P}{\partial x} - \rho_m g \frac{\partial h}{\partial x} - \left(\frac{1}{r}\right) \frac{\partial (r\tau_{rx})}{\partial r} \tag{2–67}$$

$$\frac{\partial \rho_m}{\partial t} + \frac{\partial (\rho_m v_x)}{\partial x} = 0 \tag{2–68}$$

Using Equation 2–65, and evaluating dA/dP from the expansion of the pipe, Equation 2–68 is converted to an expression involving the pressure:

$$\rho_m \frac{\partial v_x}{\partial x} + \left(\frac{1}{a^2}\right) \frac{\partial P}{\partial t} = 0 \tag{2–69}$$

where

$$\left(\frac{1}{\rho_m a^2}\right) = B\left(1 + \frac{D}{EBe}\right) \tag{2–70}$$

B is the combined compressibility of the mixture, E is the modulus of elasticity of the pipe wall, D is the pipe diameter, and e is the wall thickness. Solving Equations 2–67 and 2–69 simultaneously requires the shear stress to be predicted as a function of time.

Chapter 3
Homogeneous Slurries

3.1 HOMOGENEITY

When applied to slurries, the adjective homogeneous has so many disadvantages that one is strongly tempted to use another term. Of course, no slurry is ever truly homogeneous because it consists of distinct phases. Nevertheless, there are situations in which particular slurries can be described satisfactorily by single-phase models. However, the conditions that establish these situations include the properties of the flow as well as those of the mixture. This difficulty has led some workers to use the term *nonsettling* because it indicates that the situation is important, as well as the mixture. This introduces a secondary problem in that it becomes necessary to define nonsettling in a way that allows for the fact that the single-phase equations can often be applied to slurries that show a small but finite tendency to settle when allowed to stand.

Since the term exists in conventional parlance, the safest course is to regard homogeneity as a limiting form of behavior that actual slurries approach. In this asymptotic condition, the slurry can be described by continuum models.

In this chapter, we begin by considering the continuum models that are most likely to prove useful for slurries. We describe the most common devices that are used to obtain data for use with these models. The practical considerations associated with the models and measurements are discussed in Chapter 4.

The rheology of slurries is a field of scientific investigation in its own right. In fact, the scientific effort expended in this aspect of slurry flow probably exceeds that in all of the remaining topics combined. It is often possible to explain experimental measurements quantitatively and it is usually possible to give a qualitative interpretation of them. However, it is not possible at the present time to predict the viscosity–concentration–shear rate relationship for an industrial slurry, except perhaps as a crude estimate. Predictions are reliable only for dilute neutrally buoyant model particles forming Newtonian suspensions. Otherwise measurements are necessary.

3.2 SHEAR IN PIPE FLOW

To determine the rheology of a fluid we normally use a flow which is as simple as possible, a so-called viscometric flow, for which the velocity varies with only one position coordinate. The two viscometric flows which are most important for slurries are pipe flow (Poiseuille flow) and flow between concentric cylinders (Couette flow). Viscometers to produce these flows are illustrated in Figures 3–1 and 3–2. A convenient horizontal tube viscometer has been described by Lazarus and Sive (1984).

For steady pipe flow of a constant-density fluid, the differential equation of motion in cylindrical coordinates is Equation A1–6. In this case, we neglect inertia and assume the only velocity component is v_x.

$$0 = -\frac{dP}{dx} - \rho g \frac{dh}{dx} - \left(\frac{1}{r}\right)\frac{d}{dr}(r\tau_{rx}) \tag{3–1}$$

Since the pressure gradient and the gravity term do not vary with r,

$$\tau_{rx} = C_1 r + C_2/r$$

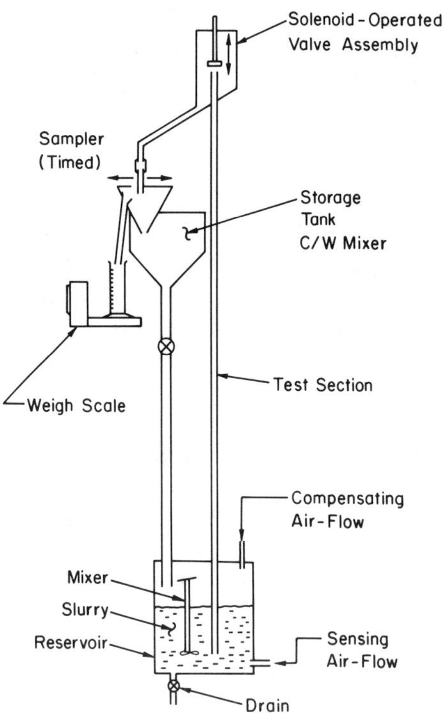

Figure 3–1. A tube viscometer. The upward slurry velocity is maintained constant by the compensating air flow as the slurry level in the reservoir falls.

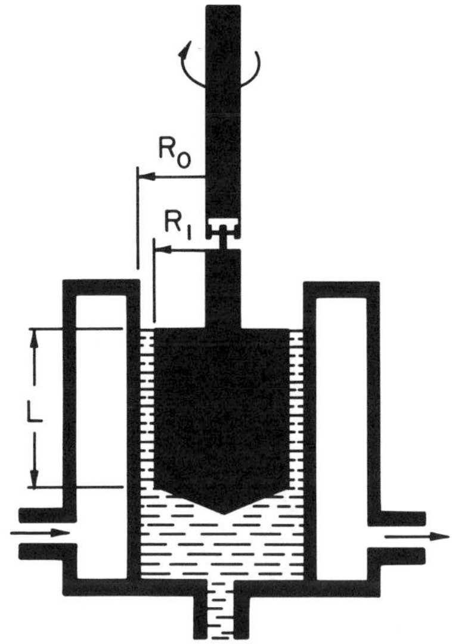

a

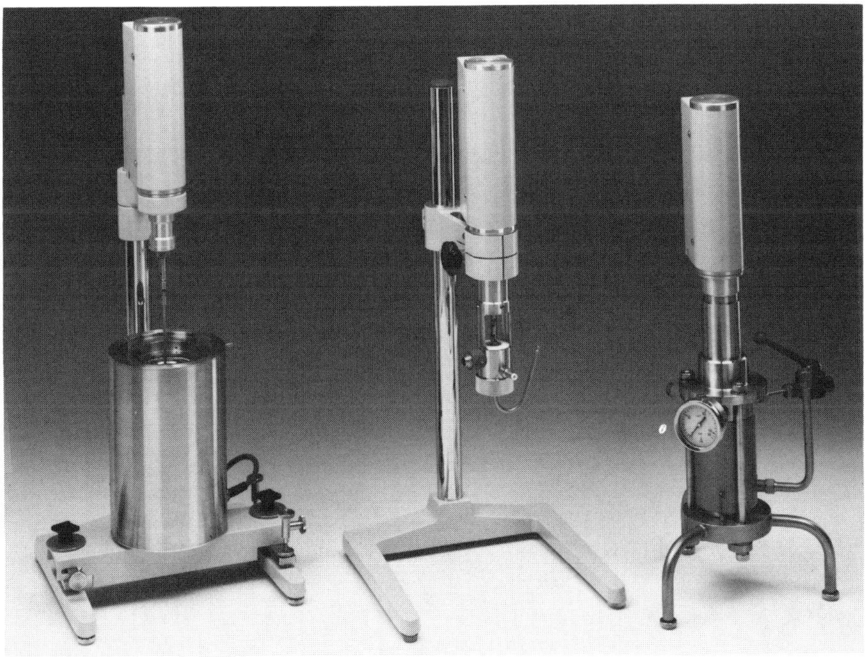

b

Figure 3–2. A Couette viscometer. (Photograph courtesy of Haake Buchler Instruments, Saddle Brook, NJ. Reprinted with permission.)

Using the boundary conditions $\tau_{rx} = 0$ at $r = 0$ and $\tau_{rx} = \tau_w$ at $r = 0.5D$ we obtain the shear stress decay law for a pipe:

$$\frac{\tau_{rx}}{\tau_w} = \frac{2r}{D} \tag{3-2}$$

To use Equation 3–2, we require the relationship between the shear stress components and the local velocity. The simplest relationship is that which applies to Newtonian fluids in laminar flow.

The Newtonian fluid relationship is expressed in terms of the stress and deformation tensors described in Equation A1–9:

$$\mathbf{T} = -\mu\boldsymbol{\Delta} \tag{3-3}$$

Values of μ for water can be calculated from Equation A3–1 in Appendix 3.

For steady laminar pipe flow, τ_{rx} is the only nonzero component of $\mathbf{T}$ and dv_x/dr is the only nonzero component of $\boldsymbol{\Delta}$, i.e., shear causes the velocity v_x to be a function of r. The shearing process in laminar pipe flow is then described by Equation 3–4.

$$\tau_{rx} = -\mu \frac{dv_x}{dr} \tag{3-4}$$

3.3 SHEAR IN AN ANNULUS

The second important viscometric flow for slurries occurs in the annular space between two concentric cylinders when one of the cylinders is rotated and the other is held stationary. The momentum equation for the tangential direction is obtained by neglecting inertia and gravity in Equation A1–5. We assume the only velocity component is tangential, that the flow is axially symmetric and independent of x. This yields

$$r^2\tau_{r\theta} = \text{constant} \tag{3-5}$$

for flow in the space between the cylinders.

For Newtonian fluids in laminar flow, the dominant components of $\mathbf{T}$ and $\boldsymbol{\Delta}$ yield

$$\tau_{r\theta} = -\mu r\left(\frac{d(v_\theta/r)}{dr}\right) \tag{3-6}$$

For non-Newtonian fluids, Equations 3–4 and 3–6 are, of course, inappropriate. To determine the general relationship between $\mathbf{T}$ and $\boldsymbol{\Delta}$, two approaches are possible: One can search for a model that represents experimental measurements satisfactorily; alternatively, one can try to determine the components of **T** and **Δ** directly from the measurements. The latter technique is used less frequently.

3.4 INTEGRATED EQUATIONS FOR VISCOMETRIC FLOWS

For pipe flow, the wall shear stress is obtained from the pressure gradient and the pipe inclination using Equation 1–3. For steady flow and constant density, the inertial and kinetic energy terms in Equation 1–3 are zero. The quantity $(-dP/dx - \rho g\, dh/dx)$ is often written $(i\rho g)$, where i is the frictional headloss with units (m fluid/m pipe length). Less precisely, it is sometimes written as $\Delta P/L$ where ΔP includes the gravitation effect. Thus, we have the important definition for steady pipe flow:

$$\frac{4\tau_w}{D} = i\rho g = \frac{\Delta P}{L} \tag{3–7}$$

The other measured quantity in tube flow is the mean velocity V defined in Equation 1–2. Combining Equations 3–2 and 3–4, integrating, and using the no-slip boundary condition at the wall, $v_x(0.5D) = 0$, we obtain the familiar parabolic velocity distribution of Poiseuille flow:

$$v_x = \left(\frac{\tau_w}{\mu D}\right)\left(\frac{D^2}{4} - r^2\right) \tag{3–8}$$

This expression may be used to obtain V by integration Equation 1–2.

For concentric cylinders, the measured variables are the angular velocity (ω) of one of the cylinders and the torque per unit height T. The latter is usually measured at the inner cylinder. Table 3–1 gives the relationship between the measured quantities for viscometric tube flow of Newtonian fluids.

Of the non-Newtonian models, the Bingham fluid is probably the most useful two-parameter relationship for slurries. According to this model, the shear stress must exceed a limiting value, the yield stress τ_y, for flow to occur at all. For tube flow, the expression corresponding to Equation 3–4 is

$$\dot{\gamma}_{rx} = -\frac{dv_x}{dr} = \frac{\tau_{rx} - \tau_y}{\mu_p} \tag{3–9a}$$

where μ_p, the plastic viscosity, is the asymptotic viscosity at infinite applied shear stresses. For concentric cylinders the corresponding expression is

$$\dot{\gamma}_{r\theta} = -r\frac{d(v_\theta/r)}{dr} = \frac{\tau_{r\theta} - \tau_y}{\mu_p} \tag{3–9b}$$

Table 3–1. Viscometric Flows

	Geometry		
	Tube	Concentric Cylinders	
		Inner Rotating	Outer Rotating
Measured variables	$\Delta P, V$	T,ω	T,ω
	$(\tau_w = D\Delta P/4L)$		
Calculated quantity	$8V/D$	ω	
Newtonian	τ_w/μ	$(T/4\pi\mu)(1/R_1^2 - 1/R_2^2)$	
Valid where	$D V\rho/\mu \leq 2100$	$^{*}R_\omega \leq 45(R_m/\delta)^{0.5}$	$^{**}R_\omega \leq 1.6 \times 10^4 (\delta/R_m)^2$
			$0.2 > \delta/R_m > 0.09$
		where $R_\omega = R_m\omega\delta\rho/\mu$; $R_m = 0.5(R_2 + R_1)$	
		$\delta = (R_2 - R_1)$	
Bingham	$(1 - 4\tau_y/3\tau_w + \tau_y^4/3\tau_w^4)\tau_w/\mu_p$	$(T/4\pi\mu_p)(1/R_1^2 - 1/R_2^2) - (\tau_y/\mu_p)\ln(R_2/R_1)$	
Power law	$[4n/(3n+1)](\tau_w/K)^{1/n}$	$(n/2)(T/2\pi K)^{1/n}(1/R_1^{2/n} - 1/R_2^{2/n})$	
Casson tube	$[1 - (16/7)(\tau_0/\tau_w)^{1/2} + (4/3)\tau_0/\tau_w - (1/21)(\tau_0/\tau_w)^4]\tau_w/\mu_\infty$		
Casson concentric cylinders		$\{(T/2\pi)(1/R_1^2 - 1/R_2^2) - 4\tau_0^{1/2}(T/2\pi)^{1/2}[(1/R_1) - (1/R_2)] + \tau_0 \ln(R_2^2/R_1^2)\}/2\mu_\infty$	

* Taylor (1936).

** Nakabayashi et al. (1982).

Somewhat less frequently used is the power law model. For tube flow,

$$\dot{\gamma}_{rx} = -\frac{dv_x}{dr} = \left(\frac{\tau_{rx}}{K}\right)^{1/n} \tag{3–10a}$$

or for concentric cylinders

$$\dot{\gamma}_{r\theta} = -r\frac{d(v_\theta/r)}{dr} = \left(\frac{\tau_{r\theta}}{K}\right)^{1/n} \tag{3–10b}$$

The Casson (1959) model is also useful for slurries. For tube flow,

$$\dot{\gamma}_{rx} = -\frac{dv_x}{dr} = \frac{(\tau_{rx}^{1/2} - \tau_0^{1/2})^2}{\mu_\infty} \tag{3–11a}$$

and for concentric cylinders,

$$\dot{\gamma}_{r\theta} = -r\frac{d(v_\theta/r)}{dr} = \frac{(\tau_{r\theta}^{1/2} - \tau_0^{1/2})^2}{\mu_\infty} \tag{3–11b}$$

There are innumerable other models and Table 3–1 has been restricted to the simplest two-parameter expressions. It must be emphasized that the models are essentially correlating devices for experimental measurements and should not be used outside the range of conditions for which their parameters are determined. For slurries, the rheological parameters depend on process variables, such as solids concentration, particle size or fluid ionic content, that influence the degree of flocculation as discussed in Chapter 1. A further compilation of models is given by Skelland (1967).

To derive expressions linking the parameters in a model to the measured quantities, the most convenient procedure is to write the fully developed shear flow relationships with stress (τ) rather than position (r) as the independent variable. For tube flow, the appropriate equation for fluids without yield stresses is

$$8V/D = \left(\frac{4}{\tau_w^3}\right)\int_0^{\tau_w} \tau^2\dot{\gamma}\, d\tau \tag{3–12}$$

For concentric cylinders, the corresponding expression is

$$\omega = \int_{\tau_2}^{\tau_1} \left(\frac{\dot{\gamma}}{2\tau}\right) d\tau \tag{3–13}$$

When using Equation 3–13 for a fluid with a yield stress, the lowest stress in the viscometer must be calculated to see if the yield criterion is exceeded. This may or may not occur. If τ_2 is less than the fluid yield stress, τ_y is used as the lower limit of integration in Equation 3–13. For tube flow, Equation 3–2 shows that there is always a region of low shear stress near the pipe axis where τ_{rx} is less than yield stress τ_y or τ_0. In this region the shear rate $\dot{\gamma}$ must be zero. Thus, the lower limit of the integration for a fluid with a yield stress is altered to τ_y or τ_0 in Equation 3–12.

For the models to apply, the flows must be laminar and one dimensional. The criteria to ensure laminar flow in these viscometers are therefore very important. For tube flows, it has been suggested (Ryan and Johnson, 1959; Hanks, 1963) that stable laminar flow requires

$$Z = \left(\frac{D\rho}{\tau_w}\right) v_x \left|\frac{\partial v_x}{\partial r}\right| < 1616 \tag{3–14}$$

everywhere in the pipe cross section. The transition to turbulent flow for non-Newtonian pipe flow is discussed in Chapter 4.

The transition criterion for a Newtonian fluid in an annulus in which the inner cylinder rotates has been obtained theoretically and experimentally (Taylor, 1936). When the outer cylinder rotates, the criterion is less definite (Nakabayashi et al., 1982).

Figure 3–3 illustrates some of the relationships of Table 3–1. The Bingham and power law models may be regarded as special cases of the three-parameter yield power law model given in Equation A1–11 of Appendix 1. The integrated form of the equation for pipe flow of a yield power law fluid is

$$\frac{8V}{D} = 4\left(\frac{\tau_w}{K}\right)^{1/n}\left[\left(\frac{\xi^2}{a}\right)(1-\xi)^a + \left(\frac{2\xi}{b}\right)(1-\xi)^b + \left(\frac{1}{c}\right)(1-\xi)^c\right] \tag{3–15}$$

where

$$\xi = \frac{\tau_y}{\tau_w}, \quad a = 1 + \left(\frac{1}{n}\right), \quad b = 2 + \left(\frac{1}{n}\right), \quad c = 3 + \left(\frac{1}{n}\right)$$

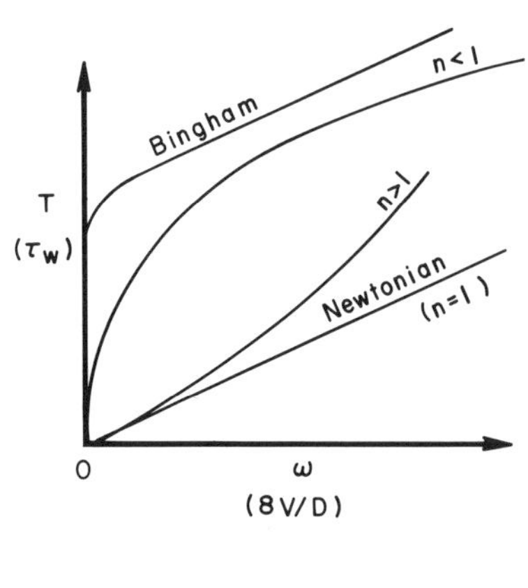

Figure 3–3. Idealized viscometric flow results.

An implicit restriction when using any empirical model is that the shear stresses (or shear rates) should be close to those for which the parameters in the model were determined, unless some justification is available for extending the range beyond the test conditions.

Table 3–1 presents the relationships with which experimental measurements of V and ΔP or T and ω may be interpreted.

If a Couette viscometer is used to obtain the $\mathbf{T} - \mathbf{\Delta}$ relationship, the component $\dot{\gamma}_{r\theta}$ of $-\mathbf{\Delta}$ at $r = R_1$ is calculated from the equation of Krieger (1968):

$$\dot{\gamma}_{r\theta} = [2\omega\alpha^{2/m}/m(\alpha^{2/m} - 1)][1 + m^2 m' f(2 \ln \alpha/m)] \qquad (3\text{–}16)$$

where

$$\alpha = \frac{R_2}{R_1}$$

$$m = \frac{d \log T}{d \log \omega}$$

$$m' = -\left(\frac{1}{m}\right)\frac{d \log m}{d \log T}$$

and the function f is

$$f(t) = \frac{t[e^t(t - 2) + t + 2]}{2(e^t - 1)^2}$$

The shear stress at the radius of the inner cylinder (τ_1 in Equation 3–13) is obtained from the measured torque T per unit height of the cylinder as

$$\tau_{r\theta} = \frac{T}{2\pi R_1^2} \qquad (3\text{–}17)$$

Although viscometer manufacturers often supply "coefficients" which purport to relate ω to $\dot{\gamma}_{r\theta}$, these are valid only for Newtonian fluids or for viscometers with extremely narrow gaps. The latter are, of course, unsuitable for many slurries.

3.5 NEWTONIAN SLURRIES

Slurries that settle slowly can be tested viscometrically. Low settling velocities can be achieved with very fine particles and/or high solids concentrations, low relative densities, and/or high fluid viscosities. If the particles are comparatively large, slurries are often Newtonian. The viscosity of the suspension is often reported as the relative viscosity μ_r, the viscosity of the mixture normalized with that of the suspending fluid.

$$\mu_r = \frac{\mu_m}{\mu_L} \tag{3–18}$$

μ_r varies with a comparatively large number of parameters. These include (not necessarily in order of importance):

1. the volume fraction of the dispersed phase c,
2. particle size and size distribution,
3. particle shape,
4. surface properties: potential or charge and Hamaker constant,
5. fluid electrolyte concentration and ionic charge,
6. temperature.

For non-Newtonian fluids, shear strain rate and, possibly, shear history are also important.

A theoretical equation relating μ_r to c was derived for very dilute suspensions of spheres by Einstein in 1906.

$$\mu_r = 1 + 2.5\,c \tag{3–19}$$

The derivation can be used to help define the conditions in which suspensions are Newtonian. It assumes, among other things, that the particles are uniform and large enough for electrokinetic effects to be insignificant. However, the particles are assumed to be small compared to the size of the domain of flow.

To extend the relationship for μ_r to higher concentrations, further simplifications are required. In a typical model, the particle is surrounded by a shell of fluid with "slurry average" boundary conditions applying at the outer surface of the shell. These model calculations suggest suitable functional forms for the μ_r versus c relationship. Useful two-parameter relationships are

$$\mu_r = \frac{\exp(Bc)}{1 - c/c_{\max}} \tag{3–20}$$

and

$$\mu_r = \left(1 - \frac{c}{c_{\max}}\right)^{-Bc_{\max}} \tag{3–21}$$

The theoretical and experimental origins of these two equations are described by Krieger (1972).

The limiting value of $(\mu_r - 1)/c$ at low concentrations, which defines the low concentration behavior of a slurry, is called the *intrinsic viscosity*. It is often represented by $[\mu]$. For spheres, Einstein's equation suggests it should be 2.5.

Both Equations 3–20 and 3–21 reflect the observation that μ_r rises very rapidly as c approaches its maximum value. The parameter c_{max} is best determined from experimental μ_r versus c measurements.

Two other relationships which may be useful are Eilers's equation

$$\mu_r = \left[1 + \left(\frac{0.5Bc}{1 - c/c_{max}}\right)\right]^2 \tag{3–22}$$

and the three-parameter expression of Vocadlo (1976),

$$\mu_r = \frac{\exp[(B - n/c_{max})c]}{(1 - c/c_{max})^n} \tag{3–23}$$

The two-parameter relative viscosity expressions are compared in Figure 3–4, taking B as 2.5 and c_{max} as 0.63. D.G. Thomas's (1965) equation (Equation 3–24) is also shown since this is a useful average of results from a number of experimental studies using deflocculated monosize spheres:

$$\mu_r = 1 + 2.5c + 10.05c^2 + 0.00273\exp(16.6c) \tag{3–24}$$

Many other equations relating μ_r to c have been proposed, including several with theoretical origins (Frankel and Acrivos, 1967; Sengun and Probstein, 1989). If laboratory data for a range of concentrations are to be fitted, one of those shown earlier will probably suffice. Of course, since concentration is not the only relevant parameter, there is no *a priori* "best" functional form of the relationship.

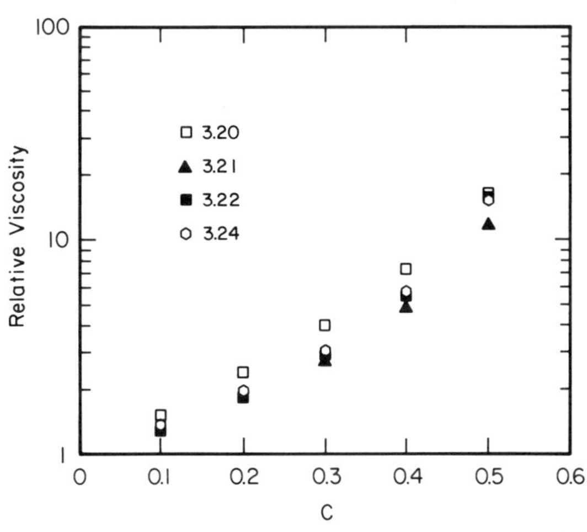

Figure 3–4. Comparison of relative viscosity equations for $B = 2.5$, $c_{max} = 0.63$.

At low concentrations, the slurry viscosity increase is primarily an excluded volume effect. That is, the particles represent regions where the continuous deformation of shear cannot occur. A nominal rate of shear is established in the form of an observable relative motion of the mixture relative to the boundaries. Because of the regions occupied by the particles, the effective shear rate in the fluid itself is increased and this increase produces higher shear stresses at the boundaries. The increase in the boundary shear stresses is interpreted as an increase in "viscosity" at the nominal shear rate.

It will be recalled that the excluded volume effect also influenced the drag force experienced by a sedimenting particle. It has been suggested (e.g., Nir and Acrivos, 1989) that $|v_s|/V_\infty$ is approximately $(1 - c)/\mu_r$ for fine particles. This relationship is derived by assuming that a particle settling in a slurry experiences a buoyant force dependent on the slurry density and is resisted by a viscous force proportional to the slurry viscosity μ_m. Further discussion of this concept is given by Happel and Brenner (1973).

3.6 DISTRIBUTION EFFECTS

Since a range of particle diameters will be found in a mixture, the effect of the breadth of the size distribution is of interest. Experiments show that as the distribution widens, the relative viscosity tends to fall. This is consistent with Equations 3–20 to 3–23 since the maximum concentration $c_{\max}$ increases as the size distribution broadens. The viscosity change can be explained by representing the actual distribution as a bimodal one in which the volume fractions of the fine and coarse particles are c_1 and c_2, respectively. The finer particles form a pseudo-continuum of viscosity $\mu_L \cdot \mu_r[c_1/(1 - c_2)]$. When the coarse particles are dispersed in this fluid, the combined viscosity is then $\mu_L \mu_r[c_1/(1 - c_2)]\,\mu_r(c_2)$. If the suspension were unimodal, the viscosity would be $\mu_L \mu_r(c_1 + c_2)$. For the functional relationships which apply at high concentrations, $\mu_r[c_1/(1 - c_2)]\,\mu_r(c_2)$ is less than $\mu_r(c_1 + c_2)$. A further discussion of this effect, including non-Newtonian (shear thinning) slurries is given by Probstein and Sengun (1987).

3.7 HIGH SOLIDS CONCENTRATIONS

As the concentration increases, interparticle contacts and lubrication interactions during shear flow become more probable. The higher rate of energy dissipation from these contacts contributes to a more rapid increase of μ_r with c. However, contact is less likely if the particles are very different in diameter (Manley and Mason, 1955) and this contributes to the lower shear resistance of mixtures with broad size distributions, compared to those with narrow ones.

Very high concentrations can probably only be achieved with broad size distributions and comparatively coarse particles (or those in which any flocculation tendency has been eliminated). As the concentration approaches the close-packed limit, the shear mechanism seems to change (Cheng and Richmond, 1978; Cheng,

1984) toward that displayed by soils and powders. This so-called *granulo-viscous* regime of flow is characterized by shear stresses and normal stresses that are weakly dependent on shear rate but depend on the concentration and the dimensions of the particles and the domain of shear.

3.8 PARTICLE SHAPE

Particle shape can be shown to affect slurry viscosity in theoretical calculations using either the dilute suspension simplification or particle and fluid shell model configurations. Model calculations show that the viscosity can increase (Happel and Brenner, 1965) as the shape departs from spherical. Indeed, for deflocculated dilute suspensions, the intrinsic viscosity defined in Section 3.5 can be considered to be an indication of particle shape. These model calculations also show that the effect of orientation on shear resistance is very strong.

To illustrate the particle shape effect, consider prolate spheroids of (major axis/minor axis) ratio 10:1. Such particles are fair approximations of cylinders or needle-shaped crystals. Depending upon orientation, the intrinsic viscosity varies between 2.01 and 4.49. For oblate spheroids with the same axis ratio, the corresponding limits of $[\mu]$ are 2.12 and 9.96 (Jeffrey, 1922). With comparatively large particles, orientation is determined only by the flow and there is a tendency toward orientations which give comparatively low flow resistance. However, as the size decreases, Brownian motion and electroviscous effects begin to become important and agglomeration becomes more likely. Brownian motion randomizes particle orientations with respect to the shear field, preventing particles from adopting orientations that minimize energy dissipation. Thus, the viscous resistance to shear of fine-particle slurries is increased by Brownian motion. Brownian motion can also contribute to non-Newtonian behavior.

3.9 ELECTROVISCOUS AND SURFACE EFFECTS

There are at least two electroviscous contributions to the shear resistance of slurries. One effect results from the distortion of the charge cloud around the particles during flow. Booth's (1950) calculation showed that this effect tended to increase the intrinsic viscosity. In addition, at higher concentrations, the charge clouds interact as the particles move past each other during shear. This interaction also produces an increase in shear resistance.

The importance of particle agglomeration in slurry rheology is well established. Although a distinction between coagulation and flocculation should probably be made, both agglomeration processes are often called flocculation. Since the aggregation tendency of fine particles is strongly influenced by surface charge effects, a number of factors influence shear resistance, including the Hamaker constant of the surfaces, the charge density or zeta potential, and the double-layer thicknesses. The latter depends on the ionic strength of the continuous phase as well as its dielectric constant. Figure 3–5 shows the effect of changes in zeta

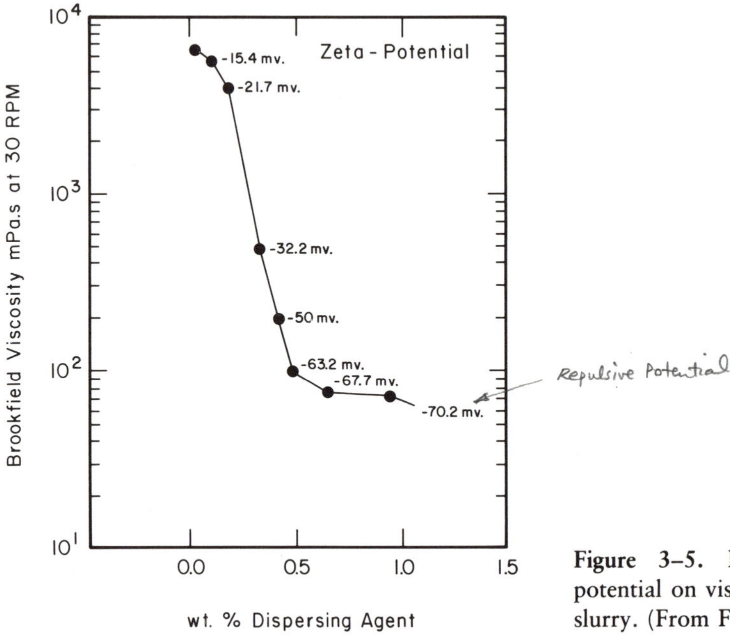

Figure 3–5. Effect of zeta potential on viscosity of a coal slurry. (From Funk, 1981.)

potential upon slurry viscosity for a coal slurry (Funk, 1981). As the repulsive force between particles increases, aggregation becomes less likely. The shear resistance then falls because the intrinsic viscosity is lower for the dispersed particle or because fluid is no longer immobilized in the aggregates. Slurries of fully dispersed particles are usually Newtonian until the solids concentration becomes high enough to reach the granulo-viscous region, or until Bagnold's inertial contact mechanism becomes important.

In a sheared slurry, the aggregate size represents an equilibrium between the association process, dependent on the attractive and repulsive forces between particles, and dispersion or separation by shear. If the aggregate size decreases with increasing shear rate, shear thinning occurs, in which the effective viscosity η decreases as Δ increases. Thus, low zeta potentials are often associated with shear thinning. However, particles with strong mutual repulsive forces can also form shear thinning suspensions. In this case flocculation in the secondary minimum of the potential energy versus separation diagram (Figure 1–12) is considered to produce the association which is destroyed by shear.

3.10 YIELD STRESSES

The strength of the floc structure is usually considered to be the source of the yield stress in Bingham fluids. Additives, such as polyvalent electrolytes, which alter surface charge or double-layer thickness, have profound effects upon yield stresses for slurries (Horsley, 1982). Kao et al. (1975) formed particle agglomerates using an immiscible bridging liquid and showed qualitatively how the non-Newtonian character of the slurries was related to agglomeration. Particle shape is known to determine the yield stresses of waxy crude petroleum slurries. If elongated needle-like crystals form, their interparticle linkages can produce structures of considerable strength. Additives which reduce the surface area by forming larger, more nearly isometric, crystals cause substantial reductions in yield stresses (Price, 1971).

Flocculation effects were first examined for clay–water mixtures, primarily because of their importance in determining the properties of soils (Lambe and Whitman, 1969; Migniot, 1968). Typical clay particles are lenticular (plates) with their large surfaces exposing sheets of atoms available for interaction with the fluid and formation of electrical double layers. Conditions at the edges of the plates are different than those on the larger surfaces and electrostatic forces between the edges and the surfaces of particles allow flocs with a house of cards structure to form. Such a structure resists shear distortion and this resistance can explain the existence of yield stresses in flocculated slurries. It can also explain thixotropy since time and previous shear history would affect the structure.

In coal slurries, the particles are more nearly isometric and the aggregates form rings and chains. However, it is the strength of the particle–particle bonds in the aggregates which are assumed to account for yield stresses in these slurries.

Particle size and concentration are, of course, extremely important since these determine the total amount of surface available for floc formation. D.G. Thomas (1963) found that the yield stress varied approximately as c^3/d^2 for his slurries, which had particle diameters between 0.4 and 17 μm.

Illustrations of these effects in industrial slurries are given in Figures 3–6 and 3–7. These show plastic viscosities and yield stresses as functions of concentration for slurries (fine coal and limestone) with comparatively coarse d_{50} values but which were homogeneous. The coal particles were all finer than 300 μm, with 33% (mass) below 44 μm. The limestone was finer than 200 μm with 67% below 44 μm.

For the coal, the broad size distribution opposes any effect of the nonspherical particle shape so that below about 35% solids by volume, where yield stresses are low, relative viscosities are fairly close to the predictions of expressions such as Equation 3–6. For the limestone, surface effects are evidently much stronger because the relative viscosities and the yield stresses are much higher at similar volume fractions.

Flocculation is known to be influenced strongly by the ionic content of the fluid phase and there have been many efforts to influence or control slurry rheology in this way. For negatively charged particles, the Schulze–Hardy rule predicts that the effectiveness of cations in inducing flocculation should increase strongly with increasing charge. Sadler and Bethany (1987) showed that they could manipulate the viscosity of slurries of hydrophilic lignite particles dramatically by vary-

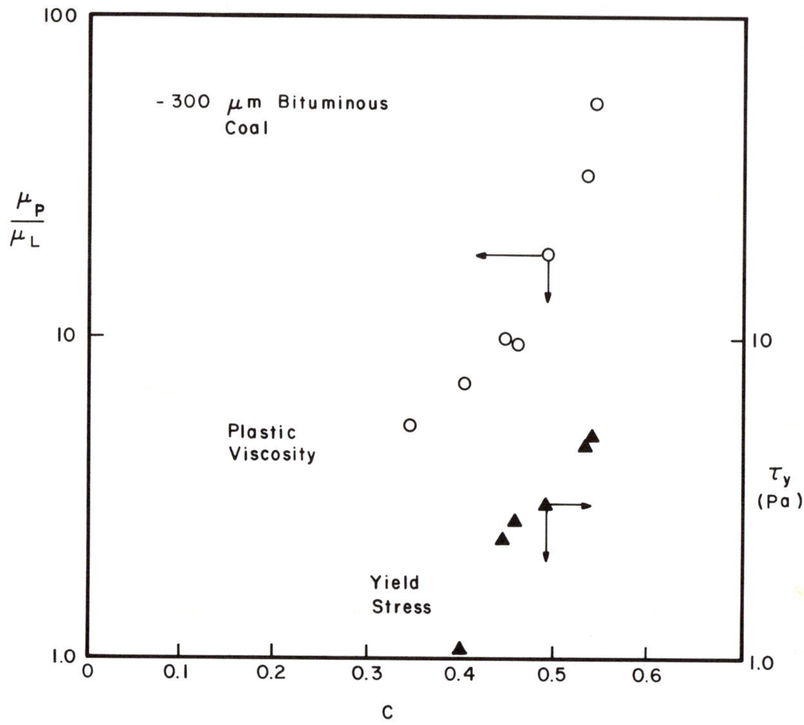

Figure 3–6. Bingham fluid parameters as functions of solids concentration for a coal slurry of the "long distance pipeline" type.

ing the ionic content of the fluid. Repeated washing or ion exchange was shown to reduce the shear resistance by reducing the concentration of cations in solution. The flocculation effectiveness of the cations decreased in the order: trivalent > divalent > monovalent as expected from the Schulze–Hardy rule. If polyvalent cations are removed from the solution by precipitation or sequestering, the shear resistance also decreases (Laapas and Aarnio, 1988).

Many solutes have been shown to act as dispersants for coal particles (Tadros, 1985) including anionic, cationic, zwitterionic, and nonionic surfactants (Laapas and Aarnio, 1988). Evidence for the importance of particle surface composition and charge in coal slurries is given by Leong et al. (1987), Boger et al. (1987), and Kanamori et al. (1990).

In terms of correlating these effects, Wildemuth and Williams (1985) have shown that variable yield stresses and viscosities can be correlated in terms of stress-dependent values of B and c_{max} in Equation 3–22. As the shear stress increases, deflocculation causes c_{max} to increase and B to decrease. Mechanistic explanations of yield stresses in flocculated slurries have also been derived (Goodwin, 1982) in terms of solids concentration and particle size.

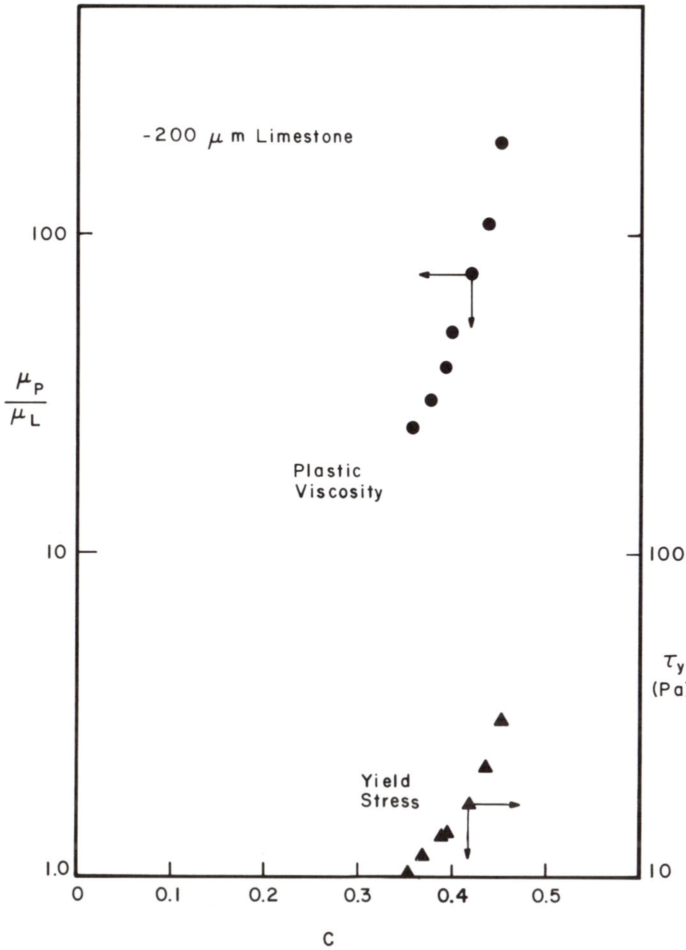

Figure 3–7. Bingham fluid parameters as functions of solids content for a fine limestone slurry.

Although flocculation provides an explanation for the existence of yield stresses in water slurries of fine particles, it cannot be the only mechanism. Slurries of quartz and of coal, dispersed in neutrally buoyant organic liquids, displayed yield stresses and shear thinning behavior although electrochemical double layers were very unlikely to have been present (Tangsathitkulchai and Austin, 1988). Using particles with Rosin–Rammler size distributions, the effect of particle size (expressed as d_{50}) was found to be strong, but the effect of the modulus m was minor, in the range $0.4 < m < 0.9$. This implies that the role played by the finest particles was not dominant because mixtures with such different m values contain considerably different quantities of fines when c and d_{50} are fixed.

3.11 SHEAR THINNING

Shear thinning flow resistance in the absence of a yield stress is sometimes called pseudoplasticity. In terms of the power law model, such slurries have n values less than 1. A variety of approaches can be used to derive a mechanistic explanation of shear thinning. In one theory, interparticle contact is considered to be a mechanism of momentum transfer which differs from liquid shear flow in its (macroscopic) shear rate dependence. Since momentum is transferred between layers by a combination of fluid shear and interparticle contact, the net effect is to give the mixture non-Newtonian flow properties.

Many shear thinning slurries are approximately Newtonian at very low and very high shear rates. The rate of interparticle contact and the duration of contact can both be predicted theoretically. Using these, a three-parameter shear thinning flow model was derived by Gay et al. (1969):

$$\frac{\tau}{\dot{\gamma}} = \mu_\infty + \frac{(\mu_0 - \mu_\infty)}{1 + (\mu_0 - \mu_\infty)\dot{\gamma}/H} \tag{3–25}$$

μ_0 and μ_∞ are the limiting "viscosities" at low and high shear rates and the flow parameter H is a measure of the deviation from Newtonian behavior at intermediate shear rates.

An alternative explanation of shear thinning assumes that an aggregate gives a greater contribution to slurry viscosity than completely dispersed particles. If aggregate size varies with shear rate, this explains the shear rate dependence of viscosity.

Even when the association between particles is temporary, as in the pairing of particles during shear flow (Krieger and Dougherty, 1959), the fact that the aggregate makes a different contribution to viscosity than a single particle of equivalent volume leads to non-Newtonian behavior.

Kinetic models have often been used in these analyses. In the approach of Krieger and Dougherty, dealing with the case where Brownian and shear forces are dominant, the numbers of single spheres and pairs per unit volume (n_1 and n_2) are assumed to be governed by these processes.

Brownian motion can form or disperse pairs according to the mechanism

$$2n_1 \underset{k_{Bd}}{\overset{k_{Bf}}{\rightleftarrows}} n_2$$

whereas shear is presumed to produce only dispersion:

$$n_2 \overset{k_{Sd}}{\rightarrow} 2n_1$$

The overall rate of pair formation is then

$$\frac{dn_2}{dt} = k_{Bf}n_1^2 - (k_{Bd} + k_{Sd})n_2$$

At equilibrium the overall rate is zero.

$$k_{Bf} n_1^2 = (k_{BD} + k_{Sd}) n_2$$

Thus, the relative proportions of the two species, and therefore the rheology, are determined by the magnitudes of the rate constants that depend on the dynamics of the Brownian motion and shear dispersion processes. Krieger and Dougherty (1959) showed that μ_r was a function of a dimensionless group formed from the shear rate $\dot{\gamma}$ and the characteristic time of the Brownian motion.

Dimensional analysis (e.g., Krieger, 1972; Probstein and Sengun, 1987) has provided a generalized interpretation of the effects of shear (S), interparticle attraction (A), interparticle repulsion (R), and Brownian motion effects (B). The Probstein and Sengun groups that express relative magnitudes of pairs of these factors are

$$N_{SB} = \frac{\mu_L R^3 \dot{\gamma}}{kT} \qquad \text{(shear/Brownian)}$$

$$N_{SR} = \frac{\mu_L R^2 \dot{\gamma}}{kT} \qquad \text{(shear/repulsion)}$$

$$N_{SA} = \frac{\mu_L R^3 \dot{\gamma}}{A} \qquad \text{(shear/attraction)}$$

$$N_{BA} = \frac{kT}{A} \qquad \text{(Brownian/attraction)}$$

$$N_{BR} = \frac{kT}{\epsilon \epsilon_0 R \zeta^2} \qquad \text{(Brownian/repulsion)}$$

R is the radius of the particle, ϵ is the relative permittivity (dielectric constant), and ϵ_0 is the permittivity of free space (S.I.). The other parameters are explained in Sections 1.10 and 1.13.

These parameters provide the independent variables which, in addition to particle shape, size distribution and concentration, determine slurry rheology. The dimensionless number N_{SB} is a Peclet number whose importance was demonstrated by Krieger and coworkers (Krieger, 1972). Probstein and Sengun explain that agglomeration will be likely when both N_{SA} and N_{BA} are low; i.e., neither Brownian motion nor shear forces are sufficiently strong to overcome the interparticle attractive forces. Macroscopic rheological measurements support this explanation. For example, Tsai and Zammouri (1988) have determined the effect of N_{BA} on model suspensions.

If flocs form, the appropriate radius in the dimensionless group is that of the floc. It is reasonable to suppose that N_{SA} would be of order unity in a floc that could be dispersed by shear. This means that R_{floc}^3 should vary as $\dot{\gamma}^{-1}$.

In dense suspensions near the maximum packing concentration shear thinning may be attributed to a layered microstructure destroyed by high shear rates (Ma and Roco, 1990).

3.12 TIME DEPENDENCE

If the effective viscosity $\tau/\dot{\gamma}$ increases with time at a constant rate of shear, the fluid is called rheopectic or antithixotropic. If $\tau/\dot{\gamma}$ decreases with time at a constant rate of shear, the fluid is thixotropic. Both types of behavior can be observed with slurries, although thixotropy is much more common.

Time-dependent shear resistance of slurries is observed more frequently in Couette viscometers than with flows which pass through pumps. Samples of flowing slurries taken from pipelines which show no time dependence are sometimes found to be time dependent when they are tested in a viscometer. It seems likely that shear in the pump destroys the structure that can be reestablished during time at rest. Thus, time dependence is likely to be most important for long pipelines that are restarted after being shut down.

A variety of equations to model the time-dependent shear response of a slurry has been proposed. The Krieger and Dougherty (1959) model illustrates the mechanistic approach which seeks to identify the physical or chemical cause of the variation with time. With industrial slurries, investigators must often employ a quasi-mechanistic view. The model of Carleton et al. (1984) employed a generalized yield power law fluid with a yield stress that varies linearly with a structural parameter λ, which depends on shear rate and time according to a kinetic model. Moore's (1959) expression for λ is

$$\frac{\partial \lambda}{\partial t} = a(1 - \lambda) - b\lambda \left| \frac{1}{2}(\mathbf{\Delta}:\mathbf{\Delta}) \right|^{m/2} \tag{3–26}$$

where a, b, and m are empirical coefficients. As generalized further by Cawkwell and Charles (1987), the rheological equation of state (A1–11) is then

$$\eta = (\kappa + \lambda \Delta \kappa) \left| \frac{1}{2}(\mathbf{\Delta}:\mathbf{\Delta}) \right|^{(n-1)/2} + \frac{\tau_{y0} + \lambda \tau_{y1}}{\left| \frac{1}{2}(\mathbf{\Delta}:\mathbf{\Delta}) \right|^{1/2}} \tag{3–27}$$

Equation 3–27 contains five additional coefficients: κ, $\Delta\kappa$, n, τ_{y0}, and τ_{y1}. The technique for determining the coefficients in a complex model of this type is illustrated by Cawkwell and Charles (1989). Once the coefficients have been determined for a given fluid, the equation of state may be used to solve Equations 2–67 and 2–69 for transient laminar flows.

3.13 SHEAR THICKENING

Although shear thickening ($\tau/\dot{\gamma}$ increasing with increasing $\mathbf{\Delta}$) has been observed with slurries, it is much less common than shear thinning. Any viscometric flow which suggests shear thickening should be investigated very carefully to verify that the flow is laminar and not turbulent.

This type of flow (power law model with $n > 1$) is sometimes called dilatant. The term originated with Osborne Reynolds who noted that beach sand appeared to dewater partially as a result of the volume expansion produced by shear. Because some unlubricated interparticle contacts are produced by dilation, subsequent flow resistance would be expected to be higher.

The link between Reynolds's dilatant flow and shear thickening is far from clear, however. A mechanistic explanation of shear thickening could involve either shear-induced flocculation or randomization of the orientations of anisometric particles. A third possible mechanism is inertial interparticle momentum transfer of the type described by Bagnold (1954).

3.14 EMULSIONS

Emulsions resemble slurries rheologically in many respects. To a first approximation, the viscosity–concentration relationships for slurries also apply to emulsions with small droplets because these are usually spherical. Because deformation can occur, c_{max} values inferred from μ_r versus c measurements are frequently much higher than those which apply to dispersed solid particles. If the drops are large or if the dispersed phase viscosity is low, circulation will occur inside the drops and this will alter the effective viscosity of the mixture.

Taylor (1932) modified Einstein's analysis of dilute suspension viscosity to calculate the effect of the circulation, assuming laminar flow of both phases. Equation 3–19 becomes

$$\mu_r = 1 + 2.5c\left[\frac{r + 0.4}{r + 1}\right] = \frac{\mu_m}{\mu_c} \qquad (3\text{–}28)$$

where r is the ratio of the dispersed phase viscosity (μ_d) to that of the continuous phase (μ_c). As r approaches infinity, Equation 3–28 becomes identical to Equation 3–19. The bracketed correction factor in Equation 3–28 represents the effect of circulation within the drops. Circulation also alters the drag force, so that the Stokes' Law drag force (calculated from Equation 1–14) for a dispersed drop is multiplied by the Hadamard correction factor $[(3r + 2)/(3r + 3)]$.

With large drops, very high concentrations or high shear rates, distortion of the drop shape may occur during flow and this can eventually lead to rupture of the interface to form smaller droplets. Taylor noted that the pressure difference between the inside and the outside of a drop becomes of comparable magnitude to the effect of the interfacial tension σ for drops of the critical diameter d_{crit}

$$d_{crit} = \left(\frac{4\sigma}{\mu_c\dot{\gamma}}\right)\left(\frac{4r + 4}{19r + 16}\right) \qquad ((3\text{–}29)$$

Drops larger than this size would tend to break up. On the other hand, collision processes and draining and rupture of the films of the continuous phase may produce coalescence during flow. For these reasons, time effects and apparent shear rate dependence are often observed with emulsion flows.

Theoretical and experimental studies of foam rheology have been reviewed by Kraynik (1988). Most microscopic models for foams consider idealized materials with spatially periodic cell structures (Khan and Armstrong, 1989; Kraynik and Hansen, 1987).

The rheology of oil-in-water emulsions with and without the presence of added solid particles has been studied systematically because of the importance of these mixtures in oil sand processing and petroleum production. Pal and Masliyah (1990) and Yan et al. (1991) determined that the oil–water mixture was Newtonian at low concentrations and shear thinning at high concentrations. This is qualitatively similar to the behavior of many solid–liquid mixtures in which the mean particle diameter is above 10 μm. With particles of diameter at least three times that of the oil drops, the three-phase mixture could be modeled as an oil-in-water pseudo-continuum with the solids forming a dispersed phase. The relative viscosity of the mixture could then be calculated using the method described in Section 3.6.

3.15 DRAG REDUCTION

There appear to be two phenomena which can be described as drag reduction for turbulent flows. The first, the so-called Toms effect, occurs for single-phase flows when the fluid at the pipe wall is viscoelastic (McComb and Rabie, 1982). The extent of drag reduction increases considerably once a critical wall shear stress is exceeded. The second reduction effect is due to the two-phase nature of the flow.

For slurries, Radin et al. (1975) examined both rigid and fibrous particle slurries and found that reduced frictional losses could only be produced at low concentrations with slurries of fibrous particles. Although the presence of solid particles has been shown to alter the turbulence level in vertical flows (Hetsroni, 1988; Nouri et al., 1987), the effect of such changes on frictional pressure drop has not been established. Further discussion of vertical flows is given in Section 5.7.

If the slurry is non-Newtonian, the Reynolds number and relative roughness are not the only parameters to determine the friction factor in Equation 1–5. The effect of the additional fluid parameters upon the flow is sometimes invoked in theoretical predictions of turbulent friction factors. These are considered in Section 4.4.

3.16 FIBER SUSPENSIONS

The motion of spheroidal particles in shear flow was first considered by Jeffrey (1922) and a description of the motion and its effects is given by Happel and Brenner (1973). Prolate spheroids of high aspect ratio are useful approximations to rods and their behavior has been examined theoretically and experimentally by many subsequent workers.

During flow, rigid rods can rotate, sweeping out volumes proportional to the cube of their lengths, and causing significant viscosity increases at very low volume concentrations. Suspensions of flexible fibers are even more complex because the fiber can deform in response to the surface forces exerted by the fluid.

Fiber suspensions can display significant reductions in friction in pipe flow, compared to those for the pure carrier fluids if the length to diameter ratio of the fibers is greater than about 30. Drag reduction increases with increasing (L/d) ratio and with decreasing fiber diameter d. Although wood pulp fibers have been studied most thoroughly, similar drag reduction behavior has been observed with nylon (Bobkowicz and Gauvin, 1965), asbestos, rayon, and acrylic fibers (Vaseleski and Metzner, 1974).

At low velocities the τ_w versus V relationship for a flocculated fiber suspension indicates plug flow. The plug is formed of interlocked fibers and because the fiber density is very close to that of the carrier fluid, the flow is essentially axisymmetric. This type of flow is rather like that occurring in fluids with yield stresses or low n (power law exponent) values (Figure 3–8). These τ_w values increase rapidly with solids concentration ("consistency" or mass fraction of fiber is the customary concentration unit) (Duffy, Moller, and Titchener, 1972). In the low velocity region, the wall shear stress for a given pulp suspension can be described by a dimensional headloss correlation of the form

$$\tau_w = \frac{igD}{4} = KV^{\alpha}C_w^{\beta}D^{\gamma} \tag{3–30}$$

We note the resemblance to the laminar power law model described in Table 3–1, for which α and $(-\gamma)$ in Equation 3–28 would equal n. However, tests with fiber suspensions show that α and $-\gamma$ can have different values.

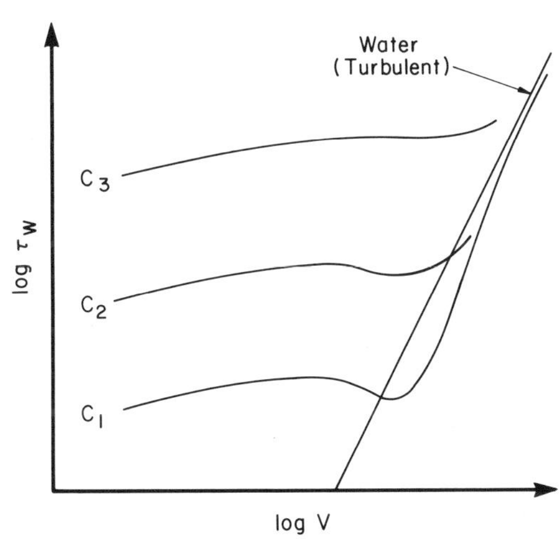

Figure 3–8. Wall shear stress as a function of mean velocity for a wood pulp suspension in pipe flow.

As the velocity increases, the headlosses or shear stresses pass through a peak and then decrease to a minimum before increasing again. The peak is associated with the onset of disruption of the central plug (Duffy and Titchener, 1974). When disruption occurs, the dislodged fiber flocs damp the turbulence of the annular region and decrease the von Karman coefficient κ in the Law of the Wall (Equation 1–8).

Because κ decreases, the friction factor for turbulent flow falls below the value for the clear carrier fluid. This reduction occurs when the diameter of the plug is still comparatively large. As the velocity increases, the core shrinks to a diameter approximately 20% of the pipe diameter at maximum drag reduction.

The reduction in κ is confirmed by velocity distribution measurements (Lee and Duffy, 1977). These distributions suggest that the viscous sublayer is not the source of the drag reduction, in contrast with the mechanism postulated by Wilson and Thomas (1985) in deriving Equations 4–11 and 4–13. Further evidence for the difference between the two drag reduction mechanisms described in Section 3.16 was given by Kale and Metzner (1974) who achieved very large drag reductions by combining a viscoelastic additive with a fiber suspension.

The most remarkable feature of the flow of fiber suspensions is that these effects occur at such low concentrations. This reflects rotation of fibers and that a large volume of fluid may move with the particles, as a consequence of fiber entanglement.

3.17 OSCILLATING AND FALLING-BALL VISCOMETRY

Besides the simple devices discussed in Sections 3.2 to 3.4, many other instruments have been used to investigate slurry rheology. An oscillating sphere viscometer measured the power required to maintain the oscillation of a sphere about its polar axis at a constant amplitude. The power is proportional to the viscosity–density product of the medium in which the sphere is immersed.

A falling-ball rheometer measures the terminal velocity of a settling sphere and relates it to the viscosity of the fluid through Equation 1–21. Since little deformation of the fluid occurs during a settling measurement, the original microstructure (and especially the particle orientation) is essentially preserved. This behavior contrasts with the particle migration effects which have been observed in Couette and tube viscometers (Leighton and Acrivos, 1987; Abbot et al., 1991; Goldsmith et al., 1966). If the falling ball is large in comparison with the dispersed particles, the variation of settling velocity from run to run is small. A device of this type has evident advantages for studying the effects of particle shape and orientation on the shear resistance of suspensions (Milliken et al., 1989; Powell et al., 1989; Powell and Walla, 1991).

Chapter 4

Calculations for Homogeneous Flows

4.1 CONCENTRIC CYLINDER VISCOMETRY

We have seen that testing of a sample of the slurry is necessary for pipeline design. Couette viscometers are convenient, economical testing devices requiring only small volumes of slurry. The measured variables with these instruments are the torque and the rotational speed. In Table 3–1 the equations relating these quantities were presented for simple two-parameter rheological models. Although more elaborate models are available and may well represent a given set of experimental data more satisfactorily, these more elaborate models are of limited use for pipeline design unless the flow remains laminar in the pipe at all conditions of practical importance.

There are a few practical problems which arise with concentric cylinder viscometry for slurries. Particle segregation in the radial direction can occur (Abbot et al., 1991). Settling of a coarse fraction can also occur, especially at low concentrations, and this removes particles from the domain of shear. Occasionally, one can overcome this problem by increasing the solids concentration beyond the range of greatest interest to take advantage of the reduced settling tendency at high concentrations. The data is then fitted to a viscosity–concentration correlation, using the condition that $\mu_r = 1.0$ at $c = 0$. The correlation is then used to obtain the desired viscosity.

Removing the coarse, rapidly settling particles (scalping) is less desirable because particle size is known to affect slurry rheology. However, if the relationship between concentration, particle size distribution (or particle surface area), and rheology is known, scalping the coarse material from the solids becomes practical. Gahlot (1987) tested coal and zinc tailings slurries and found that the rheology of scalped (s) and original (o) slurries were comparable at weight fractions related by

$$C_{ws} = C_{wo}\left(\frac{A_o}{A_s}\right)^{0.2}\left(\frac{C_{wo}}{C_{wsso}}\right)^{0.13} \quad (4\text{–}1)$$

A_o and A_s are the surface area per unit slurry volume of the original and scalped slurries and C_{wsso} is the static settled concentration (approximating $C_{w\,\max}$) of the latter. It is likely that surface and fluid properties also enter into the relationship between C_{ws} and C_{wo}.

Hanks and Hanks (1986) recommended conducting a complete set of rheological tests with solids (volume) concentration C_v and scalping fraction $(1 - \beta)$ as parameters. If β_s is the fraction of the fines in the total size distribution that produces the same rheology as the original slurry, the volumetric concentrations C_{vo} (in the original slurry) and C_{vs} (in the scalped slurry with the same rheology) for particles which do not absorb liquid on slurry preparation, are related by

$$C_{vs} = \frac{\beta_s C_{vo}}{(1 - C_{vo}) + \beta_s C_{vo}} \quad (4\text{–}2)$$

C_{vo} and C_{vs} are the volumetric concentration analogs of C_{wo} and C_{ws}. To determine β_s the rheological parameters (the relative viscosity, for example) obtained with various values of C_v and β are plotted against an equivalent (unscalped slurry) concentration computed from C_v and β. This concentration is $C_v/[C_v + \beta(1 - C_v)]$, obtained by rearranging Equation 4–2 for the general case. β_s is then the value of β for which the parameters become independent of β.

During viscometric testing it is often observed that the torque does not stabilize as quickly as it does with a Newtonian liquid. It may continue to change for some time until it finally stabilizes: the slurry is evidently time dependent. Furthermore, the torque at stabilization is often a function of previous history. In a laboratory test pipeline, the same slurry will often show no time or shear history dependence, possibly because the structure which changes slowly in the viscometer is destroyed completely by shear in passing through the pump.

A solution to this problem is to give the slurry a preliminary shear treatment at a high shear rate for a fixed period of time before attempting to characterize it. The preliminary viscometer shear should be conducted at a shear rate somewhat higher than the highest wall shear rate likely to be of interest. Although this does allow reproducible data to be obtained with the viscometer, the arbitrary nature of the preliminary shear treatment cannot be denied.

A common variant of this behavior is shown in Figure 4–1, showing shear stresses (torques) at a sequence of fixed shear rates. The fluid is shear thinning, but not time dependent, at the lower shear rates. However, at the highest shear rate a significant decrease of stress occurs with time. No time dependence was observed in pipe flow tests conducted with this slurry at the higher (wall) shear rate.

Another form of time dependence observed in concentric cylinder viscometers is due to particle diffusion from regions of high to low shear rate (Leighton and

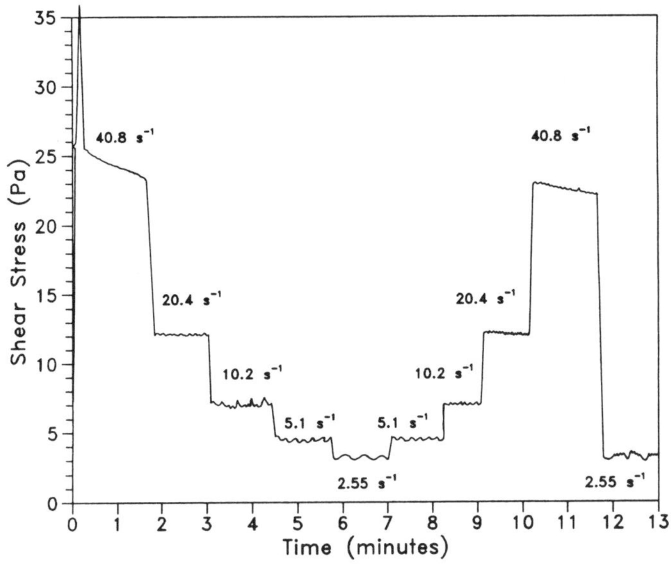

Figure 4–1. Shear stress as a function of time for a concentrated coal–water slurry tested at a series of shear rates.

Acrivos, 1987; Sengun and Probstein, 1989). In this case measurements must be made quickly to avoid concentration changes.

With non-Newtonian slurries, the region of stable laminar viscometric flow cannot be defined satisfactorily because the laminar–turbulent transition criteria have only been established for Newtonian fluids.

An advantage of concentric cylinder viscometry is that a wide range of shear rates, including the low values that apply in large pipelines, can be employed in testing. When combined with the additional considerations that only a very small sample is required, that temperature control is achieved easily, and that the instrument itself need not be expensive, it is obvious that concentric cylinder viscometry is very convenient. The method of data reduction is illustrated in the following example.

Example 4.1

A concentric cylinder viscometer with a stationary outer cylinder of radius $R_2 = 2.10$ cm and a rotating inner cylinder of $R_1 = 1.84$ cm was used to obtain data presented in Table 4–1. The data are fitted first to the Bingham model. If

Table 4–1. Couette Viscometer Data for Example 4.1

$\omega(s^{-1})$	$T\ (Nm/m)$	$\tau_1\ (Pa)$	$\tau_2\ (Pa)$
53.6	0.0404	19.0	14.6
26.8	0.0307	14.45	11.1
13.4	0.0246	11.55	8.9
6.7	0.0206	9.7	7.45
3.35	0.0168	7.9	6.06

shear strain occurs throughout the whole of the flow domain between the cylinders, then the Bingham fluid equation from Table 3–1 can be written

$$AT = \omega\mu_p + B\tau_y$$

where

$$A = \frac{1/R_1^2 - 1/R_2^2}{4\pi}$$

$$B = \ln\left(\frac{R_2}{R_1}\right)$$

Fitting the data to a linear relationship we find

$$T = 4.479 \times 10^{-4}\omega + 0.0173$$

Using A and B, we find $\mu_p = 0.024$ Pa s and $\tau_y = 7.15$ Pa. Inspecting the values of τ_2 in Table 4–1, we see that this value of τ_y implies that the last datum point should not have been included in the regression process. Repeating the fitting process with only the four higher speed points gives $\mu_p = 0.0226$ Pa s and $\tau_y = 7.7$ Pa. This now eliminates the measurement obtained at 6.7 s^{-1} from consideration.

A third trial with three data points gives $\mu_p = 0.0214$ Pa s and $\tau_y = 8.16$ Pa. In the region where τ_2 is less than the yield stress, the relationship between T and ω is obtained by integration as

$$\omega = 0.5\left[\left(\frac{T}{2\pi R_1^2}\right) - \tau_y - \tau_y \ln\left(\frac{T}{2\pi R_1^2 \tau_y}\right)\right] \Big/ \mu_p \qquad (4\text{–}3)$$

Although Equation 4–3 could be used with four data points, the fit in the region $\tau_2 < \tau_y < \tau_1$ is frequently unsatisfactory.

The measurements could also be fitted to the power law model. If this is done, one finds the empirical equation

$$T = 0.0113\omega^{0.313}$$

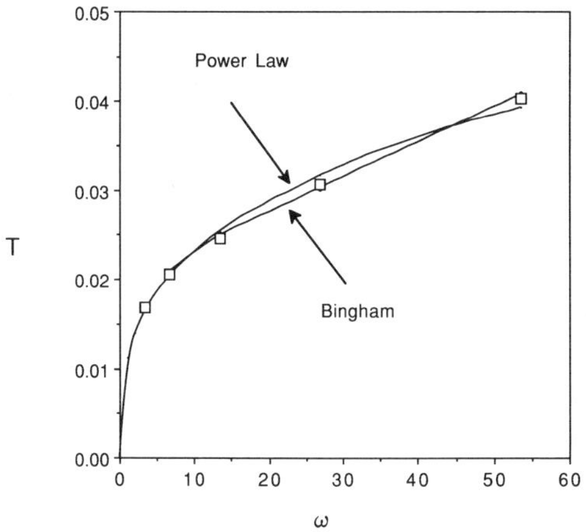

Figure 4–2. Comparison of measurements with fitted models for Example 4.1.

Using the equation from Table 2–1 we see that with $n = 0.313$

$$K = \left(\frac{0.0113}{2\pi}\right)\left(\frac{n}{2}\right)^n\left[\left(\frac{1}{R_1}\right)^{2/n} - \left(\frac{1}{R_2}\right)^{2/n}\right]^n$$

Substitution gives $K = 2.49$ Pa $s^{0.313}$. These correlations are compared with the measurements in Figure 4–2.

We note that neither model is completely satisfactory so that extrapolation beyond the experimental shear stress range would be questionable. A three-parameter combination of the two simple models, the Herschel–Bulkley yield-power law, $\tau - \tau_y = K\dot{\gamma}^n$, may also be appropriate in this case but there are not enough data points to justify it.

The example has shown how measurements can be used to provide a model which could be used for pipeline design. It is, of course, possible in principle to obtain actual values of the shear rate and shear stress for each of the viscometric data points using Equations 3–16 and 3–17. As long as the pipeline flow is laminar, the actual shear stress versus shear rate relationship can then be used to predict pressure drop as a function of mean velocity by performing the integration shown in Equation 3–12.

The example is realistic in that the number of data points which can be obtained in a particular range of shear stresses is limited. Since pipeline design requires parameters that are obtained at shear stresses near τ_w, it is often impractical to consider the more complex rheological models.

For industrial applications, the possibility of turbulent pipeline flow of the slurry must usually be considered. In these turbulent flow calculations, it will probably be desirable to employ Bingham, power law, or yield-power law models because the literature for their turbulent flows is more extensive than for slurries described by more complex models.

4.2 TUBE VISCOMETRY

Equation 1–3 (or Equation 3–7) provides a method for determination of the wall shear stress τ_w from measurements of the pressure gradient and the flowing density ρ:

$$\frac{4\tau_w}{D} = \frac{\Delta P}{L} = -\frac{dP}{dx} - \rho g \frac{dh}{dx} \tag{3–7}$$

If the flow is vertical, the gravity term is large so that the density and the pressure gradient must both be known accurately. The flow should be fully developed, with the velocity distribution independent of x. This may require a substantial entry length before the first pressure measurement position. This is achieved more easily for horizontal flows than for vertical ones.

Entry lengths for laminar flows are calculated by solving the axial momentum equation to obtain the velocity distribution as a function of position from the pipe inlet. The usual entry length criterion is a centerline velocity of 98% or 99% of the fully developed value. For Newtonian fluids

$$\frac{L_e}{D} = 0.062\text{Re} \tag{4–4}$$

For shear thinning, non-Newtonian fluids one would expect entry lengths to be lower than those of Newtonian fluids because the fully developed velocity distributions are flatter. However, there is the question of how the Reynolds numbers should be defined to permit a valid comparison to be made. For Bingham fluids, Chen et al. (1970) found that L_e/D decreases in a nearly linear manner to about 0.01Re as the ratio τ_y/τ_w increases from 0 to 0.55. In this case, the Bingham fluid Reynolds number is used for Re:

$$\text{Re}_B = \frac{DV\rho}{\mu_p} \tag{4–5}$$

For shear thinning power law fluids, the computations of Collins and Schowalter (1963) show shorter entry lengths than those of Equation 4–4 if the Reynolds number is $(D^n V^{2-n}\rho/K)$.

If segregation and/or particle deposition occurs in a horizontal tube viscometer, this will affect the ΔP versus V relationship, altering it significantly from that applying to axisymmetric (vertical) flow. Unfortunately, there are no guidelines to be used to predict whether or not segregation or deposition is likely in a laminar horizontal flow. If deposition is detected, test flow results should be fitted to a more complex flow model, such as the two-layer model described in Chapter 6.

Gamma-ray absorption measurements can be used to detect segregation in horizontal tubes, but this is more easily done with a large tube. In addition to providing information about segregation or deposition, a larger tube can provide information about the laminar-turbulent transition. This can provide a useful verification of the rheological model used for the laminar flow region because criteria have been proposed to predict the transition (Hanks, 1963).

A 50-mm I.D. test pipe is often a reasonable compromise. Smaller diameter tubes can be used if the flow is unlikely ever to be turbulent and if deposition will not occur.

Test data, in the form of plots of i or τ_w versus V or $8V/D$, should first be inspected for evidence of turbulent flow. The laminar flow data can then be fitted to the appropriate equation from Table 3–1.

Alternatively, the method of Metzner and Reed can be used for the laminar flow data. Equation 3–12 shows that τ_w is a function of $8V/D$ for any time-independent non-Newtonian fluid. This means that for a given fluid in laminar pipe flow, values of τ_w for a large pipe can be obtained from tests conducted at the same value of $8V/D$ in a tube viscometer. This simple scale-up procedure is equivalent to fitting the tube viscometer data to equation 4–6.

$$\tau_w = K'\left(\frac{8V}{D}\right)^{n'} \tag{4–6}$$

It is worthwhile to distinguish between Equation 4–6 and the expression that applies for a power law fluid. For a power law fluid, Table 3–1 gives the relationship between τ_w and $8V/D$ as

$$\tau_w = K\left(\frac{3n + 1}{4n}\right)^n \left(\frac{8V}{D}\right)^n \tag{4–7}$$

Comparing Equations 4–6 and 4–7 we see that $n' = n$ if the fluid obeys the power law model exactly. In such cases, K' can be replaced by $K[(3n + 1)/4n]^n$.

Equation 4–6 shows that it is the value of τ_w in the full-scale pipeline which is used to select the range of experimental velocities to be used in a test to characterize a particular slurry. The following example illustrates how tube flow data can be used.

Example 4.2

A tailings slurry of density 1540 kg/m^3 was tested in a pipeline (D = 0.1076 m). The data is shown in Table 4–2 and plotted in Figure 4–3. No deposition was observed at the velocities recorded here.

The highest velocities are suspected of being turbulent so that only the five lowest ones are used in modeling. A power law curve fit gives the relationship in Equation 4–8.

$$\tau_w = 5.21\left(\frac{8V}{D}\right)^{0.213} \tag{4–8}$$

These values of K' and n' can be used in any pipe as long as the flow is laminar and the wall shear stress is within the limits imposed by the data.

Table 4–2. Tube Flow Data for Example 4.2

V *(m/s)*	*Wall Shear Stress* *(Pa)*	*8V/D* (s^{-1})
0.305	10.1	22.7
0.61	11.7	45.4
0.915	12.8	68.0
1.22	13.6	90.7
1.49	14.2	111
1.86	15.7	138
2.14	16.7	159
2.47	23.7	184
2.73	27.4	203
2.75	28.8	204
3.05	31.7	227
3.05	32.7	227

The Bingham model requires a little extra effort to be fitted to these results when the ratio of τ_y to τ_w is not low enough for the term $0.33(\tau_y/\tau_w)^4$ to be neglected. We first construct two expressions from the equation in Table 3–1 to use with the experimental data. There are five data points ($N = 5$).

$$\mu_p \sum \frac{8V}{D} = \sum \tau_w - \frac{4}{3} N\tau_y + \frac{1}{3}\tau_y^4 \sum \tau_w^{-3}$$

$$\mu_p \sum \tau_w \frac{8V}{D} = \sum \tau_w^2 - \frac{4}{3}\tau_y \sum \tau_w + \frac{1}{3}\tau_y^4 \sum \tau_w^{-2}$$

These provide two equations in μ_p and τ_y. With the definitions below, μ_p is eliminated to give a fourth degree equation for τ_y.

$$(s_4 s_5 - s_1 s_8)\tau_y^4 + (s_1 s_7 - s_3 s_5)\tau_y + (s_2 s_5 - s_6 s_1) = 0 \tag{4–9}$$

where

$$s_1 = \sum \frac{8V}{D}, \quad s_2 = \sum \tau_w, \quad s_3 = \frac{4}{3}N, \quad s_4 = \frac{1}{3}\sum \tau_w^{-3}$$

$$s_5 = \sum \tau_w \left(\frac{8V}{D}\right), \quad s_6 = \sum \tau_w^2, \quad s_7 = \frac{4}{3}\sum \tau_w, \quad s_8 = \frac{1}{3}\sum \tau_w^{-2}$$

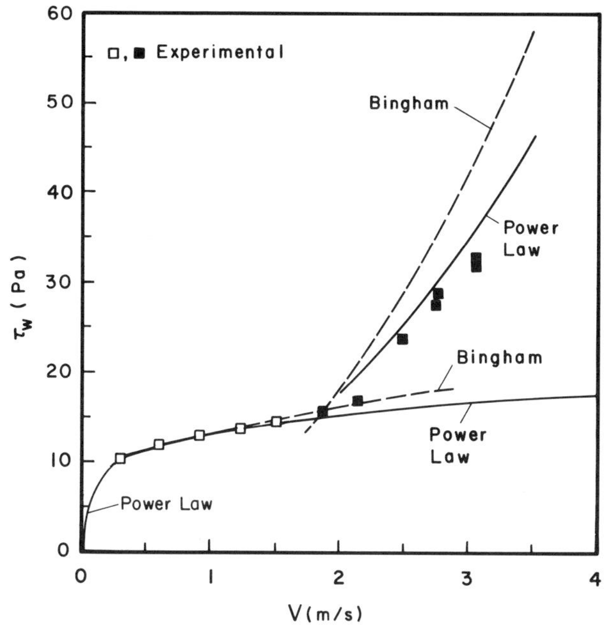

Figure 4–3. Wall shear stress as a function of velocity for Example 4.2. The lines are calculated using the Bingham fluid and power law models.

Using the data in Table 4–2 and Newton's method, we find $\tau_y = 8.13$ Pa and $\mu_p = 36.5$ mPa. The two fitted laminar flow relationships are compared with the data in Figure 4–3.

The turbulent region predictions will be discussed in Section 4.4.

4.3 WALL SLIP AND NONHOMOGENEOUS FLOW

In tests of the type described in Section 4.2, it is desirable to use more than one tube diameter. The results should be inspected for any difference in the pseudo-shear (τ_w versus $8V/D$) relationship for the slurry. If such a difference occurs and if there is no indication of turbulent flow at high velocities ($8V/D$ values), this must indicate failure of the model to apply in either or both cases. (We exclude experimental artifacts such as a small difference in concentration between the slurries used in the two tubes.)

There are at least two possible causes of model failure: wall slip and nonhomogeneous flow. We recall that the no-slip boundary condition $v_x = 0$ at $r = 0.5D$ is a feature of the continuum argument which relates τ_w and $8V/D$. A finite wall slip velocity would introduce a new parameter into the relationship between these two quantities.

A physical origin of the wall slip phenomenon can be found in the fact that conditions are necessarily different in the region between the wall and the center of the closest particle (Figure 1–5). Since the effective solids concentration in this region is low, the velocity gradient may be steeper than would be predicted using the properties of the bulk of the slurry (a continuum model extrapolation is illustrated in Figure 4–4). Any tendency toward particle migration from the pipe wall augments this effect. For particles in contact with the wall, the sliding motion may not be described adequately by a continuum model.

Repeating the integrations which provided the relationships of Table 3–1, and including a wall slip velocity as a boundary condition, one finds that V is replaced by $(V - V_{\text{slip}})$. If the slip effect corresponds to the presence of a lubricating layer of thickness x and viscosity μ, V_{slip} is $\tau_w x/\mu$ when the wall region viscosity μ is much lower than that of the slurry. Methods for correcting tube flow data for wall slip are given by Skelland (1967) and Jastrzebski (1967).

Nonhomogeneous flow is a second possible source of inadequacy in representing pipe flow results with a homogeneous non-Newtonian fluid model. Even small quantities of coarse material can affect the relationship between τ_w and V to a drastic extent, as we shall see in our examination of coarse particle flows. It is unfortunate that both laminar and turbulent coarse particle i versus V data can be misinterpreted as laminar flow of a shear thinning $(n' < 1)$ fluid.

When the size distribution of the particles is very broad, one cannot be certain that the homogeneous model is appropriate. The coarser particles can concentrate near the bottom of the pipe and increase the flow resistance considerably. This may have happened in the case of the data shown in Figure 4–5. The laminar flow data are fitted satisfactorily by either model, but the turbulent region is predicted very inadequately. This discrepancy is discussed in Section 4.5.

Further experiments, such as concentration profile measurements, viscometry, and/or use of another pipe diameter would be required to settle the matter.

Slip is also a possible complication with Couette viscometry (Mannheimer, 1985; Hanks and Hanks, 1986). Use of two different viscometer gaps (i.e., two different $R_2 - R_1$ values) allows the magnitude of the slip to be detected. The shear stress τ_m at the arithmetic mean radius $R_m = 0.5(R_1 + R_2)$ is calculated from the torque per unit height T as $(T/2\pi)R_m^2$. If $\Delta R = (R_2 - R_1)$ is small, then the mean shear rate in the viscometer gap is $(R_1\omega - V_{\text{slip}})/\Delta R$, where V_{slip} is a function of τ_m.

To determine V_{slip}, τ_m is plotted versus $R_1\omega/\Delta R$ for each of the two viscometer gaps, denoted here as a and b. At a given value of τ_m, the values of $(R_1\omega/\Delta R)_a$ and $(R_1\omega/\Delta R)_b$ are obtained by interpolation. V_{slip} is then

$$V_{\text{slip}} = \left[\left(\frac{R_1\omega}{\Delta R}\right)_a - \left(\frac{R_1\omega}{\Delta R}\right)_b\right] \bigg/ \left[\left(\frac{1}{\Delta R}\right)_a - \left(\frac{1}{\Delta R}\right)_b\right] \tag{4–10}$$

Once V_{slip} is known, the mean shear rate for this particular value of τ_m may be determined.

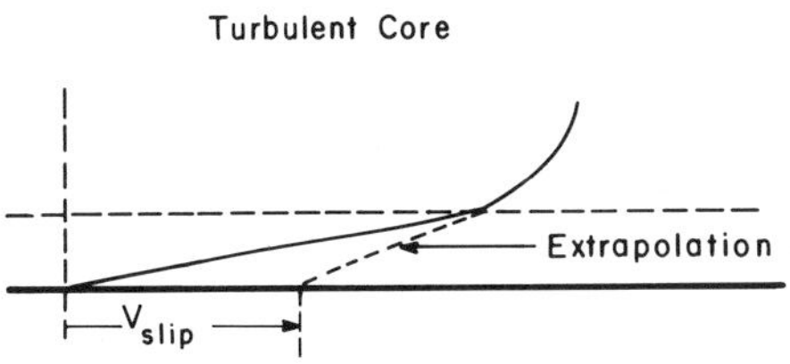

Figure 4–4. Slip at the wall of a pipe carrying a slurry.

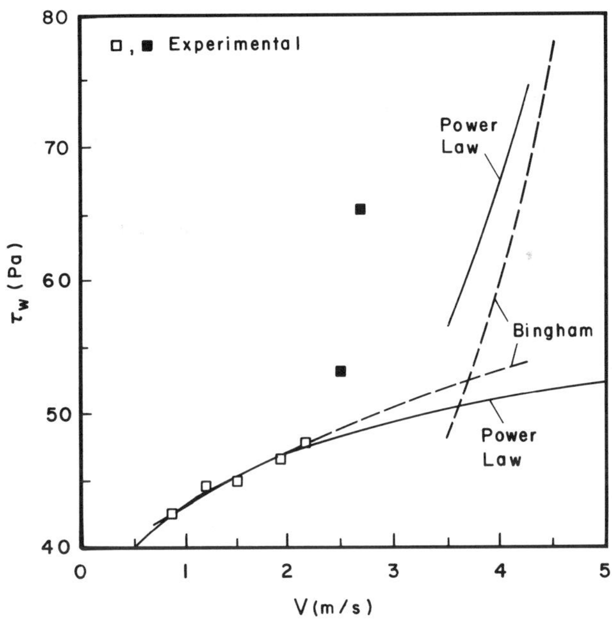

Figure 4–5. Wall shear stress as a function of mean velocity for a limestone slurry containing coarse particles.

4.4 TURBULENT FLOW

Predicting turbulent flow pressure drops is one of the most important problems encountered with homogeneous slurries. In principle, either Couette or tube viscometers can be used to test slurries for purposes of design, as long as the appropriate shear stress range is used for the test. However, differences between the results obtained with the two types of instrument have been observed with shear thinning slurries: Lazarus and Slatter (1986) show these deviations for some industrial slurries. These discrepancies may be due to differences in stress-induced deflocculation, flow-induced microstructure anisotropy, or to slip and particle migration effects in the viscometers. Abbot et al. (1991) measured this effect with Newtonian model suspensions of spherical and rod-like particles. Experimental errors can also occur when samples are removed from the flows. In the case of turbulent pipe flows of shear thinning slurries, there have been few experimental investigations which used Couette viscometers to determine the rheological parameters of the slurries. Instead, pipe flow in the laminar region has usually been used to determine the fluid properties to predict the turbulent flow behavior.

There is a fairly substantial body of experimental data available, most of it in the form of pressure drops and bulk velocities for pipe and tube viscometer flows. A number of derivations have been published, all of them plausible extrapolations of single-phase flow analyses of the near-wall and turbulent core regions of pipe flow. Since their predictions differ somewhat, in combination they give an indication of the uncertainty of a pressure drop calculation.

Of course, for a fluid with more than one rheological parameter, a Reynolds number does not necessarily have the same meaning that it does in Newtonian fluids. However, experimental measurements show that turbulent friction factors can be significantly lower than those of Newtonian fluids at comparable Reynolds numbers, if the fluid is shear thinning. Wilson and Thomas (1985) and Thomas and Wilson (1987) attributed this change primarily to an increase in the thickness of the viscous sublayer. At a given wall shear stress, this increase was considered to cause a higher mean velocity than would be observed in a Newtonian fluid with the same effective viscosity $\tau_w/\dot{\gamma}_w$. A secondary effect, according to these authors, was an altered turbulent velocity distribution in the core region. Shear thinning fluids would display somewhat blunter profiles than those of Newtonian fluids.

In the method of Wilson and Thomas, the velocity V of the slurry is related to the velocity V_N of a Newtonian fluid flowing at the same shear stress. For a Bingham fluid, Wilson and Thomas obtain

$$V = V_N + u_*\left[2.5\ \ln\left(\frac{1-\xi}{1+\xi}\right) + \xi(14.1 + 1.25\xi)\right] \qquad (4\text{–}11)$$

where $\xi = \tau_y/\tau_w$ and $u_* = (\tau_w/\rho)^{0.5}$. For smooth pipes, V_N is only a function of Reynolds number. The version of Equation 4–11 for smooth pipes will be a useful approximation for rough ones:

$$V = u_*\left\{2.5\ \ln\left(\frac{Du_*\rho}{\mu_p}\right) + 2.5\ \ln\left(\frac{(1-\xi)^2}{1+\xi}\right) + \xi(14.1 + 1.25\xi)\right\} \qquad (4\text{–}12)$$

The power law expression of Wilson and Thomas is

$$V = V_N + u_*\left[2.5\ \ln\left(\frac{n+1}{2}\right) + 11.6\left(\frac{1-n}{n+1}\right)\right] \qquad (4\text{–}13)$$

Equation 4–13 gives predictions which are similar to those of Dodge and Metzner (1959). Dodge and Metzner employed the Reynolds number $\mathrm{Re}_{n'}$ of Metzner and Reed (1955):

$$\mathrm{Re}_{n'} = \frac{D^{n'} V^{2-n'} \rho}{8^{n'-1} K'} \qquad (4\text{–}14)$$

Their power law expression is therefore suitable, in principle, for any non-Newtonian fluid. The Dodge–Metzner expression for the Fanning friction factor is

$$\frac{1}{\sqrt{f}} = 4n'^{0.75} \log\left(\mathrm{Re}_{n'} f^{(1-0.5n')}\right) - 0.4n'^{-1.2} \qquad (4\text{–}15)$$

An alternative formula for shear thinning fluids in smooth pipes has been proposed for the transition region by Kemblowski and Kolodziejski (1973).

$$f = 2.225 \times 10^{-3} \exp(3.57n'^2) \frac{\Phi^{1/\mathrm{Re}_{n'}}}{\mathrm{Re}_{n'}^{m}} \qquad (4\text{–}16)$$

where

$$\Phi = \left\{\exp\left[0.572\left(\frac{1 - n'^{4.2}}{n'^{0.435}}\right)\right]\right\}^{1000}$$

and

$$m = 0.314n'^{2.3} - 0.064$$

The transition region is that for which $3.16 \times 10^4 n'^{-0.435} > \mathrm{Re}_{n'} > 3000$. At higher Reynolds numbers, the normal smooth pipe friction factor is recommended.

Torrance (1963) generalized Clapp's (1961) formula for shear thinning fluids to include a wall shear stress. Thus, in principle, Torrance's equations can be used for power law, Bingham, or yield power law fluids. Torrance's Reynolds number Re_n is different from that of Metzner and Reed:

$$\mathrm{Re}_n = \frac{D^n V^{2-n} \rho}{8^{n-1} K}$$

For smooth pipes the Torrance formula is

$$\frac{1}{\sqrt{f}} = \frac{2.687}{n} - 2.949 + \left(\frac{1.966}{n}\right) \ln\left[(1 - \xi)\mathrm{Re}_n f^{1-n/2}\right] + \left(\frac{0.682}{n}\right)(5n - 8) \qquad (4\text{–}17)$$

For fully rough pipes, in which for single-phase flows $(k/D)\mathrm{Re}\left(\sqrt{f/2}\right) > 70$, the Fanning friction factor is, according to Torrance,

$$\frac{1}{\sqrt{f}} = 1.767 \ln\left(\frac{D}{2k}\right) + 6 - \frac{2.65}{n} \qquad (4\text{–}18)$$

For rough pipes, Szilas, Bobok, and Navratil (1981) derived

$$\frac{1}{\sqrt{f}} = -4 \log\left[\left(\frac{10^{-0.5\beta}}{\mathrm{Re}_{n'}\,(4f)^{(2-n')/2n'}}\right) + k/3.71\ D\right] \tag{4–19}$$

where $\beta = (0.707/n' + 2.12) - 4.015/n' - 1.057$. For yield power law fluids, the Thomas and Wilson relationship is

$$V = V_N + u_*[11.6(\alpha - 1) - 2.5 \ln \alpha + 2.5 \ln(1 - \xi) + 2.5\xi(1 + 0.5\xi)] \tag{4–20}$$

where $\alpha = 2(1 + n\xi)/(1 + n)$. V_N is the velocity for a Newtonian fluid of viscosity $(\tau_w/\dot{\gamma})$ flowing in the particular pipe at wall shear stress τ_w. The Newtonian viscosity $(\tau_w/\dot{\gamma})$ is evaluated from the slurry rheogram $(\tau, \dot{\gamma})$ as the slope of a line joining the origin to the point $(\tau_w, \dot{\gamma})$.

We are now able to predict headlosses for the turbulent region of Example 4.2. For turbulent flow calculations, the pipe roughness is required. Water flow experiments in the same pipe showed this to be 5 μm.

Example 4.2 Continued

Bingham Fluid (Wilson and Thomas Method)

When the fluid has a yield stress, it is easier to calculate a series of velocities for a set of assumed wall stresses than to calculate τ_w for a given V. Assuming a wall shear stress τ_w of 65 Pa, ξ is $(8.13/65) = 0.125$ and u_* is $(65/1540)^{0.5} = 0.2054$. The approximate velocity, from Equation 4–12, is 3.68 m/s. To evaluate V_N at this shear stress, we use Equation 1–5 in the form

$$V_N = \left(\frac{2\tau_w}{\rho f}\right)^{0.5}$$

f in this expression is evaluated at the Reynolds number $D\rho V_N/\mu_p$ so that iteration is required. After iteration, V_N is found to be 3.51 m/s and V is 3.74 m/s.

Power Law Fluid (Kemblowski and Kolodziejski Method)

For power law fluids, one can calculate τ_w for a given V. We recall that for this slurry K' was 5.213 Pa $s^{0.213}$ and n' (or n) was 0.213. At a velocity of 2.0 m/s, $\mathrm{Re}_{n'}$ is given by Equation 4–14 as 3257. The factor $\Phi^{1/\mathrm{Re}_{n'}}$ is 1.410 and m is (-0.055). The value of the Fanning friction factor is found to be 0.00576. τ_w is calculated from Equation 1–5 as $[0.5(0.00576)(2.0)^2(1540)]$ or 17.7 Pa.

Figure 4–3 shows that the power law predictions are somewhat better than the Bingham fluid ones in this case.

4.5 SLURRIES CONTAINING COARSE PARTICLES

It must be emphasized that the models we have considered here should only be used for homogeneous slurries. Coarse particles in a slurry can lead to serious errors of interpretation of test data. This probably occurred in the case of the limestone slurry of Figure 4–5. Although this limestone slurry contained some coarse particles, no deposition was observed in the experiments. In fitting the experimental data, turbulence was suspected at the two higher velocities so that only the five lower velocity values were used to determine the parameter pairs (τ_y, μ_p) and (K', n').

The agreement of the turbulent region calculations with the measurements for this slurry is not satisfactory and it is likely that this flow was not homogeneous. Inhomogeneity in industrial slurry flows is often difficult to determine visually because high concentrations of fine particles make the slurry opaque. The layer model described in Chapter 6 shows that coarse particles in a slurry contribute a frictional resistance which could be misinterpreted as a yield stress effect if a homogeneous model were used inappropriately. Measurement of the velocity distribution or concentration distribution is necessary to ensure that the flow is axisymmetric, as it must be if the models discussed in this chapter are to be used.

4.6 LAMINAR–TURBULENT TRANSITION

The apparent transition from laminar flow at a velocity of 2.5 m/s in Figure 4–4 can be considered with the assistance of Hanks's (1963) criterion. For transition from laminar to turbulent flow of a Bingham fluid, the criterion can be expressed in terms of the Hedstrom number He $= \rho D^2 \tau_y / \mu_p^2$ as

$$\left(\frac{DV\rho}{\mu_p}\right)_{\text{trans}} = 2100\left[1 - \frac{4}{3}\alpha_c + \frac{1}{3}\alpha_c^3\right](1 - \alpha_c)^{-3} \qquad (4\text{–}21)$$

where $\alpha_c/(1 - \alpha_c)^3 = \text{He}/16800$. The predicted value in this case $V_{\text{trans}} = 2.5$ m/s, which is indeed close to the apparent change of regime.

The Metzner–Reed Reynolds number $\text{Re}_{n'}$ in Equation 4–14 was defined so that in laminar flow

$$f = 16/\text{Re}_{n'} \qquad (4\text{–}22)$$

Experiments show that the transition region usually occurs between $\text{Re}_{n'} =$ 2100 and 3000. The uncertainty resulting from the width of this range is usually not a serious problem, provided the rheological parameters of the fluid are known.

For Newtonian fluids, the friction factor increases abruptly from the laminar flow value when the flow becomes turbulent. With non-Newtonian slurries, no discontinuity is observed in the friction factor versus Reynolds number (or headloss versus velocity) relationship although the transition is clearly defined, especially with Bingham fluids. A practical solution to the transition prediction problem is to calculate the headlosses from laminar and turbulent flow models for a range of velocities and take the higher of the two predictions.

A more serious problem arising from the laminar to turbulent flow transition is the difficulty of producing laminar flow at a desired wall shear stress in a tube viscometer. Consider the slurry of Example 4.2 and suppose that a tube viscometer of diameter 10 mm had to be used. To design a 0.1047-m pipeline with a velocity near 3 m/s, τ_w values near 32 Pa would be required. If n' and K' have the values given in Equation 4–8, one cannot achieve a wall shear stress of 32 Pa in laminar flow with the 10-mm tube. This means that one may be forced to use τ_w values below the range of greatest importance to characterize a fluid with a tube viscometer.

4.7 SCALEUP USING TURBULENT FLOW DATA

Bowen's (1961) method is based upon the assumption that in turbulent flow the friction factor for a given slurry will depend on the pipe diameter and the velocity. This means that

$$\tau_w = \frac{K'' V^w}{D^m} \qquad (4\text{–}23)$$

where w and m are positive exponents and K'' is a fluid parameter. On dimensional grounds, $w + m$ should equal 2. With tests conducted using small tubes, the functional relationship in Equation 4–23 is established and, in principle, this may be applied to any tube in turbulent flow. However, Kenchington (1972) has shown that if Equation 4–23 is used to predict behavior in a pipe very much larger than those used in the tests, an unrealistic degree of precision is required of the small tube data.

The examples have shown that the turbulent region is still not as well defined as one might wish. The discrepancies between predictions, combined with the practical difficulties mentioned previously (strong dependence of rheology on concentration and slurry composition, time dependence observed during concentric cylinder viscometry) often lead to use of safety factors in design. For major projects, testing of the slurry at close to full pipe scale is often more economical than use of these allowances for ignorance. In other cases, turbulent flow tests with a small pipe are often preferable to viscometry.

A technique for point-by-point scaleup of turbulent flow data has been proposed by Wilson (1989) using the concepts employed in deriving Equations 4–11 and 4–13. At a given wall shear stress, the velocity V_1 of flow in a pipe of diameter D_1 may be used to compute the velocity V_2 in a pipe of diameter D_2 using the relationship

$$V_2 = V_1 + 2.5u_* \ln\left(\frac{D_2}{D_1}\right) \tag{4–24}$$

where $u_* = (\tau_w/\rho)^{0.5}$. The technique may also be used for solutions in which the Toms effect produces drag reduction.

Chapter 5
Correlations for Nonhomogeneous Slurries

5.1 INTRODUCTION

For slurries which show a settling tendency, pipeline design requires prediction of deposit velocities and frictional headlosses. The complexity of these slurry flows is evident from a list of the independent variables: pipe diameter and roughness, pipe inclination, carrier fluid density and viscosity, particle density, particle shape, particle size and size distribution, particle concentration, coefficient of particle-wall friction, and the interparticle angle of internal friction ϕ. To these we must add all the factors which determine the rheological properties of the mixture and, in the case of headlosses, the mean velocity.

In dealing with complex phenomena the engineer is often forced to use correlations that summarize the results of experimental investigations. Unfortunately, some of the relevant variables have not been recorded in these investigations. This explains why correlations must be regarded as useful approximations.

The situation is not completely bleak because circumstances often only require estimations of pressure drops and deposition velocities. Moreover, practical experience with a given type of slurry will reduce dependence on a correlation. The value of this experience is greatest in dealing with slurry rheology and pipe-size effects because the vast majority of the well-documented data in the literature refer to small pipes (75 mm I.D. and below) with water-based slurries of particles with narrow size distributions and temperatures in the vicinity of 20 °C. With large pipes, nonaqueous slurries, or aqueous slurries with sufficient fines to alter the rheology, a laboratory test program will often be justified for major projects.

Since this book is intended as a guide to the designer and plant engineer rather than a historical survey of the field, the correlations presented in this chapter represent only those we consider to be the most useful. As the data base for slurry flow continues to grow, especially for large pipes, these correlations will gradually diminish in importance compared to mechanistic models.

5.2 DEPOSITION VELOCITY

Probably the simplest indication of nonhomogeneous behavior is the formation of a stationary deposit in a pipe at low velocities. The velocity at which this occurs is clearly of considerable importance because a stationary deposit will obstruct a portion of the pipe cross section and the roughness of its upper surface will normally exceed that of the bare pipe. For these reasons, headlosses would be expected to increase at V below the limiting case. Consequently, the optimum velocity, from an energy standpoint at least, should lie near V_c.

Some of the attempts to obtain a correlation for V_c have considered it to be associated with one of the following:

1. a change, such as a minimum, in the headloss versus velocity relationship (Bain and Bonnington, 1970). Such minima have been reported in some experimental studies;
2. fluctuating solid velocities, as measured by electrical, ultrasonic, or heated sensors (Ercolani et al., 1979);
3. the equilibrium of forces on the first particle layer to form (Raudkivi, 1976; Hanks and Sloan, 1981);
4. the velocity distribution of solid particles near the bottom of the pipe. An inflection point characterizes the critical flow regime (Roco and Shook, 1985b).

The first of these approaches is unsatisfactory because the minima are often poorly defined and the correlations used in obtaining the minimum are only approximate. The third approach will be discussed in Chapter 6 dealing with the two-layer model for slurry flows. The fourth approach will be discussed in Chapter 7.

Numerous correlations have been proposed for deposit velocity and the disagreement between them has been well documented (Carleton and Cheng, 1974; Oroskar and Turian, 1980).

The importance of the deposit velocity increases markedly with the size of the pipe. Wilson's (1979) nomogram, Figure 5–1, proposed originally for the highest value of V_c (at any concentration), is a useful simple method for estimating the conditions where deposition becomes likely. To predict the velocity at which a deposit forms in a pipe of diameter D, a line is drawn from the D-axis through the appropriate particle diameter. The extrapolated line intersects the central vertical axis at the value of V_c for particles of density ratio $S_s = 2.65$. For particles of other S_s values, a line is drawn from the central ($S_s = 2.65$) axis through the appropriate S_s value to intersect with the right-hand vertical axis.

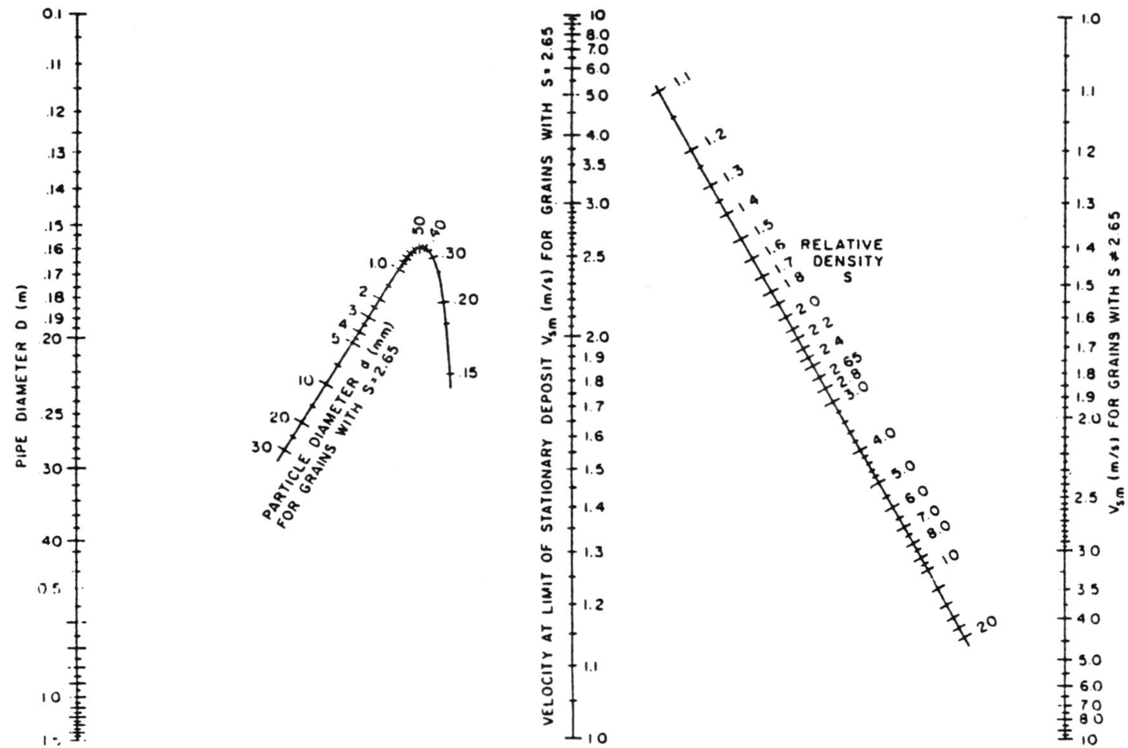

Figure 5–1. Nomogram for estimating deposition velocities. (From Wilson, 1979. *Proc. Hydrotransport 6 Conf.*, p. 11. Reprinted with permission.)

The effects of viscosity (fines content or temperature) are not included in the nomogram so that caution is required. The method does have a theoretical basis, however. Its inherent lack of precision is a realistic reflection of the uncertainty in estimating the deposition velocity from any correlation. Although slurry concentration corrections have been proposed for the nomogram, the data base to support this refinement is limited.

The nomogram summarizes important observations concerning the effects of pipe and particle diameter. For the latter, we see that V_c becomes progressively less sensitive to d and eventually shows a tendency to decrease above about 0.5 mm. The data base for large particles in large pipes is meager and some observations (such as Carleton et al., 1978) have suggested that this decrease does not always occur. This discrepancy may be a particle shape effect but the net result is to increase the uncertainty of the predictions in the coarse particle region.

Because of the neglected variables, one should not expect predictions to be much better than $\pm 20\%$ in any situation. The correlation of Oroskar and Turian (1980) was derived from data which included few results for large particles and is best used for particles of diameter below the 0.5-mm limit. Their correlation uses a hindered settling velocity of the particles, estimated from the expression

$$V_s = V_\infty (1 - C_v)^2$$

The correlation is

$$\frac{V_c}{[gd(S_s - 1)]^{0.5}} = 1.85\, C_v^{0.1536} (1 - C_v)^{0.3564} \left(\frac{D}{d}\right)^{0.378} \mathrm{Re}_p^{0.09} \mathbf{X}^{0.30} \tag{5–1}$$

where

$$\mathrm{Re}_p = \frac{D \rho_L [gd(S_s - 1)]^{0.5}}{\mu_L}$$

$$\mathbf{X} = \frac{2}{\sqrt{\pi}} \left[\frac{2}{\sqrt{\pi}} \gamma \exp\left(-\frac{4\gamma^2}{\pi}\right) + \int_\gamma^\infty \exp\left(-\frac{4\gamma^2}{\pi}\right) d\gamma \right]$$

$$S_s = \frac{\rho_S}{\rho_L}$$

$$\gamma = \frac{V_s}{V_c}$$

The correlation is implicit in V_c so that iteration is necessary. In the common situation where $V_s / V_c < 0.5$, **X** lies between 0.9 and 1.0. The factor $\mathbf{X}^{0.3}$ thus approaches unity within the accuracy of the correlation.

Gillies's correlation (Gillies and Shook, 1991) was derived from data obtained with pipes up to 0.5 m in diameter and aqueous slurries with viscosities between 0.5 and 5 centipoise. It employs an equivalent fluid density ρ_f calculated from the mean in situ total solids concentration C_r and the concentration of -74-μm particles, C_f.

C_f is defined as $C_r \times$ (volume of -74-μm solids/total volume of solids). The equivalent fluid density is

$$\rho_f = \frac{\rho_s C_f + \rho_L(1 - C_r)}{1 - C_r + C_f} \tag{5-2}$$

The deposit velocity is expressed in terms of the factor F, first used by Durand and Condolios (1952),

$$\frac{V_c}{[gD(\rho_s - \rho_f)/\rho_f]^{0.5}} = F \tag{5-3}$$

F depends upon the drag coefficient of the median (d_{50}) diameter of the $+74$-μm fraction, settling in a "fluid" of density ρ_f and viscosity μ_f. The viscosity μ_f should be measured but it may be necessary to estimate it from a correlation if data are not available.

F is obtained from the equation

$$F = \exp(0.51 - 0.0073 C_D - 12.5 K_2) \tag{5-4}$$

where $K_2 = (K_1 - 0.14)^2$. The factor K_1 contains the viscosity and density of the carrier fluid:

$$K_1 = \frac{(\mu_L/\rho_L)^{2/3}}{g^{1/3} d_{50}}$$

Gillies's correlation includes more coarse particle and large pipe data than that of Oroskar and Turian and this may explain the significant difference between the exponents of D in the two cases.

Although correlations for very fine particles have been proposed (for example, A.D. Thomas, 1979), Etchells (1986) suggests that a lower bound can be estimated from

$$V_c = 0.3[2gD(S_s - 1)]^{0.5} \tag{5-5}$$

Example 5.1

A slurry of sand ($\rho_s = 2650$ kg/m^3, $d_{50} = 0.34$ mm) was transported in a pipe of diameter 0.263 m/s at an in situ concentration of 15% by volume. The fluid viscosity was 1.1 centipoise and the density was 999 kg/m^3. The sand contained 1.6% of -74-μm material. The viscosity of the mixture of water and -74-μm solids was found to be 1.3 centipoise. Calculate V_c.

Solution

Oroskar and Turian Correlation:
The drag coefficient is calculated first:

$$\mathrm{Ar} = 1.33(0.34)^3(10^{-9})(2650 - 999)(999)9.8/(0.0011)^2 = 698$$

From Equation 1–24 and Table 1–2, the single-particle drag coefficient is found to be 3.60. This gives a single-particle settling velocity $V_\infty = 0.045$ m/s. The velocity V_s is then $0.045(1 - 0.15)^2 = 0.033$ m/s. The factor X is taken as unity initially. Substituting

$$\sqrt{gd(S_s - 1)} = \left[(9.8)(0.00034)\left(\frac{2650}{999} - 1\right)\right]^{0.5} = 0.0742$$

$$\mathrm{Re}_p = \frac{(0.263)(999)(0.0742)}{0.0011} = 1.77 \times 10^4$$

$$\left(\frac{D}{d}\right)^{0.378} = \left(\frac{0.263}{0.00034}\right)^{0.378} = 12.36$$

Since C_v is unknown, the difference between C_v and C_r is neglected in evaluating the concentration term.

$$C_v^{0.1536}(1 - C_v)^{0.3564} = (0.15)^{0.1536}(0.85)^{0.3564} = 0.705$$

Substituting these values, the predicted deposition velocity is

$$V_c = 1.85(0.0742)(0.705)(12.36)(1.77 \times 10^4)^{0.09}(1) = 2.88 \text{ m/s.}$$

Using V_s and this value of V_c, we find $\gamma = 0.033/2.88 = 0.0114$. This is very low so the value of X used above was reasonable. Although C_v at deposition was not known exactly in this experiment, it was likely to have been greater than 0.075. For $C_v = 0.075$, the concentration term would be 0.653 compared to 0.705, and the predicted V_c would be 2.67 m/s.

Gillies's Correlation:
The density of the equivalent fluid is

$$\rho_f = \frac{[(2650)(0.016)(0.15) + (999)(1 - 0.15)]}{1 - 0.15 + (0.016)(0.15)} = 1004 \text{ kg/m}^3$$

This density is used in Ar to determine the drag coefficient. d_{50} for the $+74$-μm solids is very nearly 0.34 mm so that

$$\text{Ar} = \frac{1.33(0.34)^3(10^{-9})(2650 - 1004)(1004)(9.8)}{(0.0013)^2} = 501$$

From the correlation in Equation 1–24 and the coefficients in Table 1–1, C_D = 4.22.

$$K_1 = \frac{(0.0011/999)^{2/3}}{(9.8)^{1/3}(0.00034)} = 0.1467$$

$$12.5K_2 = 12.5(0.1467 - 0.14)^2 = 5.554 \times 10^{-4}$$

$$\exp[0.51 - 0.0073(4.22) - 12.5K_2] = 1.614$$

$$V_c = 1.614\left\{9.8(0.263)\left[\left(\frac{2650}{1004}\right) - 1\right]\right\}^{0.5} = 3.3 \text{ m/s}$$

The observed value was 3.3 m/s.

The two correlations are complementary. For example, with a slurry of 0.19 mm sand particles conveyed in a high viscosity fluid (μ_L = 38.1 centipoise, ρ_L = 1133 kg/m^3) in a pipe of diameter 0.0524 m, at C_r = 0.24, the Oroskar–Turian correlation predicts V_c = 0.6 m/s, which is almost identical to the observed value. Although this flow was laminar, the Oroskar–Turian correlation was derived from data which included laminar flows. As a rule, the Oroskar–Turian correlation is less reliable for large particles in large pipes where, as Wilson's nomogram shows, V_c does not continue to increase with d_{50}. Gillies's correlation is not applicable at this high viscosity.

5.3 HEADLOSS CORRELATIONS FOR HORIZONTAL FLOW

A number of correlations have been proposed to assist in the estimation of headlosses for nonhomogeneous flows. Many of these have been restricted to particular flow regimes, but these regimes are poorly defined because they are based upon visual observations of particle motion in small laboratory pipelines.

A useful dependent variable in headloss correlations is the dimensionless excess headloss Φ. This is defined in terms of the headlosses:

i: slurry frictional headloss (meters of fluid per meter of pipe)
i_L: fluid frictional headloss in the same pipe at the same V value

$$\Phi = \frac{i - i_L}{C_v i_L} \quad \text{or} \quad i = i_L(1 + \Phi C_v) \tag{5-6}$$

i_L can be related to the wall shear stress in terms of a friction factor since the flow of the pure fluid would be axisymmetric:

$$i_L \rho_L g = \frac{4\tau_{wf}}{D}$$

A truly homogeneous flow would also be axisymmetric so that a wall shear stress τ_w could be defined

$$i \rho_L g = \frac{4\tau_w}{D}$$

C_v would be the solid concentration in a homogeneous slurry and the mixture density would be $\rho_L[1 + C_v(S_s - 1)$. Equation 1–5 shows that the wall shear stress for homogeneous fluid varies as the product of the friction factor and the density. Using the definitions of i and i_L we see that for a truly homogeneous flow, Φ approaches $S_s - 1$ if the friction factor for a slurry is the same as that for the fluid flowing at the same velocity in the same pipeline.

The approximation $\Phi = S_s - 1$ is sometimes called the *pseudo-homogeneous flow* approximation. It will be justified if the particle relaxation time is small in comparison with the time scale of the turbulence, and if the mixture viscosity approaches that of the carrier fluid. It should be noted that as we define the term, *pseudo-homogeneity* is a property of a flowing slurry. Flows which are neither homogeneous nor pseudo-homogeneous are sometimes called *heterogeneous,* although the term is sometimes used for a subset of the group (see Section 5.6 for an example of this).

One of the most frequently used groups of independent variables is Ψ defined in the Durand–Condolios equation (1952)

$$\Psi = \frac{V^2\sqrt{C_D}}{gD(S_s - 1)}$$

A common version of the correlation is

$$\Phi = 81\Psi^{-3/2} \tag{5-7}$$

Since the density correction in Ψ was inserted by later workers, somewhat different versions of this correlation have been used.

Example 5.2

A slurry of sand ($\rho_s = 2650$ kg/m^3) in water ($\mu_L = 0.00112$ Pa s, $\rho_L = 999$ kg/m^3) is transported in a pipe of $D = 0.263$ m at 3.06 m/s. d_{50} for the mixture is 0.168 mm and C_v is 0.259. i_L was 0.0222 m water/m pipe. Calculate i.

Solution

Ar for the particles is 81.5 and, from the correlation in Chapter 1, $C_D = 10.0$. Substituting

$$\Psi = \frac{(3.06)^2\sqrt{10}}{9.8(0.263)[(2650/999) - 1]} = 6.95$$

$$\Phi = 81(6.95)^{-3/2} = 4.42$$

and from Equation 5–6

$$i = 0.0222[1 + 4.42(0.259)] = 0.0476 \text{ m water/m pipe}$$

Equation 5–7 does not approach the $S_s - 1$ limit as $\Psi \to \infty$ and there have been a number of attempts to rectify this deficiency. Charles (1970) suggested

$$\Phi = 120\Psi^{-3/2} + (S_s - 1) \tag{5–8}$$

and Zandi and Govatos (1967) proposed two expressions of the form

$$\Phi = K\Psi^{-n} \tag{5–9}$$

where $n = 0.354$ and $K = 6.3$ for $\Psi > 10$ and $n = 1.93$ and $K = 280$ for $\Psi < 10$. Equation 5–9 was restricted to the region $\Psi/C_v < 40$.

Although other correlations have been proposed, those produced before 1970 or thereabouts were severely hampered by lack of data for large pipes.

A two-constant equation which will be seen to have much in common with theoretical predictions is that of Gaessler (see Prettin and Gaessler, 1976). Written in terms of our notation, this would be

$$\Phi = \frac{K_1(S_s - 1)gD}{f_L V^2} + \left(\frac{f_L}{f_s} - 1\right) \tag{5–10}$$

K_1 is a dimensionless coefficient and f_L is the fluid friction factor at the conditions of the flow. f_s is a "particle friction" coefficient. Gaessler's correlation is described in detail by Govier and Aziz (1972).

We note that Equations 5–9 and 5–10 also deviate from the asymptotic pseudo-homogeneous flow condition as Ψ approaches infinity. We also observe:

1. there is no explicit slurry viscosity term. Absence of this parameter from correlations may reflect the low concentrations used in most early laboratory studies. It may also reflect the comparatively large diameter of the particles for which the correlations were developed;
2. the correlations contain no terms dependent on a coefficient of particle-wall friction;
3. a method must be devised to deal with broad size distributions.

For moderately broad size distributions, the empirical weighting method of Equation 5–11 seems to be as good as any other. The size distribution is considered to be divided into a number of intervals. The ith interval contains a volume fraction x_i of particles, where $\Sigma x_i = 1.0$. Particles in the ith size interval settle (at infinite dilution) in the carrier fluid with drag coefficient C_{Di}. The weighted mean drag coefficient for the mixture is computed as

$$\frac{1}{C_{Dm}} = \Sigma\left(\frac{x_i}{C_{Di}}\right) \tag{5-11}$$

Correlations based upon Ψ usually overestimate headlosses as the pipe diameter increases. This can be seen in Figure 5–2 which compares headlosses, expressed as Φ and Ψ values, for a sand with a narrow size distribution. Any $\Phi - \Psi$ correlation developed from small pipe measurements would overestimate Φ for large ones.

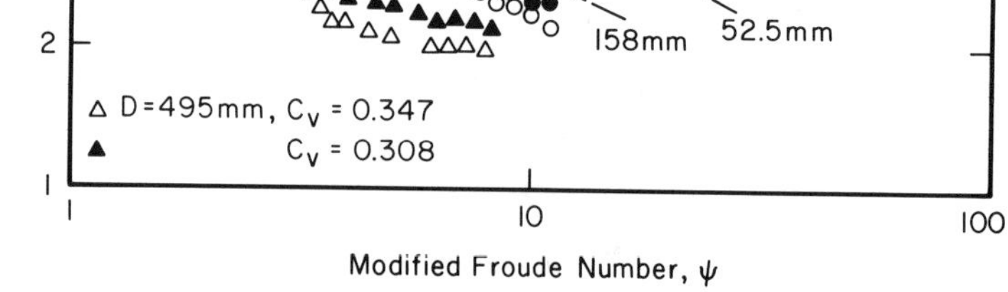

Figure 5–2. Experimental data obtained with a 0.18-mm sand, plotted as Durand–Condolios parameters. The dashed lines show the loci of the results obtained with small pipes. (From Shook et al., 1982a. *J. Pipelines* 3: p. 18. Adapted with permission.)

5.4 BROAD SIZE DISTRIBUTIONS

A number of modifications have been proposed to take account of broad particle size distributions without resorting to the empirical stratagem of Equation 5–11. These modifications are based upon the assumption that the particle mixture can be divided into "homogeneous" and "heterogeneous" fractions (Wasp et al., 1970):

$$C_v = C_{v,\text{hom}} + C_{v,\text{het}} \tag{5-12}$$

The homogeneous fraction is considered to increase the viscosity and the density of the "equivalent liquid vehicle," compared to the properties of the carrier fluid:

$$\mu_{L,\text{eq}}/\mu_L = f(C_{v,\text{hom}})$$

$$\rho_{L,\text{eq}}/\rho_L = 1 + C_{v,\text{hom}}(S_s - 1) = S_{L,\text{eq}}$$

In this approach, i_L in Φ is replaced by $i_{L,\text{eq}}$ expressed in meters of fluid of density $\rho_{L,\text{eq}}$. C_v is replaced by $C_{v,\text{het}}$. $(S_s - 1)$ and C_D in Ψ are also modified because the carrier fluid properties are considered to have been altered by the presence of the particles.

Various methods for making these corrections have been proposed. One method is to assume that particles smaller than a particular size constitute $C_{v,\text{hom}}$. Faddick (1982) suggested that particles which settle with values of C_D greater than 24 contribute to the "equivalent liquid" vehicle. We will employ this suggestion in the sample calculation shown below.

For the "heterogeneous" particles, Ψ is estimated from an expression such as

$$\Psi = \left(\frac{V^2}{gD(S_s/S_{L,\text{eq}} - 1)}\right)\left(\frac{1}{\Sigma x_{i,\text{het}}}\right)\Sigma(x_i\sqrt{C_{Di}})_{\text{het}} \tag{5-13}$$

The summations in Equation 5–13 are performed for those fractions of the original mixture which contribute to $C_{v,\text{het}}$. C_{Di} values are calculated from the Archimedes number $C_D\text{Re}^2$ and a correlation such as those shown in Figures 1–8 and 1–9. The density $\rho_{L,\text{eq}}$ and the viscosity $\mu_{L,\text{eq}}$ are used in $C_D\text{Re}^2$.

To illustrate the method, Example 5.2 will be reconsidered using the complete size distribution tabulated below.

Example 5.2 continued

We begin by assuming that the material finer than 105 microns will comprise the vehicle. The equivalent liquid thus contains 36.2% of the solids. Its density is

$$(999)[1 + 0.362(0.259)(1.652)] = 1154 \text{ kg/m}^3 = \rho_{L,\text{eq}}$$

Table 5–1. Size Distribution and Calculated Quantities in Example 5.2

Size (μm)	x_i	$d_i(\mu m)$	Ar	C_D	$(x_i\sqrt{C_{Di}})_{het}$
−2380 + 1680	0.002	2000	78060	1.09	0.002
−1680 + 1190	0.011	1414	27600	1.19	0.012
−1190 + 841	0.018	1000	9775	1.46	0.022
−841 + 596	0.077	708	3465	1.79	0.103
−596 + 420	0.112	500	1223	2.76	0.186
−420 + 297	0.076	353	430	4.54	0.162
−297 + 210	0.087	250	152	7.44	0.237
−210 + 149	0.169	177	54.0	12.16	0.589
−149 + 105	0.086	125	19.1	19.92	0.384
−105 + 74	0.030				
−74 + 44	0.048				
−44	0.284				

The concentration of the vehicle is low (0.362 × 25.9%) and we can probably estimate its viscosity from a correlation. In the absence of specific information, we assume that this fraction produces a viscosity corresponding to that of deflocculated monodisperse spheres. Using Equation 3–24, the viscosity of the equivalent liquid is estimated to be 1.34 times that of the water or 1.5 mPa. This allows the value of $C_D\,Re^2$ to be calculated for each interval. d_i is taken as the geometric mean of the upper and lower limits of the interval. The correlation given in Equation 1–24 and Table 1–2 is used to estimate the drag coefficients. The table suggests that the cutoff size for the homogeneous load fraction was approximately correct.

From Table 5–1 we find $\Sigma(x_i\sqrt{C_{Di}})_{het} = 1.697$,

$$\frac{V^2}{gD(S_s/S_{L,eq} - 1)} = \frac{(3.06)^2}{(9.8)(0.263)[(2650/1154) - 1]} = 2.807,$$

$C_{v,het}$ is $0.259(1 - 0.362) = 0.165$ and $\Sigma x_{i,het}$ is 0.638.

$$\Psi = [2.807(1.697)]\left(\frac{1}{0.638}\right) = 7.466$$

and

$$\Phi = 81(7.466)^{-1.5} = 3.97$$

Thus,

$$i = i_{L,eq}[1 + 0.165(3.97)] = 1.66\, i_{L,eq}$$

We know from Equations 1–3 and 1–5 that the pressure gradient scales as the fluid density very nearly so that $i_{L,\text{eq}}$ should be $0.0222 S_{L,\text{eq}}$ or 0.0256 m(water)/m(pipe). This gives a predicted value of 0.0425 m(water)/m(pipe) for i. This is higher than the observed value of 0.0381 m/m, probably because the Durand–Condolios equation tends to overpredict Φ for large pipes.

The predicted headloss is not very sensitive to the cutoff point location. For example, if the "heterogeneous" fraction had begun at 149 μm, the predicted value of i would have been 0.042 m/m. We note that the broad size distribution produces a lower headloss than a narrow one, for a given value of d_{50}.

In this example, use of Equation 5–11 to obtain C_D would not have produced an improved prediction, compared to the Durand–Condolios equation. Equations 5–8 and 5–9 would also have given inferior predictions. These conclusions cannot be regarded as general rules, however.

The modification of the Durand–Condolios equation shown above is a simplified version of the procedure suggested by Wasp et al. (1970). In terms of the symbols used above, their procedure is equivalent to

$$\frac{x_{i,\text{hom}}}{x_i} = \exp\left(-\frac{bV_{i\infty}}{u_*}\right)$$

where $u_* = (igD/4)^{0.5}$, $V_{i\infty}$ is the terminal velocity of species i in the "equivalent liquid" and b is a dimensionless constant. For this particular example, the measured headloss indicates that b is 4.5. However, because of the aforementioned unsatisfactory performance of the Durand–Condolios equation for large pipes, the method should be used with caution.

An advantage of the procedure described above is that it can be extended to non-Newtonian carrier fluids. Of course, an experimental verification of the predictions would be necessary, particularly for large pipes. A useful data source for the flow of coarse particles in Newtonian and non-Newtonian carrier fluids is the study of Chhabra and Richardson (1985).

For rough pipes, the limited amount of data available at present suggests that correlations which assume i to be proportional to i_L are reasonably satisfactory. Figures 5–3 and 5–4 display headlosses for a fine sand (d = 0.18 mm) in rough and smooth 495-mm pipelines (Shook et al., 1982a,b). The rough pipe measurements were conducted at progressively increasing concentrations; water friction measurements were made before and after the slurry runs to detect changes in wall roughness. The final water friction test shows that only a small change in wall roughness occurred because of the slurry flows.

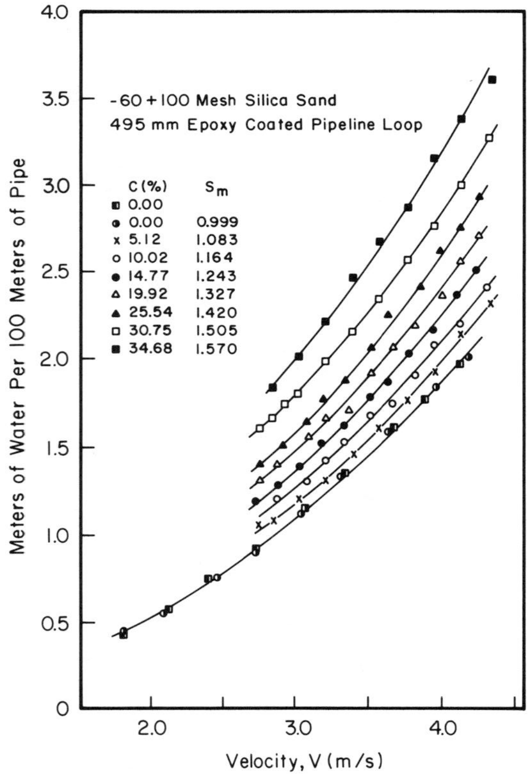

Figure 5–3. Headlosses for a sand–water slurry in a smooth pipe. (Adapted from Shook et al., 1982b.)

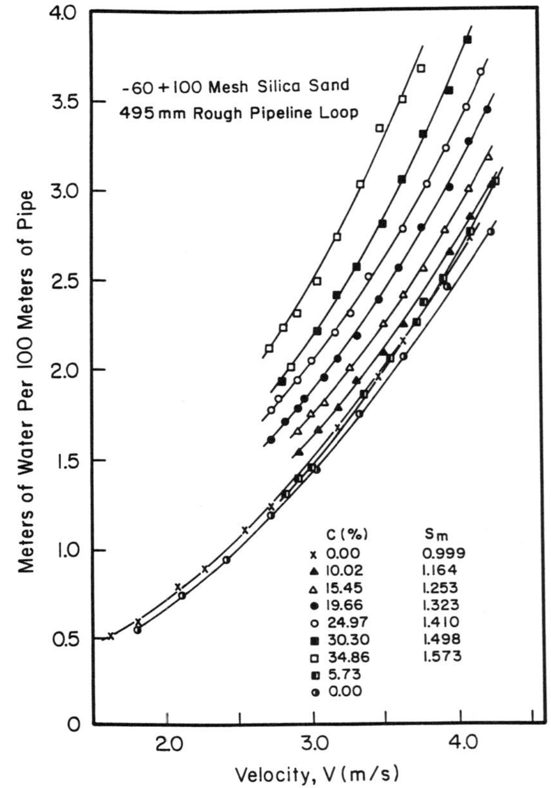

Figure 5–4. Headlosses for the sand–water slurry of Figure 5–3 in a rough pipe. (Adapted from Shook et al., 1982b.)

5.5 REGIME-SPECIFIC CORRELATIONS

For slurries of coarse particles, or for slurries in which friction is dominated by the coarse fraction of the mixture, modified versions of the Durand–Condolios equation are not recommended. Instead, recourse can be made to the two-layer model described in Chapter 6. A precursor to this model was the equation of Newitt et al. (1955). This expression could be regarded as an upper limit to the headloss:

$$\Phi = \frac{66gD(S_s - 1)}{V^2} \tag{5–14}$$

which in terms of i is approximately

$$i = i_L + 0.8C_v(S_s - 1)$$

We note the resemblance to the first term of Gaessler's equation and to other empirical expressions containing Ψ. Since the C_D values for coarse particles would be size insensitive, the right-hand side of Equation 5–14 could be expressed in terms of Ψ.

Newitt's equation was described as applying to "flow with saltation or with a sliding bed." Two other flow regimes, "heterogeneous flow" and "homogeneous flow," were also defined and separate equations were proposed for each of these. From the standpoint of data correlation, this approach is logical: one employs a set of criteria for defining the flow regime and a regime-specific set of equations.

The disadvantage of this method is that subsequent research has shown additional variables to be of importance. Furthermore, visual observations of the flow in large pipes show that flow regimes are by no means as distinct as these definitions imply. One can identify flow with a stationary deposit (bed flow) and truly homogeneous flow of nonsettling slurries. However, saltation flow and heterogeneous flow are very difficult to distinguish and sliding bed flow is extremely rare. In recent years, emphasis has shifted to mechanistic modeling, in contrast with use of regime-specific correlations.

5.6 TURIAN–YUEN CORRELATION

Because of the scope of its data base, the correlation of Turian and Yuen (1977) is likely to remain of interest. Most of the data appear to have been obtained with small ($D < 160$ mm) pipes and aqueous slurries, but the particle diameter and density ranges were broad. Four flow regimes were defined and coded as shown:

Regime	*Code*
Stationary bed	0
Saltation	1
Heterogeneous	2
Homogeneous	3

Note that "homogeneous" according to these authors includes both of the regimes we have described previously as "pseudo-homogeneous" and "homogeneous."

The slurry headloss is expressed in terms of the carrier liquid density ρ_L and a friction factor f:

$$i\rho_L g = \frac{2fV^2\rho_L}{D} \tag{5-15}$$

f is related to the carrier fluid friction factor in the same pipe at the same velocity, f_L, by the correlation:

$$f - f_L = KC_v^{\alpha} f_L^{\beta} C_D^{\gamma} \left(\frac{V^2}{gD(S_s - 1)} \right)^{\delta} \tag{5-16}$$

The set of coefficients is given in Table 5–2.

To select the appropriate regime, a set of regime numbers must be calculated from the definition

$$R_{ij} = \frac{V^2}{K1 C_v^{\alpha 1} f_L^{\beta 1} C_D^{\gamma 1} gD(S_s - 1)} \tag{5-17}$$

The regime number coefficients are given in Table 5–3.

To establish the flow regime, one computes $(R_{01} - 1)$, $(R_{12} - 1)$, and $(R_{23} - 1)$. Table 5–4 is then inspected to see if the regime is determined. If an ambiguity arises, it is resolved by computing $(R_{13} - 1)$, $(R_{02} - 1)$, or $(R_{03} - 1)$.

The drag coefficients used in these equations were computed from a correlation expressed in terms of $C_D^{0.5}\text{Re}$. However, it would be reasonable to use measured coefficients if these were available.

Table 5–2. Coefficients in the Yuen–Turian Correlation (Equation 5–16)

Code	K	α	β	γ	δ
0	0.4036	0.7389	0.7717	−0.4054	−1.096
1	0.9857	1.018	1.046	−0.4213	−1.354
2	0.5513	0.8687	1.200	−0.1677	−0.6938
3	0.8444	0.5024	1.428	0.1516	−0.3531

Table 5–3. Coefficients Determining Regime Number (Equation 5–17)

Regime Number	$K1$	$\alpha 1$	$\beta 1$	$\gamma 1$
R_{01}	31.93	1.083	1.064	−0.0616
R_{12}	2.411	0.2263	−0.2334	−0.3840
R_{23}	0.2859	1.075	−0.67	−0.9375
R_{13}	1.167	0.5153	−0.382	−0.5724
R_{02}	0.4608	−0.3225	−1.065	−0.5906
R_{03}	0.3703	0.3183	−0.8837	−0.7496

Table 5–4. Criteria for Flow Regime Definition in Yuen–Turian Correlation

$R_{01}-1$	$R_{12}-1$	$R_{23}-1$	$R_{02}-1$	$R_{03}-1$	$R_{13}-1$	*Regime Code*
neg	neg	neg				0
pos	neg	neg				1
pos	pos	neg				2
pos	pos	pos				3
neg	neg	pos		neg		0
neg	neg	pos		pos		3
neg	pos	pos		neg		0
neg	pos	pos		pos		3
neg	pos	neg	neg			0
neg	pos	neg	pos			2
pos	neg	pos			neg	1
pos	neg	pos			pos	3

Source: Turian, R.M. and Yuan, T.F., 1977. Flow of slurries in pipelines. *AICHE Journal* **23** (3): 232–243 (Table 6). Reproduced by permission of the American Institute of Chemical Engineers,

Example 5.3

A water (ρ_L = 999 kg/m^3, μ_L = 1.11 mPa s) – coal (ρ_s = 1338 kg/m^3) slurry of concentration 38% by volume was transported in a pipeline 0.4953 m in diameter at 3.18 m/s. The mean particle diameter d_{50} was 0.74 mm and the drag coefficient C_D was 4.35. The pipe roughness was 9 μm. The observed head-loss i was 0.0175 m/m, corresponding to f = 0.0042. We will use the correlations to calculate a value of f for comparison to this measurement.

Solution

From the velocity, pipe diameter, carrier fluid viscosity and density, the carrier fluid Reynolds number is calculated. The pipe roughness allows k/D to be found. Using the Churchill correlation, we find f_L = 0.00287.

Equation 5–17 is used to calculate the regime coefficients. These are $R_{01} = 309$, $R_{12} = 1.42$, and $R_{23} = 4.76$. Table 5–4 then defines the flow regime as homogeneous. This cannot be disputed since visual observation of the slurry showed no segregation. However, segregation in a coal slurry in a large pipe is just about impossible to detect visually. In fact, deposition occurred at 2.7 m/s, and at 3.18 m/s the i vs. V relationship was similar to that called saltation flow.

Equation 5–16 is then used to predict f. This gives $f = 0.00295$, which is considerably different from the actual value. The conditions of this example were entirely within the ranges set by the data used in the correlation. However, as we noted earlier, the correlation was derived primarily from small-pipe experimental data. It is worth remarking that the heterogeneous flow and saltation flow correlations do not give better predictions in this case, so that the problem is not one of regime definition.

The example illustrates the dangers inherent in a nonmechanistic correlation. In addition to the uncertainty for large pipes, the correlation should be used with caution when a mixture contains significant quantities of fine particles that can flocculate.

5.7 VERTICAL FLOWS

Potential energy effects are usually the major component of the pressure drops which occur in vertical slurry flows. This is also the case in single phase flows for which the total pressure drop is given by integrating Equation 1–3:

$$-dP = \rho g dh + 4\left(\frac{\tau_w}{D}\right)dx \tag{5–18}$$

With inclined and vertical slurry flows, we have the complication that the appropriate density in the potential energy term should be that of the mixture, calculated from the mean in situ or spatial concentration. However, it is often the delivered concentration C_v that is specified in a design situation. For very large particles at low concentrations, the difference between the in situ and delivered concentrations may become important.

To illustrate the effect, consider sand particles ($S_s = 2.65$) of diameter 2.0 mm in water at 20 °C. The mixture velocity is 2.0 m/s and the mean delivered concentration C_v is 0.10.

We can estimate the slip velocity in upward or downward flow by neglecting inertia and the wall friction effects in Equations 2–16 and 2–17. This gives us a version of Equation 2–28:

$$\frac{f_{sL}}{1 - c} = (\rho_s - \rho_L)g\frac{dh}{dx}$$

For upward flow ($dh/dx = +1$) if we substitute for the drag coefficient, we find the slip velocity to be

$$v_L - v_s = \left(\frac{4gd(S_s - 1)}{3C_{Ds}}\right)^{0.5} (1 - c)^{1.35}$$

In the absence of information about the drag coefficient in a turbulent fluid, we will take C_{Ds} as the value of C_D from the correlation of Figure 1–9 for a single particle falling at its terminal velocity. Using the properties of the particles and the fluid, we find Ar = 1.73×10^5. Thus C_D is 1.09 and the slip velocity is $0.20(1 - c)^{1.35}$ m/s.

This expression must be used with the mass balances for solids and the total mixture given in Equations 2–1 to 2–10. Neglecting any variation of velocities or concentration over the cross section, Equation 2–1 implies:

$$V = cv_s + (1 - c)v_L$$

and from Equation 2–9,

$$C_v V = cv_s$$

Using these approximations we find, at $V = 2$ m/s and $C_v = 0.10$, $c = 0.108 = C_r$.

This means that the mean density ρ_m of the mixture in the pipe at any instant is 999 [1 + 0.108(1.65)] or 1177 kg/m^3. The difference between this density and that calculated from C_v is only 1.3%. This shows that particles would have to be considerably coarser than 2 mm before the effect of slip on pressure drop estimation became important. Although evidence of reduced drag coefficients in turbulent flow has been obtained, and these lower coefficients would lead to increased slip (Sellgren, 1982), the effect would not alter this conclusion dramatically.

The example suggests that potential energy effects in vertical flows can be estimated by assuming pseudo-homogeneous flow unless the particles are very coarse or V is close to V_∞. This approximation implies that Equation 5–18 can be used to calculate pressure drops with ρ_m replacing ρ. The limited evidence available at present suggests that the wall friction can also be estimated roughly from the pseudo-homogeneous flow expression

$$\Phi = (S_s - 1)$$

for vertical flows. Equation 5–6 would be used to calculate the frictional headloss i and, since the flow is axisymmetric, in Equation 5–18

$$\frac{4\tau_w}{D} = i\rho_L g$$

Evidence for near-pseudo-homogeneous friction in a 25-mm pipe at high concentrations was obtained by Brown (1980) using a range of particle diameters, for mean velocities greater than about 2 m/s. At very high concentrations and low velocities, the flow is more complex and the wall friction mechanism changes (Wilson et al., 1979).

At low and moderate (to 30% by volume) concentrations, the wall friction for coarse particle ($d > 0.5$ mm) slurries has been found to approach that of the carrier fluid (Newitt et al., 1961). Φ in this case approaches zero.

5.8 VELOCITY AND CONCENTRATION EFFECTS IN VERTICAL FLOW

The variation of solids concentration and velocity over the cross section and the presence of wall friction complicate vertical flows somewhat. Earlier authors (Govier and Aziz, 1972) considered these complications to be dominant and suggested that the terminal falling velocity of a single particle could be used for the relative velocity ($v_L - v_s$) instead of a value calculated from Equation 2–28. However, more recent work has confirmed the effect of solids concentration on the slip velocity. Nevertheless, the complications resulting from the variation of the velocity over the cross section should be considered.

If we neglect wall friction, Equation 2–28 applies at a point. The relative velocity $v_r = (v_L - v_s)$ can be expressed in terms of V_∞. For high Ar values we assume C_{Ds} is equal to C_D. Using the symbol n for the exponent on the $(1 - c)$ term ($n = 1.7$ for spheres), at high Ar numbers

$$v_r = V_\infty (1 - c)^{0.5(n+1)} \tag{5–19}$$

In the Stokes law region

$$v_r = V_\infty (1 - c)^{n+2} \tag{5–20}$$

If v is the time-average mixture velocity at a point,

$$v_s = v - v_r(1 - c) \tag{5–21}$$

c, v_s, and v_L all vary with radial position and we must use a modified form of Equation 5–21 to relate the mean velocities to the mean concentration. Since V_∞ is likely to be high only when Ar is high, we can define dimensionless factors β_1 and β_2 such that

$$V_s = \frac{V}{\beta_1} - \frac{V_\infty}{\beta_2}(1 - C_r)^{0.5(n+1)+1} \tag{5–22}$$

where β_1 and β_2 are both less than 1. The limited evidence available at present suggests β_2 is near 1 and β_1 ranges between 0.9 and 1.

If we reconsider the example in Section 5–7, using β_1 and $\beta_2 = 0.95$, the effect on C_r is very small.

5.9 MINIMUM VELOCITY FOR VERTICAL FLOW

In selecting the minimum velocity for vertical flows, Govier and Aziz (1972) suggested $V > 2V_\infty$, where V_∞ is the terminal velocity of the largest particles in the flowing mixture. This suggestion used V_∞ to estimate the maximum possible slip and included an allowance for the effect of the variation of solids concentration over the cross section.

Although it is possible to operate at such low velocities (and even lower with narrow particle size distributions), Sellgren's (1982) criterion (a factor of 6 or 7 instead of 2) seems to be a useful practical guide. If the pipeline is to have any horizontal sections, to avoid deposition in these, the velocity will be much greater than a limiting value chosen to accommodate the vertical pipe.

An upper limit to the particle size is imposed by the need to avoid bridging or jamming of particles to form an obstruction. The possibility of this occurring at low solids concentrations is directly related to the possibility of two (or perhaps three) particles being present at any point in the pipe. Thus, a useful rule to avoid jamming is to employ a pipe with D greater than three times the diameter of the largest particle.

If the particle size distribution varies with time, concentration changes can occur during vertical flow (Shook, 1989). In extreme situations, such concentration increases could cause a region of very high concentration to form. As we saw in Section 2.9, once the particles are in continuous contact, the particle-wall force is fundamentally different from that which applies in a flowing fluid, and that we normally assume to apply to slurry flow. At such very high concentrations, very high frictional forces are possible and a plug may form.

5.10 MEAN DENSITY FROM PRESSURE DROP

If we assume that vertical flows are pseudo-homogeneous, the pressure drop in a vertical section of pipe can be used to estimate the mean in situ concentration. This approach can also be used, with an additional degree of approximation, for inclined flows. The following example illustrates how the correction for wall friction can be estimated.

Example 5.4

A mixed salt in brine slurry flows upward in a pipe of I.D. 0.158 m at a velocity of 2.83 m/s. The total pressure drop over a 3.92-m section of pipe is measured as 63.09 kPa. The brine is known to have a density of 1180 kg/m^3.

When the clear brine flowed through the same pipe at a velocity of 2.89 m/s, the measured pressure difference was 47.00 kPa. Calculate C_r.

Solution

We use the clear brine measurement to determine the wall friction by subtracting the potential energy effect from the total measured pressure drop.

$$\text{Frictional pressure drop} = 47000 - 3.92(1180)(9.8) = 1670 \text{ Pa}$$

The velocity of the clear brine was close to that of the slurry so we use this wall friction in the first estimation of the in situ density.

$$\rho_m = \frac{63090 - 1670}{(3.92)(9.8)} = 1599 \text{ kg/m}^3$$

We can make a second estimate of the wall friction from the pseudo-homogeneous approximation.

Slurry frictional pressure drop $\cong 1670(2.83/2.89)^2(1599/1180)$ or 2170 Pa. The second estimate of the flowing density is

$$\rho_m = \frac{63090 - 2170}{(3.92)(9.8)} = 1586 \text{ kg/m}^3$$

The particle size would determine which of these two estimates is preferable. If the particles are fine, the second estimate is probably better, and if the mean particle density is 2050 kg/m^3, the in situ concentration can be calculated from the rearranged form of Equation 2–5:

$$C_r = \frac{1586 - 1180}{2050 - 1180} = 0.467.$$

5.11 INCLINED PIPES

These present more difficult problems than vertical flows because deposition velocities are required. The additional independent variable (inclination) can be expected to affect both headlosses and deposit velocities but experimental data are sparse, especially for large pipes. Suitable experiments are difficult to conduct because fairly long inclined pipes are necessary to allow the disturbance produced by a bend to decay before representative measurements are made.

Deposition velocities were measured as a function of inclination and pipe diameter for typical fine, intermediate, and coarse particle slurries by Roco (1977). Figures 5–5 and 5–6 show some of the results.

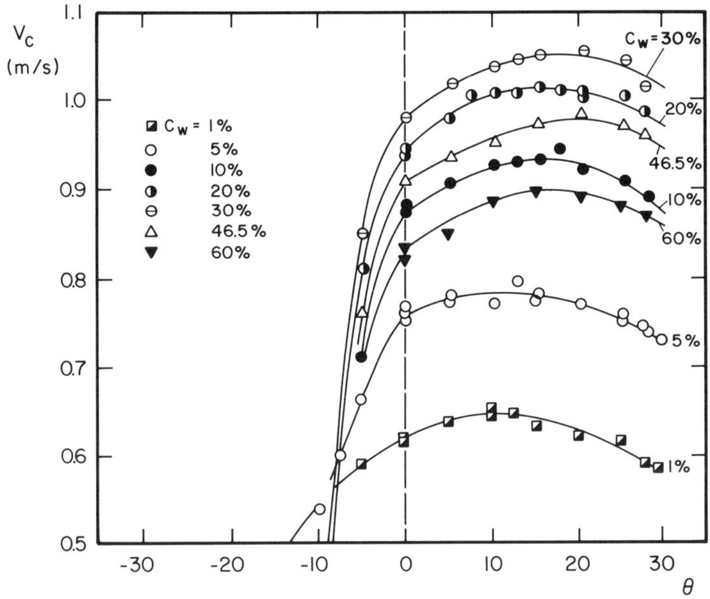

Figure 5–5. Effect of inclination on deposit velocity for a fly ash slurry (S_s = 2.3, d_{50} = 78 μm) in a 100-mm pipe. (Adapted from Roco, 1977.)

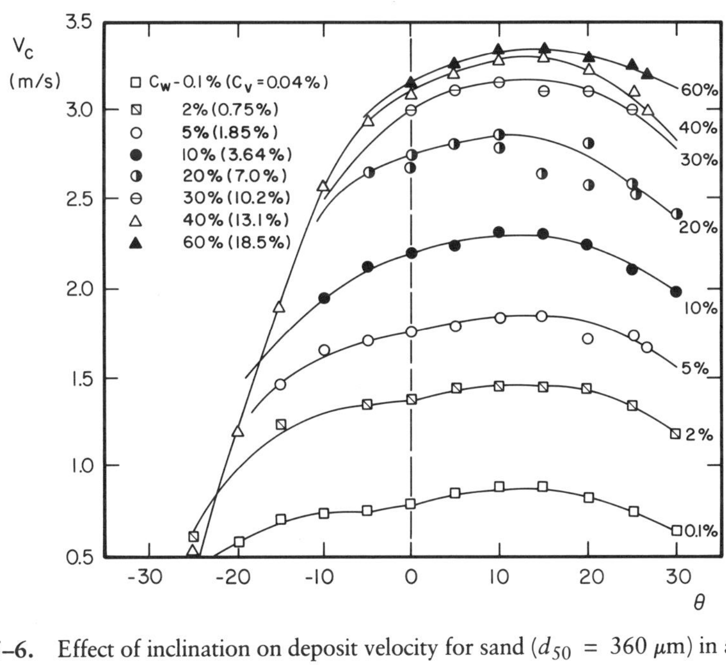

Figure 5–6. Effect of inclination on deposit velocity for sand (d_{50} = 360 μm) in a 100-mm pipe. (Adapted from Roco, 1977.)

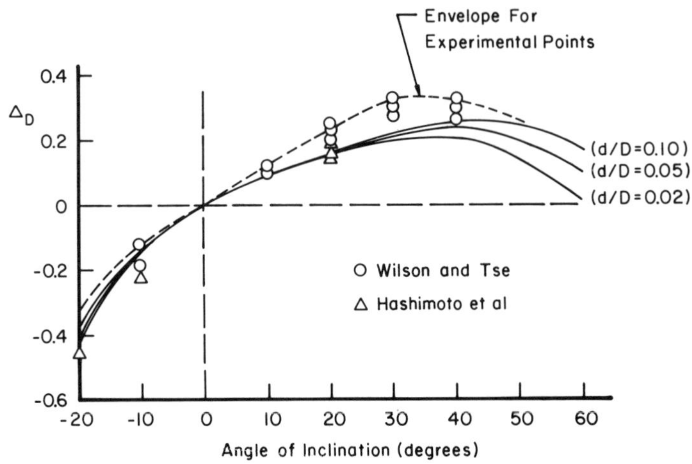

Figure 5–7. Wilson–Tse correlation for the effect of inclination on deposition velocity. (From Wilson and Tse, 1984. *Proc. Hydrotransport 9 Conf.*, p. 161. Reprinted with permission.)

We see that the deposition velocity increases slightly (of the order of 10%) for upslope flows at angles below about 15 degrees. Above this region, the deposit velocity decreases slowly with increasing angle. A more rapid decrease occurs for downward sloping pipes. The latter change is to be expected since for downward slopes the immersed weight of the particles contributes an impelling effect.

Wilson and Tse (1984) produced a correlation in the form of an increment Δ_D to the value of $V_c/\sqrt{gD(S_s - 1)}$ for horizontal pipes

$$\Delta_D = \left(\frac{V_c}{\sqrt{gD(S_s - 1)}}\right)_{\text{inclined}} - \left(\frac{V_c}{\sqrt{gD(S_s - 1)}}\right)_{\text{horizontal}} \tag{5–23}$$

Theoretical predictions of the correction were obtained from a two-layer model (slightly different from the version to be discussed in Chapter 6). Predictions and experimental results are shown in Figure 5–7.

Measurements of total pressure drop have been reported for small ($D <$ 0.16 m) inclined pipes. From these, frictional headlosses can be inferred by subtracting the potential energy effect, estimated by neglecting slip. The results are most reliable for small pipes since for these the potential energy term $\rho g\ dh/dx$ will not be dominant. No correlation has been proposed for headlosses in inclined pipes, however. For this reason one must either use a horizontal flow correlation or models, such as those to be described in Chapters 6 and 7.

In upslope flows at very low velocities, backflow can occur near the bottom of the pipe. For a calculation of this backflow, see Roco and Balakrishnan, 1985.

Chapter 6

The Two-Layer Model

6.1 ORIGIN OF THE MODEL

We saw that flow of coarse particle mixtures was described by Newitt et al. (1955) with Equation 5–14. Unlike the other early headloss expressions, this equation had a theoretical origin that suggested it should apply to pipes of other sizes. As a design equation, its utility was limited to a particular case which is rather uncommon: "flow with saltation or a sliding bed." These flows are, of course, highly stratified, but the derivation did not take this segregation into account.

The original Newitt coarse particle experiments used comparatively low concentrations for which the incremental term due to the presence of the solids was sometimes small. A significant advance occurred when Streat and Bantin (1972) transported very high concentrations of coarse particles and measured the in situ concentration as well as headloss. These high-concentration results provided an important test of the derivation of Wilson et al. (1972). The derivation was important because it was mechanistic. Furthermore, Wilson's analysis showed why the coefficient 0.8 in Newitt's equation varied with concentration for highly stratified flows.

Wilson's derivation differed profoundly from Newitt's because the stratification of the flow was its central feature. On the other hand, Gaessler's equation (Equation 5–10) can be viewed as an attempt to generalize Newitt's equation by adding terms to account for effects omitted from Equation 5–14 without taking the inherently segregated nature of nonhomogeneous horizontal flow into account.

Wilson's original statement of the model was restricted, in effect, to very coarse particles. It was interesting because it did not contain any empirical flow-derived constants. He employed a kinematic coefficient of friction conceptually similar to the coefficient of particle-wall friction η_s, defined in Equation 2–47. η_s was assumed to be a constant that could be determined from sliding friction tests using a method such as that shown in Figure 2–8, or by tilting a section of pipe and noting the point at which continuous sliding occurs. Subsequent studies have shown the model to be capable of being extended to finer particles and mixtures of coarse and fine solids, if flow-derived coefficients are employed.

The model presented here is essentially that of Wilson (1970, 1976). In addition to incorporating subsequent experimental results, it will be derived in a form which should be easier to follow or modify. The model can be used to estimate headlosses as an alternative to the correlations shown in Chapter 5. It can also be used to scale up experimental measurements to larger pipes.

6.2 THE TWO-LAYER MODEL

The slurry flowing in a horizontal or sloping pipe is visualized as forming two layers separated by a hypothetical horizontal interface. This is shown schematically in Figure 6–1a. Within each region, variations of solids concentration and velocity are neglected when computing boundary stresses and the stress at the interface. In the usual case, when the particles are denser than the fluid, the upper layer is assumed to contain only particles whose immersed weight is supported by fluid lift forces.

The mixture in the upper layer, of volumetric solids concentration C_1, behaves essentially as a liquid as far as the wall shear stress is concerned.

The lower layer (see Figure 6–1b) is assumed to have the high total solids concentration C_{lim}. The increment C_2 is assumed to consist of particles whose immersed weight is transmitted to the pipe wall by interparticle contact. Coulombic sliding friction, of the type described in Chapter 2, is assumed to occur at the contact between the wall and the lower layer.

According to the model, C_{lim} is a specified quantity. In the original formulation, C_{lim} was designated as the concentration of a loose-packed bed, which could be determined from nonflow measurements. For this reason, models of this type are sometimes described as *sliding bed* models. It should be noted that flow with an unsheared bed is rare in large pipes, but as we shall see, the nature of the flow in the lower layer is less important than the stress transmission mechanism. Fortunately, headloss predictions for coarse particles are rather insensitive to the value of C_{lim}, so that a value of 0.60 can be used unless the size distribution is very broad and the mean concentration is very high.

The model employs mass balances relating the usual design variables, V and C_v, to the quantities defined in Figure 6–1. The volumetric flow rate of mixture is

$$AV = A_1 V_1 + A_2 V_2 \tag{6–1}$$

For the solids, we neglect the local slip velocity of the particles relative to the fluid so that the volumetric flow rate of solids is

$$C_v AV = C_1 AV + C_2 A_2 V_2 \tag{6–2}$$

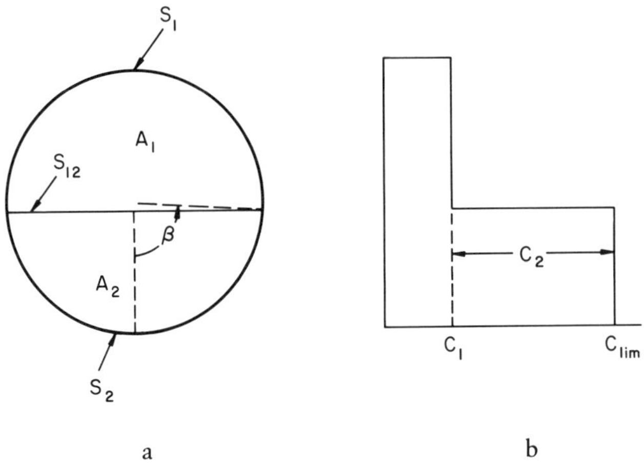

Figure 6–1. (a) Pipe cross section, as idealized in the model. (b) Concentration variation with elevation, according to the model.

C_1 was zero in the original formulation, but because we want to use the method with finer particles, we must employ a method (derived from experimental results) for predicting C_1. This is usually done by determining the quantity C_2A_2/A, sometimes described as the "contact load" of the flow. This is denoted as C_c. Physically, C_c represents the mean volumetric concentration of particles contributing Coulombic friction to the flow.

Early attempts to generalize the model sought to relate C_c to the delivered concentration C_v, defined in Equation 2–2. This was convenient but since the ratio C_c/C_v can exceed unity near deposition, it is preferable to relate C_c to the mean in situ concentration C_r, defined in Equation 2–7. In terms of the quantities shown in Figure 6–1, the definition of C_r becomes

$$C_r = C_1 + \frac{C_2A_2}{A} \tag{6–3}$$

In terms of C_c

$$C_c = \frac{C_2A_2}{A} \tag{6–4}$$

A tentative correlation for C_c is

$$\frac{C_c}{C_r} = \exp\left[-\alpha_1 \mathrm{Ar}^{\alpha_2}(V^2/gd)^{\alpha_3}\left(\frac{d}{D}\right)^{\alpha_4}(S_s - 1)^{\alpha_5}\right] \tag{6–5}$$

where Ar is the Archimedes number for a particle settling in the fluid:

$4gd^3(S_s - 1)\rho_L^2/3\mu_L^2$. For narrow size distributions, tentative values of the coefficients appear to be (Shook et al., 1986), at least for Ar $< 3 \times 10^5$,

$$\alpha_1 = 0.124$$

$$\alpha_2 = -0.061$$

$$\alpha_3 = 0.028$$

$$\alpha_4 = -0.431$$

$$\alpha_5 = -0.272$$

The model employs momentum equations for each layer. These are written in terms of the boundary and interfacial stresses. For the upper layer in steady flow, we have a modified form of Equation 1–3:

$$-\frac{d(P + \rho_1 gh)}{dx} = \frac{\tau_1 S_1 + \tau_{12} S_{12}}{A_1} \tag{6–6}$$

Equation 6–6 assumes, in effect, that the interfacial stress τ_{12} opposes motion of layer 1 in the same way that stress τ_1 does. This implies that V_2 is less than V_1. For the lower layer, the corresponding momentum equation is

$$-\frac{d(P + \rho_2 gh)}{dx} = \frac{-\tau_{12} S_{12} + \tau_2 S_2}{A_2} \tag{6–7}$$

The sign of the τ_{12} term in Equation 6–7 reflects the fact that the interfacial stress has an impelling effect on the lower layer. If the interfacial stress terms are eliminated between Equations 6–6 and 6–7, we have the force balance for the pipe as a whole:

$$-\frac{d(P + \rho_m gh)}{dx} = \frac{\tau_1 S_1 + \tau_2 S_2}{A} \tag{6–8}$$

Stress τ_1 is evaluated from Equation 1–5, using the velocity and density of the upper layer:

$$\tau_1 = \frac{1}{2} f_1 V_1 |V_1| \rho_1 \tag{6–9}$$

Strictly speaking, the friction factor should be calculated from the viscosity and density of the mixture in the upper layer and the hydraulic equivalent diameter of the region. However, for C_r values less than about 0.35 and particles larger than 0.18 mm (d_{50}), there appears to be little effect of concentration on the effective viscosity to be used in the model. Thus, the friction factor can be estimated from the properties of the fluid and the mean velocity. The latter is often specified in a design situation, i.e.,

$$f_1 = f_1\left[\frac{D\rho_L V}{\mu_L}, k/D\right]$$

and f_1 is evaluated using Equation 1–6.

Stress τ_{12} is presumed to result from the difference in velocity between the two layers. The stress is calculated from the density of the upper layer:

$$\tau_{12} = \frac{1}{2} f_{12}(V_1 - V_2)|V_1 - V_2|\rho_1 \tag{6–10}$$

The interface is visualized as a boundary whose roughness depends on the diameter of the particles. Again, an approximation eliminates the hydraulic equivalent diameter of the upper layer. f_{12} is calculated from a modified Colebrook friction factor equation:

$$f_{12} = \frac{2(1 + Y)}{[4 \log_{10}(D/d) + 3.36]^2} \tag{6–11}$$

where $Y = 0$ for $d/D < 0.0015$ and $Y = 4 + 1.42 \log_{10}(d/D)$ in the range $0.00015 < d/D < 0.15$. This expression has been deduced from data taken at $\text{Ar} < 3 \times 10^5$.

At the boundary S_2 of the lower layer, two mechanisms contribute to the frictional resistance:

$$\tau_2 S_2 = \tau_{2m} S_2 + \tau_{2s} S_2 \tag{6–12}$$

The fluid and the suspended solids contribute a stress dependent of the velocity of the lower layer. This is estimated from the friction factor f_1 and the density ρ_1:

$$\tau_{2m} = \frac{1}{2} f_1 V_2 |V_2| \rho_1 \tag{6–13}$$

The contact load solids contribute the velocity-independent resisting stress τ_{2s} at the boundary S_2. In terms of the quantities defined in Chapter 2, the boundary frictional resistance depends on the radial interparticle stress τ_{srr} shown in Figure 6–2. Note that θ and β are measured from axes displaced by 90 degrees.

$$\tau_{2s} S_2 = D\eta_s \int_{\theta = \pi/2 - \beta}^{\pi/2} \tau_{srr}\, d\theta \tag{6–14}$$

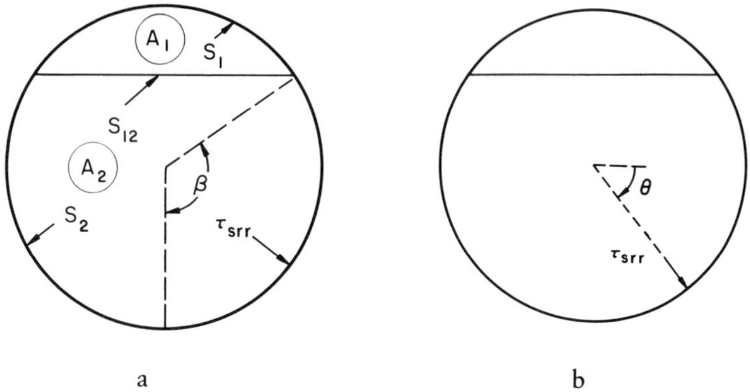

Figure 6–2. Cross section of the pipe.

Stress τ_{srr} reflects the balance of forces in the vertical direction in the lower layer for the fluid, the suspended particles, and the contact load particles. These forces are obtained from Equations 2–13 and 2–14 assuming

1. no inertial forces or vertical wall-derived forces act on the fluid or the suspended load particles;
2. no fluid drag force acts on the contact load particles.

With y measured downward, the force balances for these three components can be written

$$0 = -\frac{dP}{dy} + \rho_L g + f_{Lsy} \tag{6–15}$$

$$0 = -\frac{dP}{dy} + \rho_s g + f_{sLy} \tag{6–16}$$

$$0 = -\frac{dP}{dy} + \rho_s g + f_{ssy} \tag{6–17}$$

In Equation 6–17, the force f_{ssy} becomes f_{swy} at the pipe wall.

The interfacial forces between the fluid and the suspended load particles must balance. Equation 2–18 must be rewritten for this particular situation as

$$C_1 f_{sLy} + (1 - C_1 - C_2) f_{Lsy} = 0$$

Using this with Equations 6–15 and 6–16, we find an expression for the vertical pressure gradient. Substituting this in Equation 6–17, we find the interparticle force per unit volume:

$$f_{ssy} = -\frac{(1 - C_1 - C_2)(\rho_s - \rho_L)g}{1 - C_2} \tag{6–18}$$

Equation 6–18 shows that the effect of the suspended solids (volume fraction C_1) is to reduce the force transmitted to the wall by increasing the buoyant effect on the contact load particles.

If inertial contributions are small, the force f_{swy} can be expressed in terms of the interparticle effective stress as in Equation 2–44:

$$f_{ssy} = -\frac{1}{C_2}\left(\frac{\partial \tau_{syy}}{\partial y} + \frac{\partial \tau_{sxy}}{\partial x} + \frac{\partial \tau_{szy}}{\partial z}\right) \tag{6–19}$$

The shear stress terms in Equation 6–19 are neglected on the grounds that the layer of contact load solids is shallow compared to its width. Figure 2–7 suggests that in this case the interparticle stress increases linearly with depth. If $\tau_{syy} = 0$ at the interface $y = 0$, we have

$$\tau_{syy} = \frac{(\rho_s - \rho_L)gC_2(1 - C_1 - C_2)y}{1 - C_2} \tag{6–20}$$

The radial stress τ_{srr} is assumed to be equal to τ_{syy}. Using this in Equation 6–14 we obtain

$$\tau_{2s}S_2 = \frac{0.5D^2\eta_s g(\rho_s - \rho_L)(\sin\beta - \beta\cos\beta)C_2(1 - C_1 - C_2)}{1 - C_2} \tag{6–21}$$

The partial perimeters S_1, S_{12}, and S_2 are functions of the angle β which defines the location of the interface. An iterative solution technique can be used to solve the equations and predict the pressure drop. The BASIC program given in Appendix 4 can accomplish this. To illustrate the method and to provide a set of results for use in debugging, a sample of the calculation will be given.

6.3 SAMPLE CALCULATION: TWO-LAYER MODEL

Example 6.1

Find the pressure drop for a slurry of sand, $d_{50} = 0.5$ mm, flowing at 4.0 m/s in a pipe of I.D. 0.25 m, if the delivered concentration is 20% by volume. The pipe roughness k is 45 μm and the fluid is water at 20 °C. η_s can be taken as 0.50.

The operating velocity is first compared to the deposit velocity prediction. Using the nomogram of Figure 5–1, we find a deposit velocity of 3.4 m/s so that the proposed operating velocity of 4 m/s is reasonable. Since C_v is specified and C_r is required for the calculation, it is necessary to assume a value of C_r. We try $C_r = 0.22$.

To determine C_c we first evaluate S_s (2.65) and Ar (2694) for the particles. Using these in Equation 6–5 we find $C_c = 0.065$. Equations 6–3 and 6–4 then give $C_1 = 0.155$. Since (Figure 6–1b) C_2 is $(C_{\lim} - C_1)$ or 0.445, Equation 6–4 provides the area ratio $A_2/A = 0.1459$.

From the area ratio we can find the angle β that defines the location of the hypothetical interface. Equations 6–1, 6–7, and 6–8 can then be solved simultaneously to give $-dP/dx$ (or i), V_1, and V_2. These values are $\beta = 0.9360$, $V_1 = 4.46$ m/s, $V_2 = 1.32$ m/s, and $i = 0.096$ m/m. This allows us to calculate C_v from Equation 6–2 to compare with the specified value of 0.20. The result is $C_v = 0.177$, which indicates that the assumed value of C_r was too low.

With a new value of C_r, the calculation can be repeated. A second iteration with $C_r = 0.245$ gives $\beta = 0.990$, $V_1 = 4.52$ m/s, $V_2 = 1.43$ m/s, and $i = 0.102$ m/m. In this case C_v is 0.20.

We note that V_2 is considerably less than V_1 and this reflects the fact that the operating velocity is fairly close to deposition. The deposition condition would not be expected to correspond to the situation where $V_2 = 0$ since the lower layer is normally much thicker than one or two particle diameters, the thickness of the first stationary deposit. Although the model (in the form shown here) does not predict deposition, V_2 values do indicate where deposition becomes more likely.

Thus, for example, as the concentration increases, the model suggests that the deposition velocity should fall. In Figure 6–3, the low-velocity terminations of the parametric curves cannot be fixed precisely but the model does indicate a decrease in V_c. This agrees with experimental observations for this slurry.

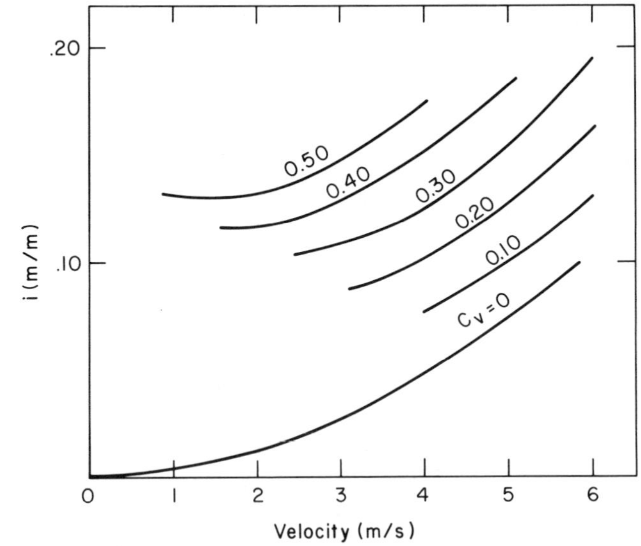

Figure 6–3. Headlosses predicted for sand slurries ($d = 0.5$ mm, $D = 0.25$ m).

6.4 DEVELOPMENTS IN THE MODEL

We must expect Equations 6–5 and 6–11 to be improved as the data base expands. In a later version of Equation 6–5, developed for sand slurries with 0.15 mm $< d_{50} <$ 2.5 mm and $C_r < 0.35$, the parameter $(V^2/gd)^{\alpha_3}$ has been replaced by a group consisting of the velocity V and the deposit velocity $(V/V_c)^{\alpha_6}$. The new coefficients are

$$\alpha_1 = 0.122 \qquad \alpha_2 = -0.12 \qquad \alpha_4 = -0.51$$

$$\alpha_5 = -0.255 \qquad \alpha_6 = 0.30$$

If the contact load concentration in a slurry is very small, convergence of the interface-locating subroutine in the program in Appendix 4 will be very slow. If convergence is very slow, or if no convergence is obtained, one can conclude that the two-layer model is not the appropriate method to predict headlosses in this particular case. Instead, a homogeneous model (if the rheology can be determined) or the pseudo-homogeneous approximation should be used.

Figure 6–3 shows a shallow minimum in the headloss–velocity relationship at the higher concentrations. We note that the minimum occurs well above the deposition velocity. Such minima have been reported in experimental studies. The model suggests that differentiation of a headloss versus velocity relationship is an unsatisfactory method to predict where deposition occurs.

Reconsidering the equations that comprise the model, we see that Equations 6–5 and 6–11 contain the results of slurry flow experiments. The assumed concentration of the lower layer is physically unreasonable for fine-particle slurries and the correlations in Equations 6–5 and 6–11 must compensate for this inadequacy. If better models are to be produced, we require methods for predicting concentration distributions in slurries. With very coarse particles, it would be desirable to allow for slip in the lower layer.

It is important to emphasize that the model presented here has been developed from experiments conducted using particles with fairly narrow size distributions. Many industrial slurries contain significant quantities of coarse particles and of fines that would affect the fluid viscosity. To employ the model with such broad particle size distributions, one might separate the total solids content *a priori* into two fractions: the fines and the remaining solids.

The finest particles would contribute to fluid viscosity and density as in the Wasp or Faddick extension of the Durand–Condolios equation described in Chapter 5. The remainder of the solids would contribute C_1 and C_2 to the two-layer model. Experimental verification of headloss predictions made with this type of approximation would be desirable, however.

Notwithstanding these deficiencies and recognizing the need for further experimental work with slurries of fine particles, of very coarse particles, and those with very broad size distributions, the model is useful for predictions or scaleup. If predictions are required for circumstances which have not been covered in experiments, the model is probably preferable to empirical correlations.

We note that many similar models have been produced since Wilson's original derivation. In one variant, the contact load solids are assumed to comprise a complete "sliding bed" with the suspended ones above it [e.g., Doron's model, (Doron et al., 1987)]. In another version, some of the contact load solids are considered to be present in the upper layer. The quantity of these solids depends upon the stress τ_{12} (Wilson, 1988). In such cases, the interface has some physical meaning, in contrast with the model we have described. To reiterate, the interface is essentially hypothetical in the model shown here.

It is important to emphasize that coefficients (and equations using them) derived from flow data for use with the model shown here cannot be used with these other models.

6.5 EFFECTS OF PARTICLE DIAMETER AND FLUID VISCOSITY

Figure 6–4 has been prepared from the model to illustrate the effect of particle diameter on slurry headloss. It applies to slurries delivering 20% sand in a pipe of I.D. 0.25 m. The model suggests that headlosses increase continuously with particle diameter, eventually approaching limiting values as C_c approaches C_r. There is some difficulty in comparing these predictions with experimental data for the flow of coarse particles. This results from the fact that experimental conditions have often been documented inadequately. Besides failing to determine η_s, investigators have rarely reported complete particle size distributions. Using particles with narrow size distributions, Carleton et al. (1978) found headlosses to decrease with particle diameter above 5 mm. This could reflect the influence of an additional lift force such as the Magnus force described in Section 1.8.

The model shows that the effect of temperature (fluid viscosity) is small for coarse particle slurries. To illustrate this, consider the effect of increasing the temperature of the water, in Example 6.1, from 20 °C to 60 °C.

The density and viscosity of the fluid are altered and the other quantities remain unchanged. With $C_r = 0.25$, we find $i = 0.109$ m/m, $V_1 = 4.62$ m/s, $V_2 = 1.36$ m/s, and $C_v = 0.196$. The effect on i is very small.

With fine particles, the effect of fluid viscosity on headlosses is substantial, especially in large pipes. The small contact loads increase with the terminal settling velocity V_∞ and their effect in large pipes is often considerable.

Slurries with very high concentrations and broad particle size distributions may contain sufficient fine particles to require the mixture in the upper layer to be represented as a non-Newtonian fluid. An alternative to the two-layer model in such slurries considers the coarse particles to flow as an unsheared plug or core which may be either concentric (Brown, 1988) or eccentric (Tatsis, 1990). Shear takes place in the carrier fluid in the space between the core and the pipe wall. The flow of the carrier fluid is assumed to be laminar. The headloss–velocity relationships for these mixtures are rather similar to those shown in Figure 6–3, but without the shallow minimum.

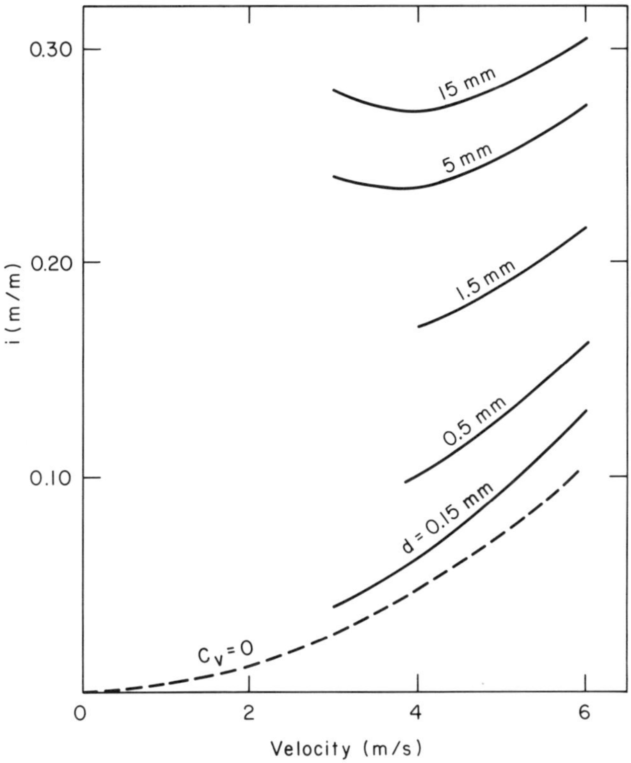

Figure 6–4. Predicted effect of particle diameter on headloss ($C_v = 0.20$, $D = 0.25$m).

6.6 INCLINED FLOWS

Continuing Example 6.1, we consider the effect of pipe inclination at a temperature of 20°C. The results are given in tabular form below for a constant value of $C_r = 0.25$.

Angle (degrees) of Inclination	V_1 *(m/s)*	V_2 *(m/s)*	*Frictional Headloss (m/m)*	C_v
+10	4.65	0.90	0.104	0.193
+5	4.60	1.16	0.103	0.198
0	4.54	1.45	0.103	0.203
−5	4.47	1.78	0.102	0.209
−10	4.39	2.17	0.102	0.216

These predictions are consistent with the trends predicted by the correlation described in Chapter 5. The effect of inclination on deposition velocity, shown in Figures 5–5, 5–6, and 5–7, is reflected in the V_2 values.

6.7 INCLINED PIPES AT SHUTDOWN

If flow ceases in a pipeline containing a slurry, settling of the particles begins to take place. A region of low concentration and low density forms at the top of the pipe while a region of high concentration and density forms at the bottom. If the pipeline is inclined, free convective flow results. This flow transports solid particles down the incline at low angles of inclination. It would be desirable to be able to predict the behavior of this free convective flow.

We can make an approximate analysis by assuming the region containing the flowing slurry is homogeneous and of the same concentration as the original slurry. Its density is ρ_2 and the mean velocity is V_2. The upper layer is assumed to be clear fluid (density ρ_1) with mean velocity V_1. The upslope direction is that of positive x. A diagram of the idealized flow is shown in Figure 11–9.

The area of the upper layer varies because of axial flow and the settling that occurs at the interface (we use the symbols defined previously in Figure 6–1a).

The mass conservation condition for layer 1 is

$$\frac{\partial A_1}{\partial t} + \frac{\partial (A_1 V_1)}{\partial x} = S_{12} V_{12} \cos\theta \qquad (6\text{–}22)$$

If the settling and flow processes proceed independently, the interface velocity V_{12} can be estimated or measured in batch settling tests. For the moving slurry layer (layer 2) the corresponding equation is

$$\frac{\partial A_2}{\partial t} + \frac{\partial (A_2 V_2)}{\partial x} = -(S_{23} V_{23} + S_{12} V_{12}) \cos\theta \qquad (6\text{–}23)$$

where S_{23} defines the interface between layer 2 and stationary sediment.

If compaction occurs in the sediment layer, it will be described by Equation 2–60. However, we neglect these concentration changes.

The net volumetric flow is zero:

$$A_1 V_1 + A_2 V_2 = 0 \qquad (6\text{–}24)$$

The momentum equations for the two moving phases are

$$\rho_1 \frac{DV_1}{Dt} = -\frac{\partial P}{\partial x} - \rho_1 g \sin\theta - \frac{\tau_1 S_1 + \tau_{12} S_{12}}{A_1} \qquad (6\text{–}25)$$

and

$$\rho_2 \frac{DV_2}{Dt} = -\frac{\partial P}{\partial x} - \rho_2 g \sin\theta - \frac{\tau_{23} S_{23} + \tau_2 S_2 - \tau_{12} S_{12}}{A_2} \qquad (6\text{–}26)$$

Stresses τ_1 and τ_{12} are negative, opposing velocity V_1. Stresses τ_{23} and τ_2 oppose velocity V_2 and are therefore positive. The inertial terms are likely to be minor in the two momentum equations.

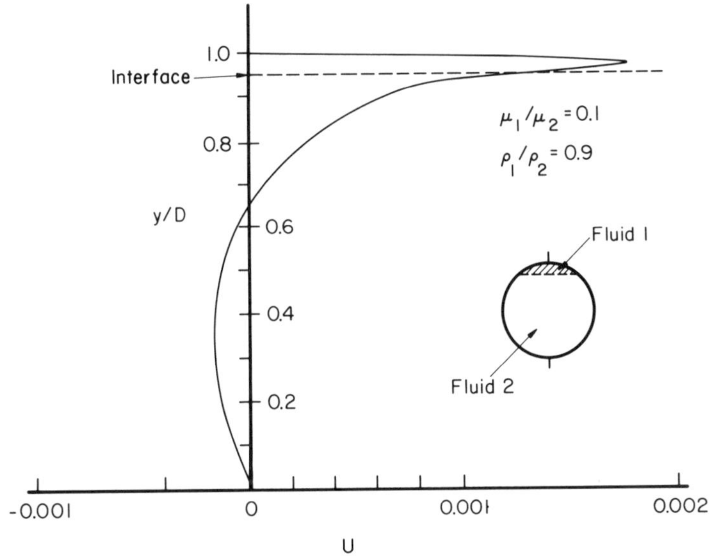

Figure 6–5. Dimensionless local velocity $(4v_x\mu_2/D^2\rho_2 g \sin\theta)$ at the midplane of a zero net flow at shutdown of an inclined pipe. (From Masliyah and Shook, 1978. *Can. J. Chem. Eng.* 56: p. 171. Reprinted with permission.)

If these equations are solved, they provide a prediction of the ultimate position of the upper interface as a function of x. Qualitatively, we see that the flow rate A_1V_1 increases with $(\rho_2 - \rho_1)\sin\theta$.

Using the predictions of Masliyah and Shook (1978) for shear stresses in Newtonian laminar flow, we also find that the flow rate varies as D^4/μ_1. This means that the phenomenon grows in importance as the pipe size increases. For high slurry concentrations, with layer 1 occupying a small fraction of the cross section, the viscosity ratio of the two layers is not an important factor. Figure 6–5 shows the velocity distributions, as functions of position, for a flow of this type in which inertial effects are negligible. The velocity is very high in the upper layer.

An investigation of the combined sedimentation and flow process for a slurry in contact with an inclined flat plane is given by Nir and Acrivos (1990).

6.8 DEPOSITION AND THE MODEL

The mechanistic model also allows us to understand the onset of deposition. Consider the three-layer model shown in Figure 6–6 and, for simplicity, restrict the analysis to horizontal flow. Equation 6–6 is unaltered, but Equation 6–7 becomes

$$-\frac{dP}{dx} = \frac{-\tau_{12}S_{12} + \tau_2 S_2 + \tau_{23}S_{23}}{A_2} \tag{6–27}$$

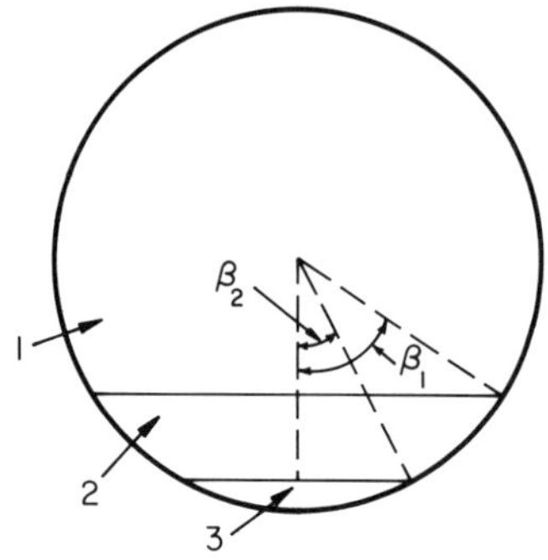

Figure 6–6. A three-layer model to interpret deposition in slurry pipeline flow. The layer concentrations are C_1, C_2, and C_3.

and for the bottom layer the corresponding momentum equation is

$$-\frac{dP}{dx} = \frac{-\tau_{23}S_{23} + \tau_3 S_3}{A_3} \tag{6–28}$$

The resisting forces per unit pipe length, $\tau_2 S_2$ and $\tau_{23}S_{23}$, can be calculated from the approach used to obtain Equation 6–21. $\tau_2 S_2$ depends on velocity V_2, concentration C_2, and angles β_1 and β_2. $\tau_{23}S_{23}$ depends on the velocity difference $(V_2 - V_3)$, C_2, β_1, and β_2. The coefficient η_s applies to friction on the partial perimeter S_2, but along the perimeter S_{23}, the appropriate coefficient should be $\tan \phi$ defined in Equation 2–50.

Since we are interested in incipient deposition, we assume that within the bottom layer there is no suspension by fluid lift forces, i.e., within layer 3

$$\frac{d\tau_{syy}}{dy} = (\rho_s - \rho_L)gC_3 \tag{6–29}$$

where y is measured downward.

Equation 6–29 can be integrated to obtain the stress distribution in the bottom layer. This, in turn, can be used to find $\tau_3 S_3$. In addition to the contact load effect, there will be a fluid-boundary stress which depends upon the velocity V_3 of the bottom layer.

To examine the case of incipient deposition, we choose a small but finite value for β_2 and allow V_3 to approach zero. Equations 6–1 and 6–2 apply, with $V = V_c$, and Equation 6–3 becomes

$$AC_r = (A_1 + A_2)C_1 + A_2 C_2 + A_3 C_3$$

If we assume that C_2, C_3 and the interfacial friction factors f_{12} and f_{23} are predictable, we see that we have one additional momentum equation (Equation 6–28) and one additional variable, V_c.

Besides indicating the importance of the concentration distribution, this layer model suggests that an additional variable (tan ϕ) should be measured if deposition velocities are to be explained mechanistically.

Since η_s is determined experimentally in a test conducted at low velocities, it is appropriate for use in an analysis of flow near deposition. The actual particle-wall interaction mechanisms at higher velocities include particle sliding, rolling, and saltation so that lubrication forces may play an important role. Any flow dependence of the actual coefficient of friction is included, so far as the model is concerned, in the expression for C_c.

Chapter 7

Microscopic Modeling of Slurry Flows

7.1 THE NEED FOR MODELS

The two-layer model described in Chapter 6 is obviously oversimplified. However, it provides useful information concerning pressure drops and, to a lesser extent, limiting velocities for nonhomogeneous slurries at low and moderate concentrations. Better headloss predictions and deposition velocity predictions require the velocity distribution. In other words, we need a Law of the Wall and constitutive relationships for slurry flows. Such expressions would link the local velocity to the pressure gradient. The mean velocity could then be obtained by integrating the local velocities over the pipe cross section.

The equations of motion which were described in Chapter 2 suggest that the local solids concentration affects the velocities of the individual phases. Consequently, a method for predicting solids concentration must be part of the strategy of a comprehensive model.

It is worth remarking that even for single-phase turbulent flows, simple models are not entirely satisfactory for predicting velocity distributions. The complexity of modern single-phase models has grown as the variety of experimental information about the flows has increased.

For slurries, because we have more variables, the problem is more complex. However, the approach would be similar to that used in other fluid mechanics problems:

1. using the equations of motion to provide as much information as possible,
2. inferring the form of the appropriate physical laws for constitutive relationships and boundary conditions from experimental measurements.

Model development is essentially a research activity and since these models are still under development, the designer and plant engineer could be excused for regarding the subject as academic. However, it is desirable to understand the direction in which improved models can be expected to evolve. Since we can expect continuous evolution of models, we concentrate upon the evidence that should be considered in developing them. In so doing, we hope to minimize the obsolescence of this chapter and provide a framework for further research.

7.2 CONCENTRATION DISTRIBUTIONS IN A CLOSED CHANNEL

Figure 7–1 illustrates the type of concentration variation observed for slurries in horizontal flow. These measurements were taken for slurries of 0.53-mm sand in isothermal flow (17 °C) in a closed rectangular channel with a breadth four times the depth (Daniel, 1965). Gamma ray absorption was used for these measurements.

Although the pressure drops showed that the contact load was fairly high for these flows, the profiles are considerably different from those postulated in the two-layer model. In particular, we note that no "sliding bed" forms in the physical sense, although the two-layer model would be an appropriate method with which to correlate headlosses.

Inspecting these results, we also observe

1. the concentration decays with height in an approximately exponential manner in the upper region,
2. the decay is not exponential at high concentrations or near the bottom of the channel. In fact, a reversal in the C profile occurs in some cases.

With lower headlosses (i values) or finer particles, the reversal was less evident.

If we consider the high C_r results, the fact that the particles were narrowly sized implies that the limiting concentration for continuous shear would be of the order of 55% by volume, according to the constraints of particle packing. The region of near-constant concentration in the lower half of the channel evidently represents a fairly dense mixture.

The mixture velocity distribution was measured for the slurry with the second highest value of C_r and this showed that shear occurred in the lower half of the channel. There was no evidence of *en bloc* sliding, at least at velocities significantly above the deposition velocity.

7.3 THE DIFFUSION MODEL

Since most early studies of concentration distributions avoided high concentrations and coarse particles, it is not surprising that an exponential concentration variation with height is the best known feature of such distributions. A diffusion model which related the volumetric settling rate in a quiescent fluid to a diffusion flux was suggested by Schmidt (1925) and Rouse (1937) (independently) for open channel flows.

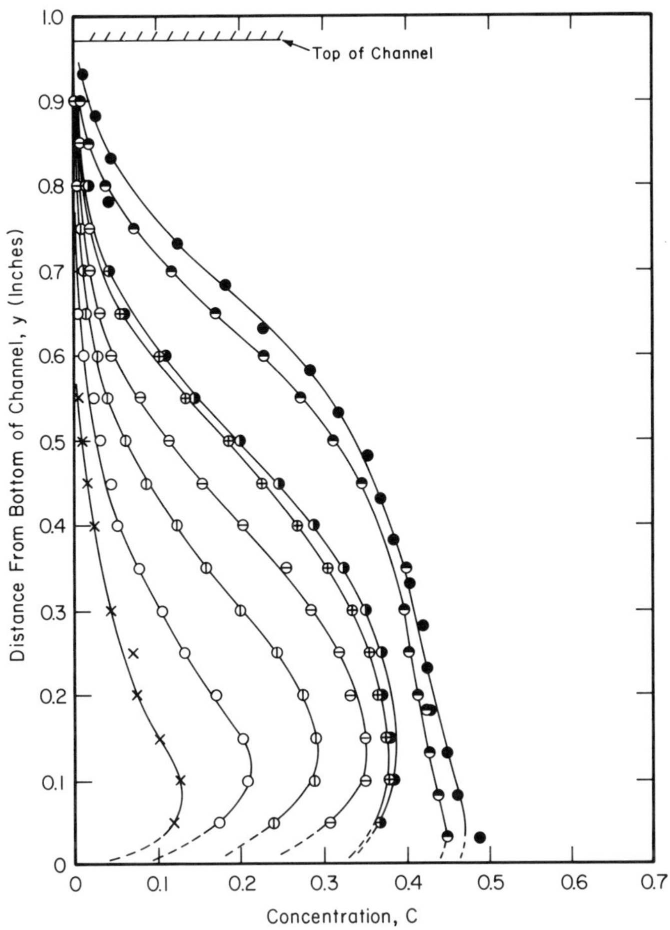

Figure 7–1. Concentration as a function of height for horizontal flow of sand (d = 0.35 mm) in a closed rectangular channel 25 mm high by 100 mm wide, i = 0.334 m/m.

With y measured downward, this model is

$$\epsilon_s \frac{dC}{dy} = V_\infty C \tag{7–1}$$

where ϵ_s is the solids diffusion coefficient.

To examine the origin of this model, consider unit area in a horizontal plane as shown in Figure 7–2. The idealized fluid velocity fluctuation v_0 is assumed to be upward in half the area and downward in the other half. In terms of a scalar slip velocity, v_r, the particle velocities in the two regions are $v_0 - v_r$ and $v_0 + v_r$,

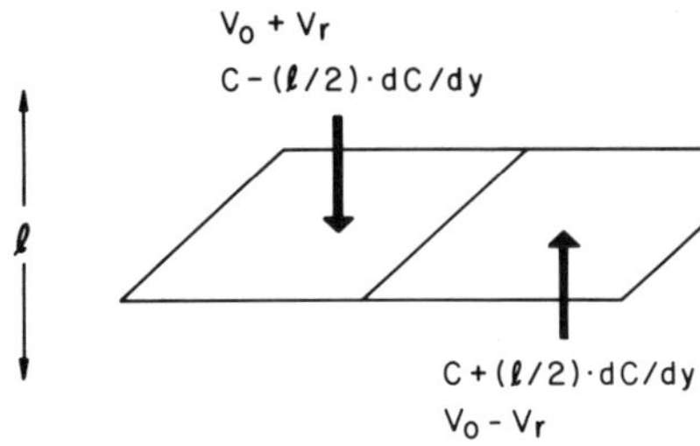

Figure 7–2. Mixing length model of particle exchange by turbulence.

neglecting any variation in v_r between the two sections. If C is the concentration at the plane and if a fluid velocity fluctuation has an effective length l, the upward volumetric flux of particles will be given by the product of the area fraction, the concentration at the bottom of the fluctuation (eddy), and the velocity, i.e.,

$$\text{Upward flux per unit area} = \frac{1}{2}\left[C + \left(\frac{l}{2}\right)\frac{dC}{dy}\right](v_0 - v_r)$$

Similarly,

$$\text{Downward flux per unit area} = \frac{1}{2}\left[C - \left(\frac{l}{2}\right)\frac{dC}{dy}\right](v_0 + v_r)$$

Equating these expressions we have

$$Cv_r - \left(\frac{l}{2}\right)v_0\frac{dC}{dy} = 0 \tag{7–2}$$

Using the mixing length theory for single-phase turbulent flows, the product $(l/2)v_{f0}$ (or ϵ_s) is identified as being related to the eddy kinematic viscosity ν_t defined in Chapter 1 for single-phase fluids.

Equation 1–11 shows that a considerable variation of ν_t over the cross section of a pipe is required to explain the turbulent velocity distribution. To obtain an exponential concentration distribution from an equation such as 7–2, however, no such a variation in ϵ_s can occur.

It is known that the transfer of heat and dissolved matter in turbulent flow can be described with a diffusion coefficient which is close to ν_t in magnitude. The situation may be different for ϵ_s because the latter describes the transfer of finite objects.

For transfer of heat or soluble matter in a flowing fluid, the turbulent diffusion coefficient (Goldstein, 1938) is

$$\epsilon_L = \left(\frac{\overline{v_L'^2}}{t}\right)\int_0^t R(t')\,dt' \tag{7–3}$$

where the correlation coefficient relates fluctuating fluid velocities at times t_1 and t_2 in terms of the time averages denoted by the overbar:

$$R(t_2 - t_1) = \frac{\overline{v'_L(t_1)v'_L(t_2)}}{\overline{v'^2_L}}$$

Friedlander (1977) shows that a similar equation can be derived for diffusion of solid particles in a turbulent flow, at least in the region at some distance from the pipe wall. However, the velocity fluctuation terms in the integral must be evaluated for fluid *in the vicinity of the particle* during its motion. Only for small particles, which can follow the fluid motion exactly, would the turbulent diffusivities be the same for finite objects and dissolved matter. Friedlander suggests that a measure of the "heaviness" of a particle would be the ratio $(\rho_s d^2/\mu_L)(\epsilon\rho_L/\mu_L)^{0.5}$, where ϵ is the time rate of energy dissipation per unit mass (m^2/s^3).

If we neglect transient effects (added mass and Basset force complications and the effect of turbulence on the drag coefficient), the relative velocity in Equation 7–2 can be estimated from the terminal velocity of the particle and the hindered settling correction. This yields the simple form

$$\epsilon_s \frac{dC}{dy} = V_\infty C(1 - C)^{m-1} \tag{7–4}$$

which reduces to Equation 7–1 at low concentrations. m is the exponent in hindered settling equations of the type

$$\frac{|v_s|}{V_\infty} = (1 - C)^m \tag{7–5}$$

Equation 7–5 is a generalized form of Equations 2–30 and 2–31.

An equation for estimating m in Equation 7–5 for spheres has been given by Wallis (1969):

$$m = \frac{4.7[1 + 0.15\ \mathrm{Re}_\infty^{0.687}]}{[1 + 0.253\ \mathrm{Re}_\infty^{0.687}]} \tag{7–6}$$

The term $(1 - C)^{m-1}$ has the same limitation in Equation 7–4 as it has in sedimentation rate expressions because $(1 - C)^{m-1}$ does not become zero when C reaches its limiting (packed-bed) value. The exponent in Equation 7–4 has usually been taken as m in previous work but since ϵ_s may be a function of concentration, it is the net exponent of $(1 - C)$ which is of interest. Better agreement with some experiments can be achieved by including $(1 - C/C_{max})$ in the model (Roco and Frasineanu, 1977).

7.4 FINE-SAND CONCENTRATION DISTRIBUTIONS

We now consider the experimental evidence for concentration distributions in pipe flows. A sand of $d_{50} = 0.18$ mm has been used in experimental studies at

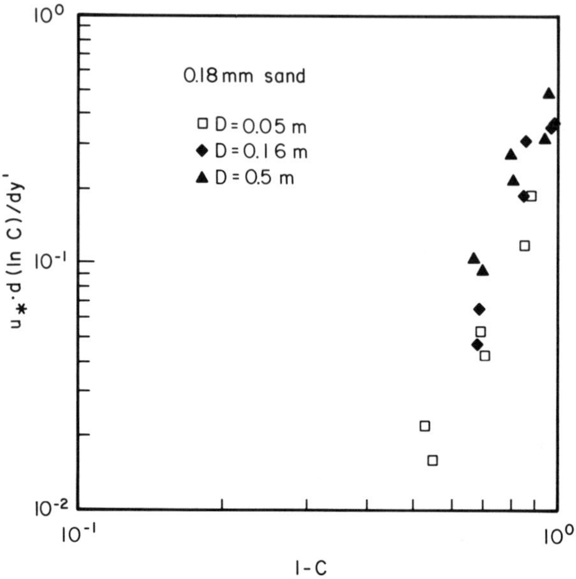

Figure 7–3. Concentration gradients observed for 0.18 mm sand at $y' = 0$ for horizontal pipe flows.

the Saskatchewan Research Council laboratory extending over many years (Gillies, 1990). Concentrations have been measured by resistivity probe, isokinetic sampling, and gamma ray absorption. Little lateral variation of concentration occurred at the midplane ($y' = 0$) of the pipes with this sand.

To test the diffusion model, chord average concentrations have been fitted to polynomials of the form

$$\ln C = b_0 + b_1 y' + b_2 y'^2 + b_3 y'^3$$

where $y' = 2y/D$ is measured downward from the pipe axis, so that values of $d \ln C/dy'$ could be obtained.

These derivatives are shown, multiplied by the friction velocity $u_* = \sqrt{igD/4}$ in Figure 7–3. The scatter in the plot results from the combined effects of small differences in temperature and particle size distribution between experiments as well as the inadequacies of the numerical differentiation procedure. However, we observe

1. The concentration gradient is a strong function of $(1 - C)$. The net exponent of $(1 - C)$ is close to m in the hindered settling equation.
2. There is evidence of a minor pipe diameter effect, with lower concentration gradients occurring in the small pipe.

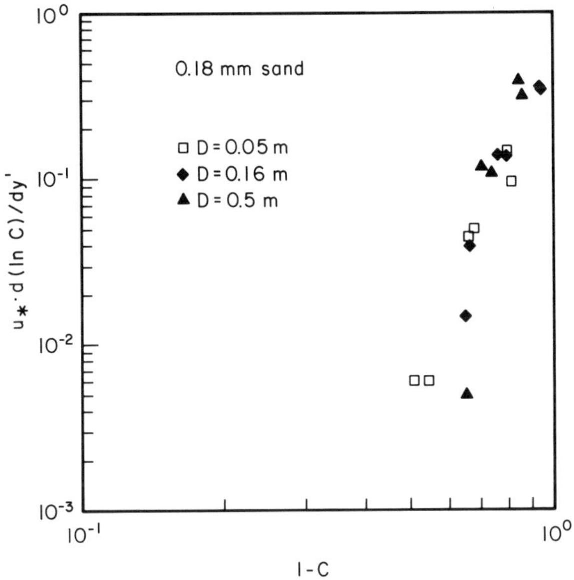

Figure 7–4. Concentration gradients observed with the sand of Figure 7–3 at $y' = 0.37$.

As one moves into the high-concentration region in the lower half of the pipe, a consistent pattern of deviation is observed. Figure 7–4 shows the corresponding results for $y' = 0.37$. Here we observe

3. The exponent on $(1 - C)$ is higher as we move toward the bottom of the pipe, i.e., the concentration gradient is smaller than expected if an equation such as Equation 7–4 is applied to the whole flow.

This trend continues as y' increases downward. However, the possibility of lateral variation of concentration makes the evidence from high y' values of chord average concentration less convincing than the evidence at $y' = 0$ or 0.37.

In terms of the two-layer model, these slurries all had significant contact loads, ranging between about 7% of C_r for $D = 0.5$ m to about 30% for $D = 0.05$ m. Observations 2 and 3 suggest that another mechanism influences the C distributions and we should consider the possibility that it is related to the contact loads that were inferred from experimental headlosses with the two-layer model.

7.5 COARSE-SAND CONCENTRATION DISTRIBUTIONS

There have been few systematic concentration distribution measurements made with narrowly sized coarse particle slurries. However, data collected by Gillies

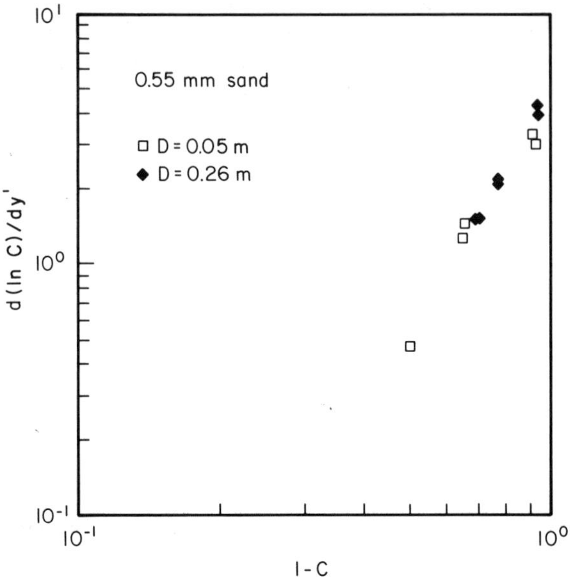

Figure 7–5. Concentration gradients observed for 0.55-mm sand at $y' = 0$.

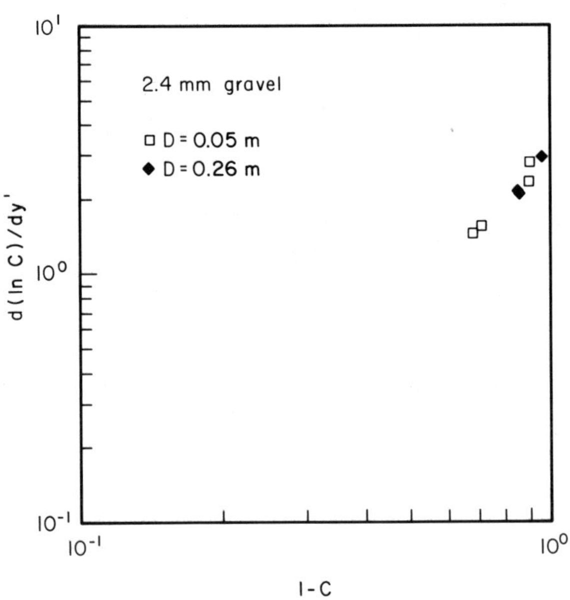

Figure 7–6. Concentration gradients observed for 2.4-mm gravel at $y' = 0$.

(1990) allows derivatives at $y' = 0$ to be obtained by the same procedure. These are shown in Figures 7–5 (0.55-mm particles) and 7–6 (2.4 mm) for two pipes. We again observe the strong dependence upon $(1 - C)$. Also,

4. The exponent on $(1 - C)$ is lower for the larger particles than for the fine ones. The exponent m in Equation 7–5 is affected similarly.
5. The friction velocity does not have the same effect upon the gradients that it does for the finer particles. (Note that u_* has been omitted from the ordinates of these figures.)

These experiments were confined to velocities of engineering interest, although, in retrospect, it would have been interesting to examine the gradients at very high velocities.

In addition to the lower exponent on $(1 - C)$, we observe that the extrapolated intercept at $C \rightarrow 0$ for the 2.4-mm particles is actually *lower* than that for the 0.55-mm sand. The net effect is that there is not much difference between the C distributions although there is a significant difference in the V_∞ values for 0.55-mm and 2.4-mm particles. Observations 3, 4, and 5 were confirmed by measurements made with the coarser particles at $y' = 0.37$.

7.6 MODIFYING THE DIFFUSION MODEL

These observations show that the concentration distribution should probably be written as

$$(1 - C)^{-m'} \frac{d \ln C}{dy} = f(V_\infty, u_*, d, D, y, \ldots) \tag{7–7}$$

where m' is not necessarily identical to $m - 1$.

We note that the tendency toward a more uniform concentration in the bottom half of the pipe (lower gradients) is not consistent with reduced particle diffusivities. Something increases the relative ability of the flow to suspend the particles when the concentration is high.

If the concentration distribution in the vertical direction displays a maximum, it cannot be explained with the simple mixing model. Furthermore, some other mechanism must act if slurries of settling particles can be transported in horizontal laminar flow when Brownian motion is not important. There must be other lifting effects to balance the immersed weight of the particles on a time-average basis.

The mixing model of Equation 7–4 was not derived from the momentum equations which we used in Chapter 6. Before discussing possible modifications, it is useful to try to reconcile the models we have already used.

In the momentum equations of Chapter 2, the forces f_{ss} and f_{sw} contained inertial turbulent (Reynolds stress) and particle interaction (i.e., Bagnold stress) effects. Both of these effects are dependent on strain rate. However, f_{ss} and f_{sw}

also contain any Coulombic (strain rate independent) forces transmitted between particles. To examine the concentration and velocity distributions, we define these individual contributions to the effective stress $\mathbf{T}_s$ using subscripts t (turbulent), B (Bagnold), and C (Coulombic), after Roco and Shook (1983):

$$\mathbf{T}_s = \mathbf{T}_{st} + \mathbf{T}_{sB} + \mathbf{T}_{sC}$$

The fluid effective stress $\mathbf{T}_L$ can be written as the sum of viscous (subscript v) and turbulent contributions:

$$\mathbf{T}_L = \mathbf{T}_{Lv} + \mathbf{T}_{Lt}$$

Eliminating the pressure gradient $\partial P/\partial y$ from Equations 2–16 and 2–17 gives the instantaneous force balance for the vertical direction. If y is measured downward, for macroscopically steady flow ($\partial/\partial t = 0$)

$$\left(\frac{1}{C}\right)[\nabla \cdot \mathbf{T}_{st}]_y - \left(\frac{1}{1 - C}\right)[\nabla \cdot (\mathbf{T}_{Lv} + \mathbf{T}_{Lt})]_y - \frac{f_{sL}}{1 - C} = (\rho_s - \rho_L)g - \left(\frac{1}{C}\right)[\nabla \cdot (\mathbf{T}_{sB} + \mathbf{T}_{sC})]_y \qquad (7\text{–}8)$$

In Chapter 2 we saw that the gravitational and drag effects combine in simple settling to determine the terminal falling velocity of particles. We can therefore interpret Equation 7–4 as an attempt to model Equation 7–8 in circumstances when the terms $\mathbf{T}_{sB}$ and $\mathbf{T}_{sC}$ are negligible. For the models to be consistent, the left-hand side of Equation 7–8 should be

$$\left(\frac{3\rho_L C_D}{4d}\right)\left(\frac{\epsilon_s^2}{(1 - C)^{2m-2}}\right)\left(\frac{d \ln C}{dy}\right)^2$$

If κ_c is the fraction of the immersed weight of the particles transmitted downward by the Coulombic interaction, then

$$[\nabla \cdot \mathbf{T}_{sC}]_y = \kappa_c C(\rho_s - \rho_L)g \qquad (7\text{–}9)$$

so that Equation 7–8 could be written, for the y direction, as

$$\left(\frac{3\rho_L C_D}{4d}\right)\left(\frac{\epsilon_s^2}{(1 - C)^{2m-2}}\right)\left(\frac{d \ln C}{dy}\right)^2 = (\rho_s - \rho_L)(1 - \kappa_c)g - \left(\frac{1}{C}\right)[\nabla \cdot \mathbf{T}_{sB}]_y \qquad (7\text{–}10)$$

It is really a matter of preference as to whether Equation 7–7 or 7–10 forms the basis for model development. The number of profiles that have been measured is unfortunately still very limited but it seems likely that the Bagnold dispersive

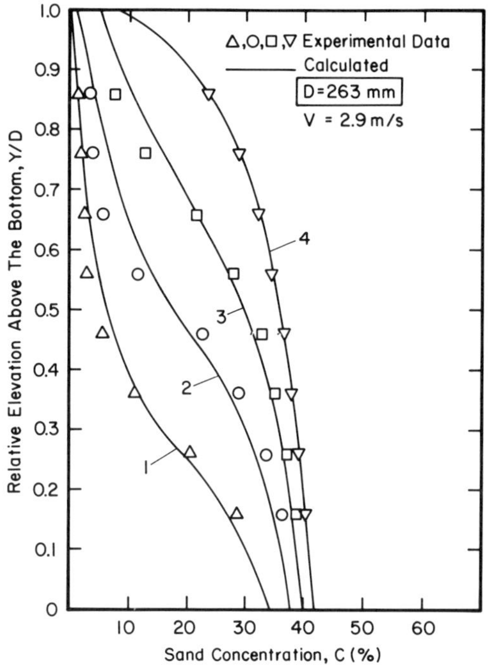

Figure 7–7. Concentration as a function of height for 0.17-mm sand particles.

stress and the Coulombic term cannot be neglected in predicting the concentration distributions. We conclude that Equation 7–4 should be regarded as an asymptotic expression useful for estimations under conditions of low concentrations, fine particles, and large pipes.

An illustration of experimental concentration distributions and predictions with a model is given in Figure 7–7 (Roco and Shook, 1981) for the fine sand of Section 7.4. The pipe diameter was 263 mm and the mean flow velocity was 2.9 m/s. As in Figure 7–1, the concentration profiles flatten as the mean concentration increases, under the combined effects of hindered settling and particle–particle interactions.

Predictions made with Equation 7–8 for several limiting flow regimes where either the particle–fluid interaction (f_{sL}), turbulent diffusion (ϵ_s), Bagnold stress (T_{sB}), or Coulombic stress (T_{sC}) is the dominant two-phase interaction mechanism were discussed by Roco (1990). The lift forces caused by particle spin and inertial effects caused by the containing wall were found to play a relatively minor role compared to ϵ_s, T_{sB}, and T_{sC} (Roco and Cader, 1990).

Evidence for the existence of what we have called T_{sB} comes from experiments with small gravity forces: those with $(\rho_s - \rho_L)g$ approaching zero. A lateral variation of concentration has been observed in experiments using nearly neutrally buoyant particles in a horizontal 50-mm pipe (Nasr-El-Din et al., 1987). This tendency increases with particle diameter but it is not yet clear that the force can be predicted using the equations which were discussed in Section 2.10.

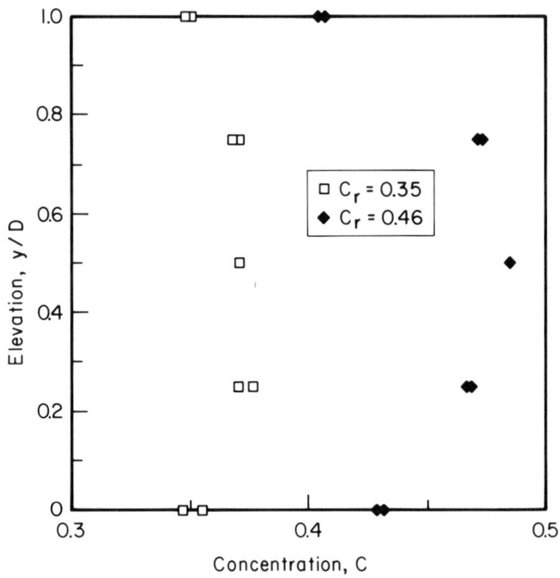

Figure 7–8. Concentration as a function of height for 0.3-mm polystyrene particles.

Figure 7–1 shows that maximum in the concentration distribution can occur and the effect is pronounced when the particle density is close to that of the fluid. Figure 7–8 shows local measurements of concentration on the vertical axis for slurries of 0.3-mm polystyrene particles in a 50-mm pipe (Sumner et al., 1989). The profiles are nearly symmetrical because the gravitational force is small.

Further research is required to quantify ϵ_s, κ_c, and $\nabla \cdot \mathbf{T}_{sB}$ in terms of flow parameters, so that concentration distributions can be predicted reliably for a wide range of conditions.

7.7 VELOCITY DISTRIBUTIONS

Figure 7–9 shows the variation of chord-average concentration with vertical position which is typical of nonhomogeneous horizontal flows. The sand particles in this experiment were coarse (d = 0.9 mm) in a pipe of 50 mm I.D. (Sumner et al., 1989).

Figure 7–10 shows the velocity distribution in the vicinity of the vertical pipe axis for this slurry. The velocities were measured with a traversable particle velocity probe described in Section 10.13 and with a similar device flush mounted in the pipe wall. We note the distortion of the velocity distribution, the elevation of the point of maximum velocity, and the finite velocities of particles closest to the wall. In the region of high solids concentration we note that $|dv_x/dy'|$ is both

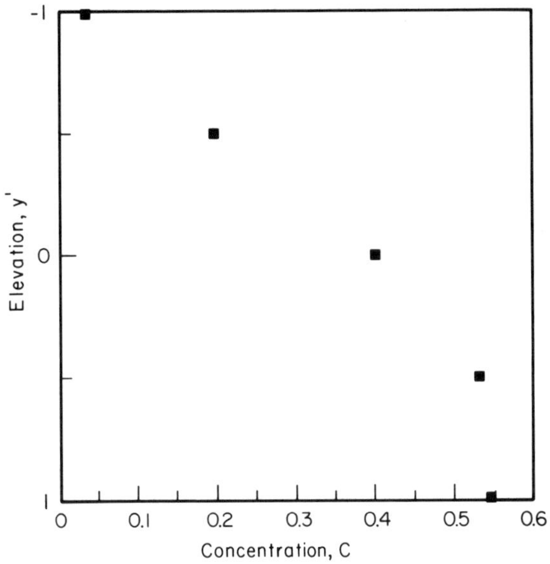

Figure 7–9. Concentration as a function of height for 0.8-mm sand particles in a horizontal pipe.

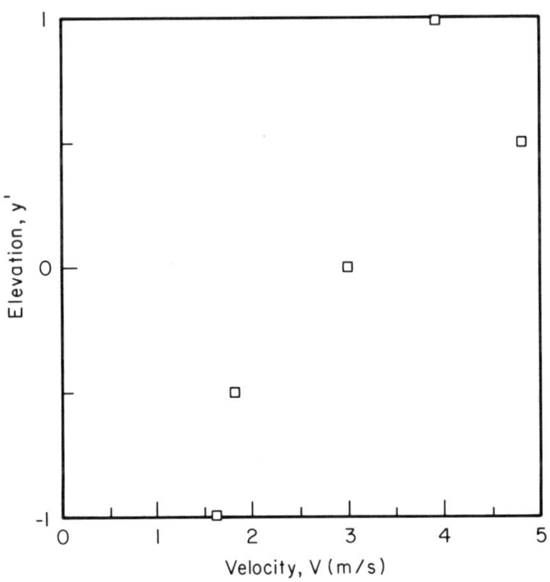

Figure 7–10. Velocity as a function of height (on the vertical axis of symmetry) for the slurry of Figure 7–9.

smaller (near the wall) and larger (near the pipe axis) than it would have been in the clear carrier fluid.

In order to demonstrate a satisfactory understanding of slurry flow, we would have to predict both Figures 7–9 and 7–10. We have already seen the complications which arise for the concentration distribution and the problem is no simpler for the velocities. We begin by examining the momentum equations for the individual phases in macroscopically steady flow.

If the interfacial drag force is eliminated between the momentum equations for the solid and fluid phases (see Appendix 2), for the axial (x) direction

$$[\nabla \cdot (\mathbf{T}_{st} + \mathbf{T}_{sB})]_x + [\nabla \cdot (\mathbf{T}_{Lv} + \mathbf{T}_{Lt})]_x = -\frac{dP}{dx} - \rho_m g \frac{dh}{dx} - [\nabla \cdot \mathbf{T}_{sC}]_x \tag{7–11}$$

The left-hand side contains all the inertial terms, including Reynolds stress and interparticle contact effects, as well as the term due to viscous fluid shear. The latter would be dominant in a thin layer near the pipe wall if the flow is turbulent.

In the turbulent core of a slurry flow, if particle–particle interactions and turbulence modification by particles are not important, the left-hand side of Equation 7–11 should approach (ρ_m/ρ_L) times the value of this term in the clear carrier fluid. If, in addition, the Coulombic term $[\nabla \cdot \mathbf{T}_{sC}]_x$ is negligible, we have the pseudo-homogeneous flow condition. To achieve this, the particles should follow the fluid motion identically and no modification of the flow structure should occur because of the presence of the particles.

In Equation 7–11

$$[\nabla \cdot \mathbf{T}_{sC}]_x = \kappa_c C(\rho_s - \rho_L) g \tan \beta \tag{7–12}$$

where $\tan \beta$ is the stress ratio $|\tau_{syxC}/\tau_{syyC}|$, a kinematic coefficient of friction in the slurry. In Equation 7–12, C is the local solids concentration.

In using Equation 7–11 we cannot merely substitute Bagnold's equation for τ_{syxB} when the flows are turbulent. In the derivations of the interparticle stress term $\mathbf{T}_{sB}$ in dry granular solids, the velocity fluctuations which produce interparticle contact were considered to be determined by the particle diameter and $\dot{\gamma}_s$, the gradient of the time-averaged particle velocity. However, for particles in a turbulent fluid, unless the particles are very heavy in the sense explained by Friedlander (1977), particle velocity fluctuations which derive from the fluid will also be important. Furthermore, particle interactions are affected by the interstitial fluid.

7.8 MODELING VELOCITY DISTRIBUTIONS

A popular model, widely used in open channel flow, relates the local velocity v_x to distance y from the wall and the von Karman coefficient κ

$$\frac{dv_x}{dy} = \frac{u_*}{\kappa y} \tag{7–13}$$

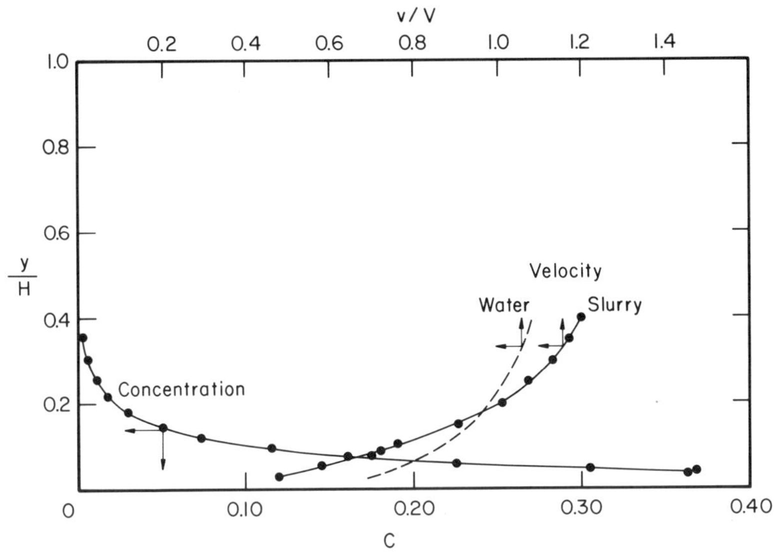

Figure 7–11. Velocity measurements of Einstein and Chien in an open channel.

The data of Einstein and Chien (1955), who used this model, are shown in Figure 7–11 for their highest slurry concentration. There are evidently reduced velocities (higher gradients) in and near the region of high solids concentration. In slurries, the von Karman constant would have to be lower in the slurry than in the clear fluid to "explain" higher velocity gradients. However, the T_{sB} and T_{sC} terms in Equation 7–11 provide an alternative interpretation of the altered velocity distributions.

If we are to use Equation 7–11 as the starting point for predicting velocity distributions, we can identify the following requirements.

1. We will require values of the product (κ_c tan β) unless T_{sC} is negligible.
2. If an effective viscosity μ_{mt} is used for the turbulent core, the left-hand side of Equation 7–11 can be written as $\nabla \cdot (-\mu_{mt}\Delta)$. However, the presence of solids (and T_{sB}) suggests that μ_{mt} should vary with concentration, d, ρ_s, and $\dot{\gamma}_s$.
3. A separate approach will be required for the near-wall region where T_{Lv} will be an important component of the left-hand side of Equation 7–11. Frictional headloss measurements suggest that the appropriate viscosity for use in the near-wall region will range between that of the carrier fluid and that of the mixture.

We see that the parameters required to describe the velocity distribution using Equation 7–11 are κ_c, tan β, μ_{mt}, and the wall region viscosity. Since tan β should resemble the coefficient tan α discussed in Section 2.10, it should also depend upon the factors which determine μ_{mt}.

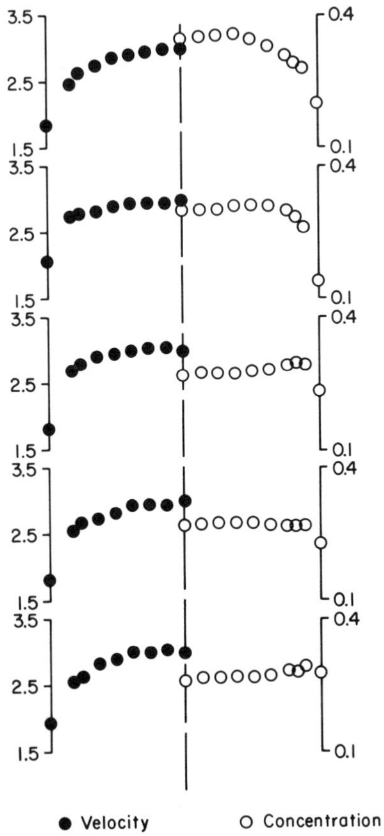

Figure 7–12. Velocity and concentration distributions measured for slurry flows in a vertical pipe (C_r = 0.30, V = 3 m/s, D = 38 mm). Particle diameters—coarse plastic: 1.4 mm; coarse sand: 0.8 mm; medium sand: 0.47 mm; fine plastic: 0.3 mm; fine sand: 0.17 mm.

In addition to these factors, there is the kinematic friction condition, used in deriving the two-layer model, to be satisfied at the pipe wall. That is, the Coulombic component of the immersed weight of the particles contributes a wall resistance proportional to the coefficient of kinematic friction η_s.

Comparing Equations 7–10 and 7–11, we observe that some of the parameters are common to the two distribution equations and this must be considered in modeling. For satisfactory model development, one requires experimental measurements of headlosses, concentration distributions, and velocity distributions.

Figure 7–12 illustrates the experimental results of Sumner et al. (1991) who measured velocity and concentration distributions for vertical flows using the devices described in Sections 10.8 and 10.13. Both distributions were symmetrical about the axis of the pipe. The left-hand diagrams show the velocities for the region between the wall and the pipe axis. The right-hand diagrams present concentrations. Points plotted on the pipe wall were measured with a flush-mounted sensor in the pipe wall, so that they describe the flow of the closest particles. The velocity distributions are somewhat flatter than would be predicted by the Law of

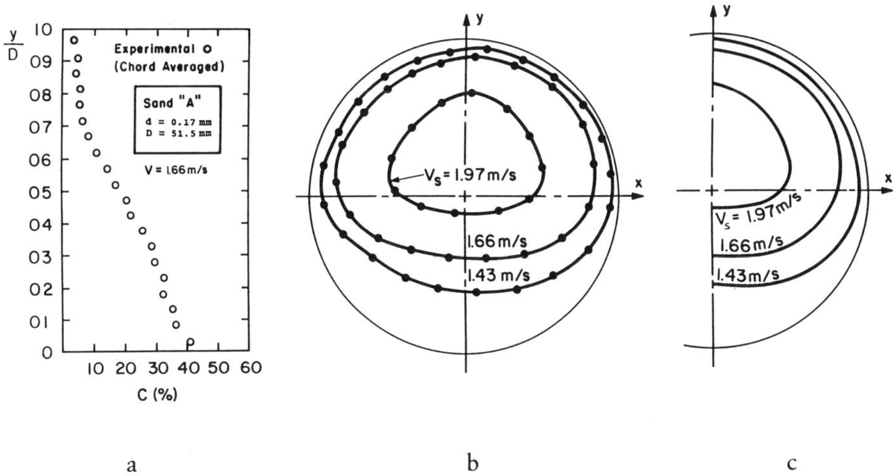

Figure 7–13. Experimental (b) and calculated (c) velocity distribution for 0.17-mm sand–water slurry, $D = 51.5$ mm.

the Wall for the *fluid* (i.e., Equation 1–9) and there is certainly no indication of a decrease in the von Karman coefficient κ.

We can also see that Equation 7–10 is qualitatively reasonable for this simple flow ($g = 0$). A region of reduced concentration occurs at the pipe wall, where the velocity gradient (see Equation 2–54) is highest.

An illustration of particle velocity distributions and predictions made with a model based on Equation 7–11 is shown in Figure 7–13 (Roco and Shook, 1985a). The chord average concentration profile is presented in Figure 7–13a; the mean flow velocity was 1.66 m/s. As in Figure 7–10, the maximum velocity is shifted upward, compared to single-phase flow. The Coulombic stress effect $[\nabla \cdot T_{sC}]$ is dominant near the pipe bottom in this stratified flow. In dilute suspension flow, the turbulent eddy–particle interaction plays an important role in velocity distribution (Abou-Arab and Roco, 1989).

Scaleup from laboratory to large scale flows using Equations 7–8 and 7–11 and some constitutive relationships was illustrated by Roco and Mahadevan (1986). Several particle-dependent coefficients in the constitutive relationships were determined from the laboratory results.

7.9 INTERPRETING EXPERIMENTAL HEADLOSSES

In the light of experimental results shown in the preceding sections, we see why one should exercise caution in interpreting pipeline headlosses using a continuum model for the whole pipe. For example, in the case of the particles whose concentration distributions were shown in Figure 7–8, the velocity distributions were

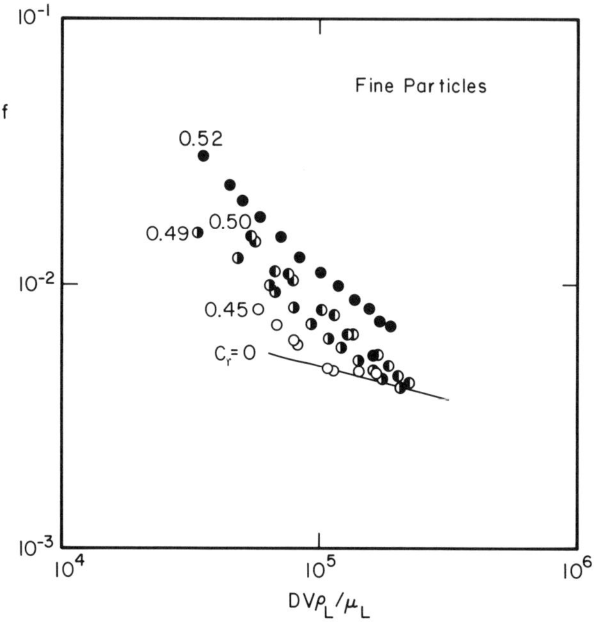

Figure 7–14. Experimental friction factors as a function of fluid Reynolds number for slurries of 0.3-mm polystyrene particles. The parameter is the mean in situ concentration C_r. (From Shook, 1985. *Can. J. Chem. Eng.* 63:865. Reprinted with permission.)

flat, suggesting either turbulent or plug-like flow. When the headlosses were expressed as friction factors using Equation 1–5,

$$f = \frac{igD\rho_L}{2V^2\rho_m}$$

the friction factors showed strong velocity and concentration dependence.

These friction factors are plotted in Figure 7–14 with the Reynolds numbers calculated from the *fluid* viscosity. The friction factors increase markedly with concentration and with decreasing Reynolds number (bulk velocity).

If the headlosses are interpreted with a continuum model, in terms of μ_m, one infers a velocity-dependent viscosity. In view of the discussion of Chapter 3, shear thinning is unlikely for suspensions of these dispersed, nearly neutrally buoyant, large rigid spheres.

Figure 7–15 shows these viscosities, normalized with the fluid viscosity, compared with viscosities measured in a Couette viscometer. We note that the viscometer measurements do not agree with either the laminar or turbulent region viscosities. Similar behavior has been observed in pipeline flow of coal–oil slurries (Frankiewicz et al., 1990, 1991), although in the latter case the pipeline laminar flow viscosity was considerably lower than the viscometer value. It seems likely

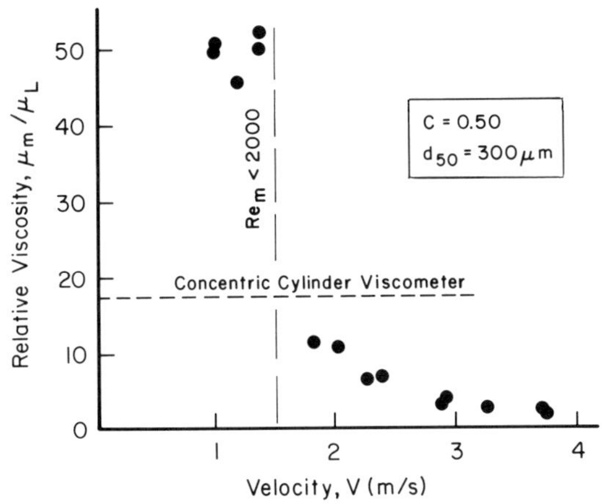

Figure 7–15. Comparison of Couette viscometer measurements t_0 with relative viscosities inferred from pipeline pressure drops.

that the solids concentration in the near-wall region is of dominant importance in these flows.

It is important to note that these slurries were apparently homogeneous, i.e., settling was very slow and viscometry was possible. These observations illustrate the limitations of a continuum model for interpreting pipeline pressure drops. Measurements of the velocity and concentration distributions are desirable for the interpretation and scaleup of headloss measurements for slurries.

7.10 INTERPRETING DEPOSITION VELOCITIES

The deposition velocity in a pipe can be identified from the velocity of the particles in the vicinity of the lower boundary. Several situations are shown in Figure 7–16. Line A corresponds to flow without deposition. Line B corresponds to a sheared

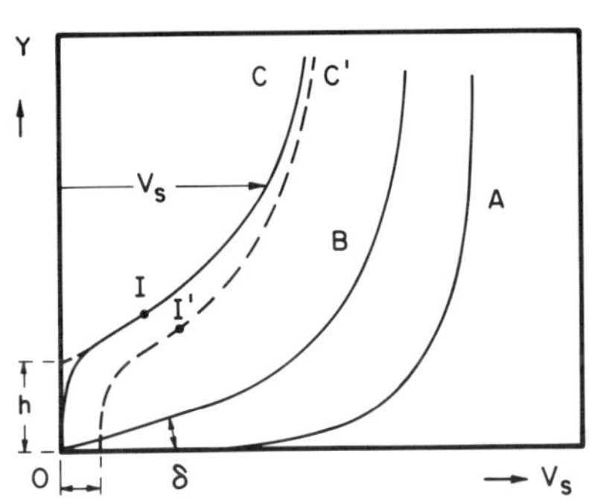

Figure 7–16. Velocity distributions of solids near lower boundary.

bed of particles and line C to a stationary unsheared bed of particles when the particle-wall friction coefficient is larger than the interparticle friction coefficient. If the interparticle friction coefficient is larger, there is a slipping velocity Δv at the bottom of the pipe (curve C′). The inflection point (I on curve C and I′ on curve C′) characterizes the critical flow regime (Roco and Shook, 1985b). This point of inflection is caused by the predominant contribution of the Coulombic stress in the region below. An unstable flow regime is expected if an inflection point occurs, with increased velocity fluctuations.

Chapter 8
Wear in Slurry Equipment

8.1 INTRODUCTION

The erosion wear of a material in contact with a flowing suspension can be defined by the loss of weight per unit area (Δw) or loss in thickness (Δs) under the dynamic action of solid particles. The loss of material is usually reported per unit time or per unit solids throughput.

Erosion wear is so important that a voluminous scientific and engineering literature on the subject already exists. The number of variables and related phenomena to be considered is very large. The task here is to summarize the basic principles of the erosion wear process. In so doing, an attempt is made to construct a framework for the plant engineer to fit his particular experience.

The large number of variables can make it difficult to interpret particular observations and this fact has influenced the view that we will take on the subject. We will attempt to follow a basic principle of scientific investigation: to give the greatest weight to evidence from experiments that can be duplicated. This has the disadvantage of emphasizing idealized laboratory wear simulations that minimize the number of variables. Observations of actual pipeline system wear have to be regarded as anecdotal unless they are completely documented. As we shall see, this documentation would have to include many properties of the eroding particles, fluid, and eroded surface, as well as the flow conditions (temperature, particle trajectories, etc.).

Excluding energy consumption, erosion wear is the principal concern in the operation of equipment handling granular materials. The dimensions and surface roughness suffer alterations in time compared to the original design. Wear affects the hydraulic and mechanical performances of slurry equipment, reduces its reliability and, most important, its operation life. Slurry erosion has similarities to gas–solid erosion in transportation systems, coal-burning boilers, etc. The engineering objectives are to predict the erosion rate and to develop techniques to reduce this rate. The first step in attaining these objectives is to identify the basic mechanisms of the wear process.

8.2 WEAR MECHANISMS

The most accepted wear theories for the material removal processes which occur at the contact zone between a particle and wall are the adhesive approximation (Burwell and Strang, 1952), the abrasive or cutting model (Richardson, 1967; Rabinovich, 1965), the delamination model (Suh, 1973), the rolling wear model (Berthe et al., 1980), fretting wear, the plastic deformation model (Hutchings, 1981), and the fatigue model (see Figure 8–1). The local mechanical deformation of the interacting particle-wall combination can be simulated numerically starting with the elasticity equation (Follansbee et al., 1980; Moore, 1975).

The *adhesive model* considers the wear process between two surfaces subjected to a normal load. Contact zones are assumed to be formed between asperities on the two surfaces (Figure 8–1a). Part of the load is transmitted to the bases of the asperities, causing their plastic deformation. As the surfaces continue to slide relative to each other, wear particles are formed and some of them adhere to the opposite sliding surface. Larger loadings would cause the number of contact zones and their area to increase, creating larger wear particles.

According to *the abrasive or cutting wear model* there are two methods by which a surface is worn. In the first method, particles remove material by cutting chips out of the abraded surface. In the second method, the abrasive particles, or more specifically the asperities of those particles, displace by cutting or ploughing some particles on the abraded softer surface to form grooves (Figure 8–1b). Both wear mechanisms impart a plastic strain to the abraded surface, causing wear by fatigue later in the wear process.

The *delamination theory* explains the wear process from the point of view of subsurface cracks (parallel to the surface), void formation, and the joining of cracks due to surface shear deformation. When a subsurface crack reaches a critical length, the material between the crack and surface will shear off (Figure 8–1c). The results are flake-like wear particles.

The *rolling wear model* was suggested for relatively smooth highly elastic materials and would explain the formation of surface damage such as micropits, sponges, and spalls. Rolled shreds are formed through the interaction of the asperities of two surfaces (Figure 8–1d). After an increase in density of the rolled shreds, the formation of macropits and spalls occur.

Fretting wear is generally produced between two hard surfaces, where the fractured materials detached from one or both surfaces form an abrasive medium (Figure 8–1e). This “third body” debris in contact with both surfaces causes the damage.

Plastic deformation is generally considered to be caused by particles impacting the surface at large angles. When the impingement energy exceeds the yield limit,

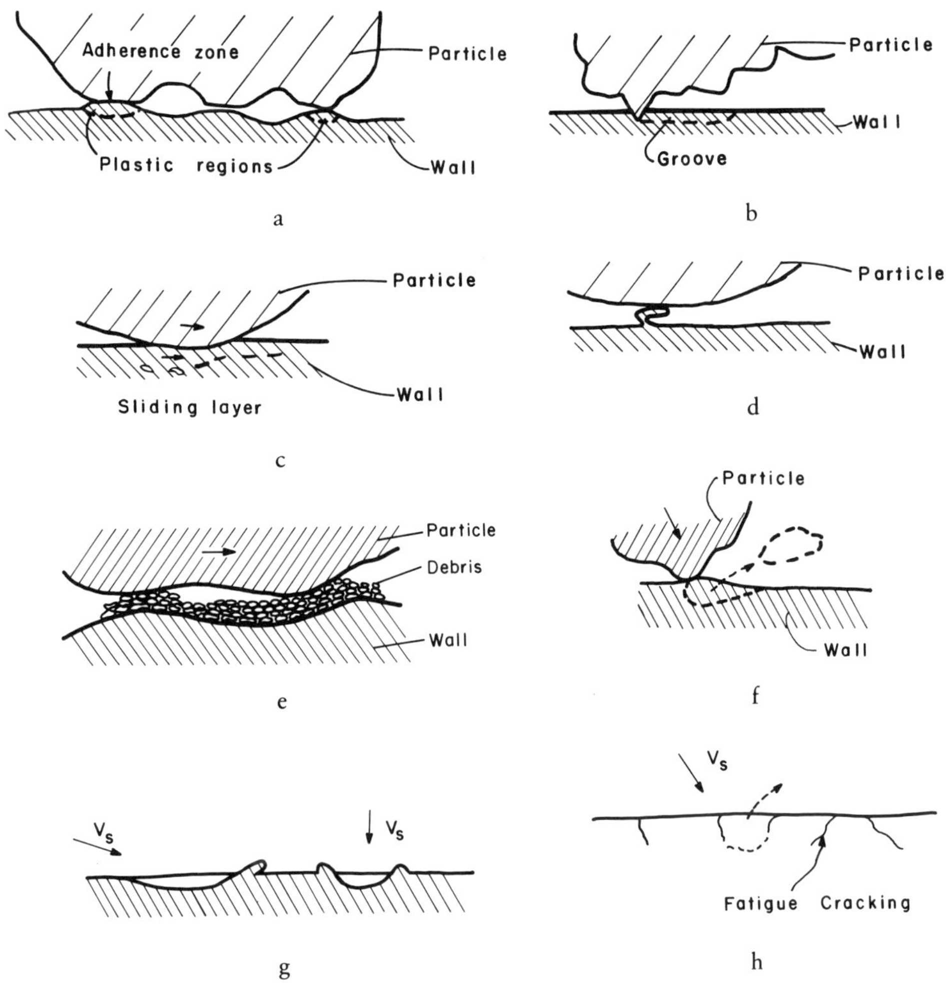

Figure 8–1. Schematic for model processes which occur at the contact zone particle-wall (micromechanical view): (a) adhesive model, (b) abrasive model, (c) delamination model, (d) rolling wear model, (e) fretting wear model, (f,g) plastic deformation model, and (h) fatigue.

the surface layer is locally deformed (Figure 8–1f). Further impacts of solid particles thereafter remove the material. At large impingement velocities some "chips" of material may be detached by a single impact (Figure 8–1g).

The particle-wall interaction is a function of particle dynamics near the wall, and therefore of fluid–particle and particle–particle interactions in that region. A *single-particle impingement* can be considered to be of dominant importance in dilute suspensions close to the wall. The particle interacts only with the carrier fluid before the impact. The fluid drag and lubrication forces play important roles

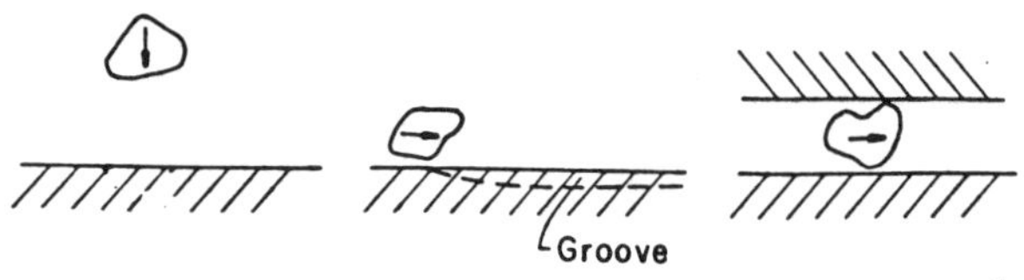

Figure 8–2. Schematic for material removal by individual particles: (a) impingement mechanism, (b) frictional mechanism, and (c) grinding wear.

in determining the particle trajectories and velocities. Single-particle impingement can produce wear by one of the following:

- *Impingement* at relatively large impact angles (for instance in Figure 8–2a), when the plastic deformation and fatigue wear are prevalent.
- *Friction,* when the adhesive, abrasive, delamination, and rolling wear develop as a function of the pair of materials for particle and wall and their surface roughness. It occurs when the particles strike the surface under small angles or effectively slide on the surface (Figure 8–2b). If the resulting local stress exceeds the limiting shear strength of the surface, a scratching or tearing damage will be produced.
- *Grinding,* when individual solid particles are entangled and pounded between two walls in relative motion (Figure 8–2c).

The impingement angle of particles on the wall varies from 0 to 90 degrees in equipment handling slurries, so that both impingement (predominantly deformation) and frictional (predominantly cutting) wear mechanisms occur simultaneously in diverse proportions. The relative rates of the two types of erosion also depend upon the material of the exposed surface. If the material is hard and brittle (white iron alloys, ceramics, etc.), deformation wear has a higher rate and the maximum occurs when solid particles strike at a normal direction (Figure 8–3b, Wilson, 1972); (Roco et al., 1984a,b). In the case of ductile and soft materials (rubber, aluminum, etc.), the cutting wear is the largest erosion component, and the critical angle is usually between 10 and 30 degrees (Figure 8–3a).

In a centrifugal pump, the impingement mechanism (deformation wear) is important at the inlet to the impeller and at the tongue of the pump shell. The frictional mechanism (cutting wear) is more significant at the impeller outlet and along the shell wall, and grinding wear occurs at the seal as well as between the impeller plates and casings.

Multiparticle impingement occurs in dense slurry flow, where the particle–particle interactions are at least of the same order of importance as the liquid–solid interactions. The particles will interact with the wall if their convective velocities are directed toward the wall, if fluctuating particle velocities are generated in the turbulent sheared layer of particles near the wall, if particles in Coulombic

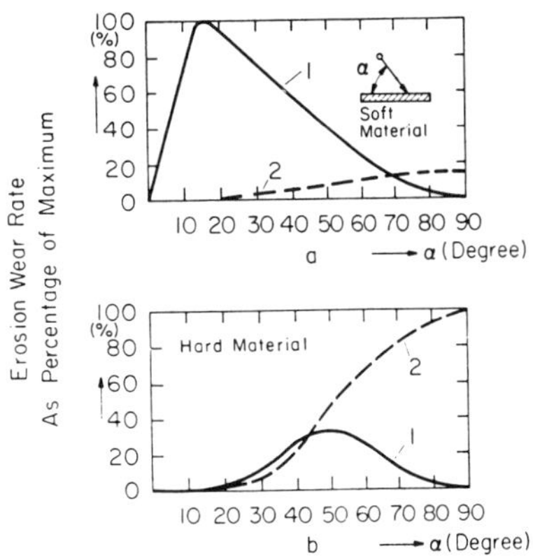

a

b

Figure 8–3. Effect of impact angle on erosion rate of different types of materials: (a) soft ductile material, (b) hard material; curve 1 = cutting wear and curve 2 = deformation wear.

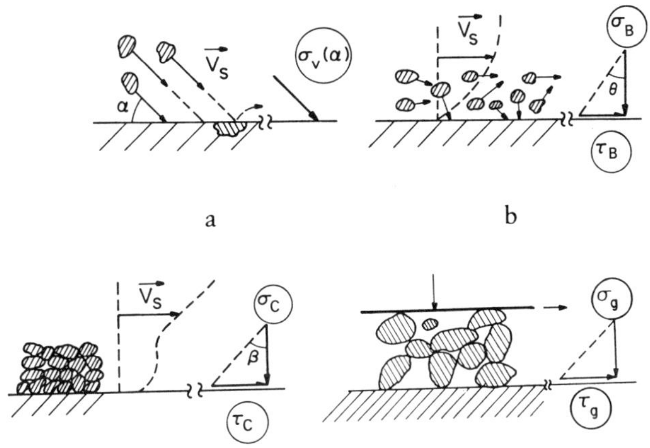

a b

c d

Figure 8–4. Schematic for material removal by dense slurries: (a) mean solid velocity (directional impingement), (b) fluctuating solid velocity (random impingement), (c) Coulombic friction, and (d) grinding wear.

contact move along the wall, or if particles are pressed against the wall by another bounding surface. We consider the erosion process in dense slurry flow to have four components (Figure 8–4a to d), caused by

a. the directional impingement of solid particles,
b. the random impingement of particles, due to the fluctuating turbulent motion,
c. the friction of a sliding bed pressing onto the wall, and
d. the friction of a layer of particles between two walls in relative motion.

These components are related to the stresses acting in dense slurry flows, which may be obtained from a fluid dynamics analysis. These stresses include

a. dynamic or impact pressure of solid particles,
b. Bagnold dispersive stress,
c. Coulombic friction, and
d. grinding normal stress.

The wear intensity and the energy dissipated by the particle–wall interaction are assumed to be correlated in these four mechanisms (Roco et al., 1984a,b). The Bagnold dispersive stress in dense slurry flow is caused by collisions and lubrication forces between solid particles in neighboring layers. The Coulombic friction is transmitted through persistent contacts.

The general wear pattern on the worn surface can be uniform (with a smooth or rough surface), quasi-uniform (wavy surface), or localized (local grooves, pitting, etc.). The mechanical removal of material from the surface may be accelerated by corrosion and cavitation, and damped by scaling. Macroscopic roughness of the worn surfaces generally is a sign of the simultaneous action of erosion and corrosion. Other particular wear patterns depend on the equipment.

8.3 EFFECT OF LOCAL PARAMETERS ON WEAR

The factors determining erosion wear at a given point can be separated into the following groups:

- flow characteristics (velocity, impingement angle, concentration of solid particles, shape of the wall surface),
- size, shape, surface roughness, and relative density of solid particles,
- properties of the carrier fluid,
- the relation between the mechanical properties of the exposed material and of the solid particles,
- the simultaneous action of corrosion, cavitation, and/or scaling phenomena.

Some of these effects are illustrated by experimental results. The intensity of erosion, the loss in weight by erosion wear per unit area and per time, increases with the *impact velocity* to a power m, varying between 1.5 and 3.5 in experiments carried out in different laboratories. The empirical exponent m averages about 2 for wear of metal plates by a water–particle jet (Finnie, 1960), between 1 and 4 in slurry pipes, and from 2.2 to 3 in centrifugal pumps (Wilson, 1972). The dispersion of the values for m can be explained by the diversity of wear mechanisms and the mean flow parameters, as well as differences in experimental methods.

In Figure 8–5 we show experimental results obtained with a water–sand jet striking on plates of various materials under various impingement angles (Wellinger and Uetz, 1955). The erosion is higher at impingement angles α = 60–90 degrees for hard materials and at angles α = 0–30 degrees for the soft ones.

The existing results concerning the effect of particle diameter are contradictory. However, it appears that erosion wear increases for some ranges of particle

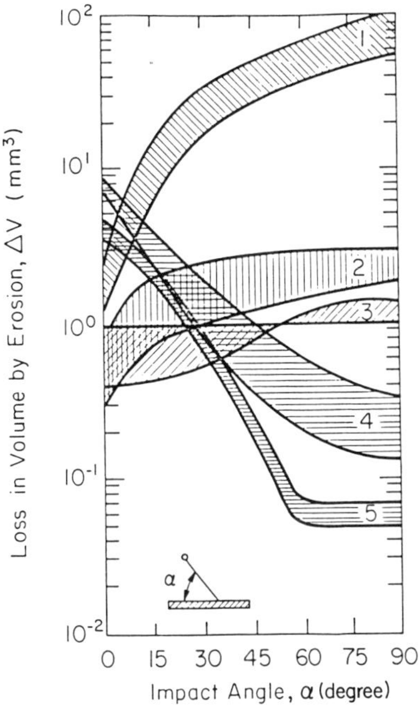

Figure 8–5. Erosion wear produced on plates under diverse impact angles: 1: basalt; 2: steel; 3: hard cast iron; 4: rubber; 5: urethane (elastomer).

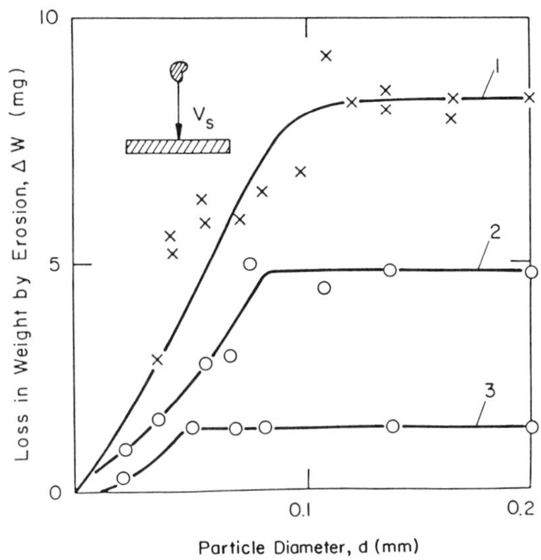

Figure 8–6. Influence of solid particle diameter on wear rate at diverse impact velocities: V_s—solid particle velocity; 1, 2, and 3—erosion curves for different V_s.

size, being thereafter insensitive to size (for example, see results with sand in Figure 8–6) (Tilly and Sage, 1970). The abrasivity increases with the density, nonsphericity, and the roughness of the lateral surface of solid particles.

The relation between the hardness of particles and that of the worn material (Brinell scale or Mohs scale) are important in the wear mechanisms. The solid particles should have at least the same hardness as the wall for the erosion to be significant. The hardness of some materials and an approximate correlation between Brinell scale (ASTM Standard) and Mohs scale (ASTM Standard) are shown in Figure 8–7 (Wilson, 1972).

8.4 EFFECT OF CORROSION

The simultaneous action of mechanical wear and electrochemical corrosion on the wetted walls of slurry equipment mutually enhance the wear process. The planes of lower resistance created by corrosion at the wall surface facilitate material removal by particle impingement. By subtracting the separate effects of erosion and corrosion from the total wear, one obtains the synergism, which may have values over 50% of total wear (Madsen, 1988).

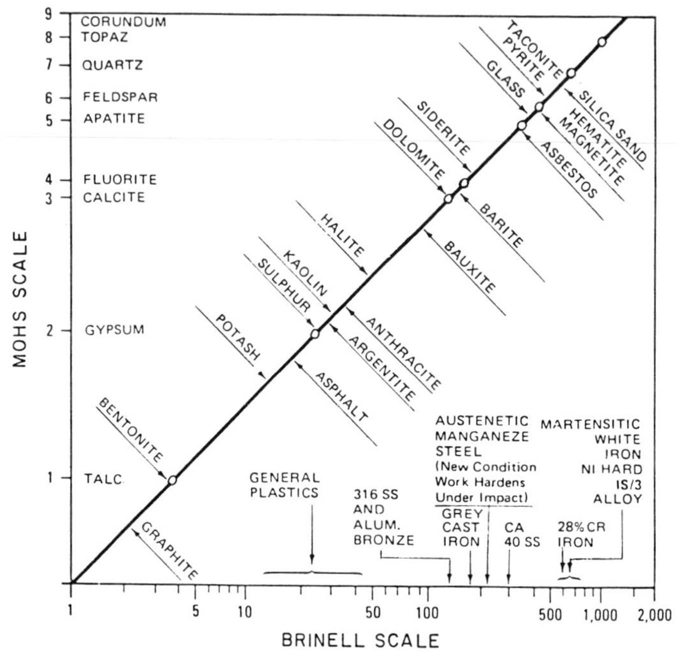

Figure 8–7. Approximate correlation between Brinell scale and Mohs scale.

The symbolic form of an overall corrosion reaction for metal M is

$$M + H_2O + 0.5O_{2(aq)} \rightarrow M(OH)_2 \tag{8–1}$$

showing that dissolved oxygen is an essential component in the liquid. When the protective film is penetrated, the corrosion rate may be increased by the exposure of fresh metal if the corrosion is not controlled by some other factor such as the diffusion of oxygen to the metal interface.

Although oxygen-scavenging species have been shown to reduce wear rates, it is rarely practical to alter the oxygen content of slurries to improve pipe life (long slurry pipelines are a possible exception, but these comprise a minute fraction of slurry transport applications). A more practical course is to select appropriate pipe wall materials or, perhaps, to use inhibitors which promote formation of a passive film at very low inhibitor concentrations. Numerous inhibitors are available for steel pipelines, including chromates, zinc compounds, polyphosphates, silicates, molybdates, and polymeric materials (Jacques and Neil, 1977; Postlethwaite, 1979). The combination of chromate and certain organic compounds was shown to be effective in controlling corrosion in coal–water slurries (Titus and DiGregorio, 1975). Postlethwaite (1981) showed that chromates could reduce erosion corrosion by sand slurries significantly.

Although coal often contains minerals (pyrite, silica) with high hardness, which produce wear, cleaning a fine coal ($d_{50} \approx 70\ \mu m$) can lead to negligible abrasive pipe wall wear by a slurry (Ferrini et al., 1982). In this case the total weight loss could be attributed to corrosion.

Methods are available for determining the corrosion tendencies of a given fluid–wall combination (ASTM Standards D2776–79, G5–87, G102–89, and G59–78, 1989). These may not require pipeline flow and usually can be conducted with small quantities of fluid or slurry. In comparison, simulations conducted with slurries in pipe flow are more cumbersome and expensive. Using bench scale pilot scale ($D = 100$ mm) measurements, Ferrini et al. (1982) showed that the polarization resistance method could be used satisfactorily if weight loss measurements were used for calibration. Postlethwaite et al. (1986) show satisfactory agreement among the diffusion current, resistance polarization, and weight loss measurements for erosion-corrosion of pipe walls by sand slurries. In a two-step process, the corrosion current i_{corr} can be estimated by first using linear polarization measurements (ASTM Standard G59–78, 1989) to establish the polarization resistance R_ρ. Cathodic and anodic polarization measurements are then made to determine the Tafel slopes b_c and b_a (ASTM Standards G59–78 and G5–87, 1989). The corrosion current, which determines the rate of weight loss, is

$$i_{corr} = \frac{b_c b_a}{2.303(b_c + b_a)R_\rho} \tag{8–2}$$

Ferrini et al. (1982) employed a series of weight loss measurements to determine how the polarization test results should be interpreted to yield b_c and b_a.

The effect of corrosion increases in slurry mixtures because of several factors:

- The protective layer or film of corroded products is removed by wear and fresh surfaces are exposed to the carrier fluid.
- The amount of oxygen available is often increased because air is entrained in solids mixing installations.
- Dissolved species from the solids may react with the liquid and the wall.

Although test methods are available, misleading results can be obtained by extrapolating the laboratory data to industrial situations, even under identical pH and other flow conditions (Madsen, 1988).

8.5 EFFECT OF MATERIAL

The mechanical properties of the eroded surface material relative to those of transported solid particles determines the wear rate, as a function of the flow situation and predominant wear mechanism(s). For instance, if the wear by friction is dominant, it is desirable for the wall material to have a greater hardness, whereas if impingement wear is dominant, one selects more compliant materials (rubber, polyurethane, etc.).

Besides the metallic materials (white cast irons, alloys), one should also consider rubber, polyurethane, high-density polyurethane, ceramic materials, fiberglass, cast basalt, and composite materials for slurry flow applications (Loadstar Publ., 1988). The surface microstructure and roughness play an important role on the rate of material removal. In addition to the wear resistance, factors to be considered in selecting material include corrosion resistance, any change in hydraulic resistance, effects of temperature, property change in time for plastic materials, and repair and replacement costs.

8.6 WEAR PREDICTION

Semiempirical formulas are useful in specific applications to relate the wear rate to mean flow parameters, such as flow rate, mean concentration, and particle size. Their degree of generality is limited by the experimental base used in their formulation, and extrapolations are not recommended.

The most suitable area of application for semiempirical formulations would seem to be pipe flow because of its relative simplicity as compared to pumps, valves, tees, and other equipment handling slurries. However, there are still no general semiempirical laws of variation of wear with the flow parameters. One of the earlier formulas is that of Bergeron (1952)

$$\Delta\dot{s} = \frac{k\rho_L(S_s - 1)d^3CV^3}{D} \tag{8–3}$$

where $\Delta\dot{s}$ (10^{-6} mm/h) is the wear rate per unit time, S_s is the solid relative density, d (m) is the mean particle size, C (by volume) is the mean solids concentration, V_m (m/s) is the mean mixture velocity, D (m) is the pipe diameter, and k is an experimental coefficient. Generally, the exponents of $(S_s - 1)$, d, C, V, and D are functions of the flow pattern in the pipe and cannot be extrapolated outside the experimental data base (Roco and Cader, 1988). For instance, the power of the mixture velocity was 3.27 in the experiments of Karabelas (1976). However, Faddick (1975) presented a correlation for industrial pipelines in which the amount of wear per unit mass of solid material transported is proportional to V_m to a power of 2. On the other hand, Shook et al. (1990) found their wear rates per unit mass of transported material, produced by quasi-spherical particles in horizontal pipe flow in the laboratory, to be nearly independent of V_m.

A micromechanical approach based on idealized representations of the erosion mechanisms can provide an alternative to semiempirical formulas using mean flow parameters. In simulations, wear is determined by the initial kinetic energy of the particles and the fluid–particle and particle–particle forces that act on the particles during their contact with the eroded surface. The kinetic energy of particles consists of both translational and rotational components. The particles may be decelerated before impingement and can bounce back into the flow, adhere to the surface, or may exhibit a multicycle attachment–detachment pattern. Since the single-particle motion, the particle–fluid interaction, and particle–particle interaction, as well as the wear mechanism itself, are not completely understood, the present models adopt some idealizations to describe a complex flow situation.

The abrasion (erosive cutting) of ductile materials is better understood. Finnie (1960) suggested a model with a single wear mechanism similar to the material removal by a machine tool, in which the particle–fluid and particle–particle interactions are neglected. With similar simplifying assumptions Neilson and Gilchrist (1968) simulated plastic deformation wear. Bitter (1963a,b) combined the cutting and plastic deformation models. The deformation wear measured by loss of volume per particles of mass m is

$$\frac{1}{2}\frac{m(V_s \sin \alpha_i - K)^2}{\phi} \tag{8–4}$$

where V_s is particle velocity and α_i is the impact angle. K is the particle velocity at incipient erosion and can be calculated from mechanical and physical properties. ϕ represents the energy required to remove a unit volume of material from the body surface and describe the plastic–elastic properties of the material.

Cutting wear occurs if particles strike the surface at a small angle and Bitter (1963b) shows that cutting wear is a function of both normal and tangential velocity components. Damage occurs only if the normal velocity component exceeds the K value corresponding to the elastic load limit. Two formulas were proposed for the situations in which (a) the particle still has a tangential velocity when it leaves the body surface and (b) the tangential velocity component becomes zero during the collision. The classical micromechanical models for wear by particle impingement were reviewed by Tilly (1979).

The Hertzian contact time for a spherical particle colliding with a semi-infinite block was determined by Goldsmith (1960). It plays an important role in the stress distribution and wave propagation in the impacted wall. For a constant impingement velocity vector at an angle between 0 and 90 degrees, the rebound velocity is described by a frequency distribution function (Matsumoto and Salto, 1970) and has an out-of-plane (third) component (Eroglu and Tabakoff, 1991).

Pourahmadi and Humphrey (1983) proposed a two-dimensional approach to predict the turbulent motion of a very dilute suspension of spherical particles and to evaluate the wall wear with the Finnie (1972) cutting model.

The energy approach makes a connection between the wear and flow indices near the wall. Wall erosion and particle attrition are complementary processes, in that both involve creation of new surface area. The energy dissipated during erosion is assumed to create new surface or produce plastic and elastic deformation of the wall and/or particles. In the energy approach, the mechanical energy dissipated by the particle–wall interactions is assumed to be proportional to the amount of material removed (Tuzson, 1982; Rocò et al., 1984a). The coefficient of proportionality depends on the wear mechanism (directional impingement, random impingement, friction, grinding) and the presence of other surface processes, such as corrosion, cavitation, or scaling.

The erosion mechanisms have a probabilistic character: particles of random size and shape slide, roll, and impact on the wall with random velocities under various impingement angles; the inhomogeneities of the exposed material have a random distribution. Therefore, the material removal is best characterized by a probabilistic function.

An energy-based predictive approach for dense slurry wear combines computational and experimental steps. A computational methodology to determine slurry wear has been tested in collaboration with several laboratories and industrial units and has been presented in the work of Roco et al. (1984a,b; 1987) and Roco and Cader (1990). In this section we summarize its main features.

First, small-scale experiments in the laboratory provide the empirical "material" correlations between the wear rate and energy dissipated for each simple wear pattern for the required pair of materials (solid particles and eroded wall), i.e., $\Delta\dot{s}_i = f(\dot{E})$ (Roco et al., 1987). Then, the concentration and velocity distributions close to the exposed walls are predicted numerically (Roco and Dehaven, 1989; Roco and Cader, 1988). The interaction energy between particles and wall corresponding to each simple wear mechanism, $\dot{E}_i(V_s, \theta, C)$ is calculated from the particle motion in the proximity of the wall.

The wear rate at any location in slurry equipment is obtained by assuming superposition of the contributions from individual wear mechanisms:

$$\Delta\dot{s} = \sum_i \Delta\dot{s}_i = \sum_i \phi_i' \cdot (\dot{E}_i - \dot{E}_{i0}) \tag{8–5}$$

where $\Delta\dot{s}$ = the loss in wall thickness per unit time, averaged in time;

i = v (directional impingement), k (random impingement caused by the kinetic energy of particles), C (frictional wear), and g (grinding wear);

ϕ_i' = $(\partial s_i/\partial E_i)_{\text{constant conditions}}$ = coefficient of the effect of the ith wear mechanism, for a given pair: (particles–worn wall) when all other surface processes are invariant;

$\dot{E}_i$ = time rate of energy dissipation per unit area by the ith wear mechanism ($\dot{E}_i$ is equal to the solid velocity at the wall multiplied by the corresponding solid stress caused by directional impact, Bagnold dispersive effect, Coulombic friction, and grinding, respectively); and

$\dot{E}_{i0}$ = the threshold energy rate for incipient wear by the ith mechanism. This value may be approximated to zero for a set of data ($\Delta\dot{s} \approx \phi_i' \dot{E}_i$) with a corresponding adjustment of ϕ_i'.

This predictive approach can be applied to two- and three-dimensional flow geometries, such as pumps, pipes, cyclones, and fittings. The mixture velocity V_M may be evaluated using turbulence models or inviscid flow simulations with slip boundary conditions when this simplification is acceptable (Roco et al., 1984b).

The contact between two bodies is made on their surface asperities. Accordingly, the lubrication forces between particles and particle and wall are functions of the surface roughness. Their effect is lower for the irregularly shaped particles encountered in many industrial applications because the lubrication force is approximately inversely proportional to the distance between surfaces (Frankel and Acrivos, 1967). However, the lubrication effect is significant for spherical beads encountered in some industrial processes and frequently used in laboratory studies. The role of lubrication forces complicates an energy approach for predicting or correlating erosion rates since the increased energy dissipation in slurry flow need not involve actual particle-wall contact (Levy and Hickey, 1987; Shook et al., 1990).

8.7 EXPERIMENTAL AND COMPUTATIONAL ASPECTS

Laboratory methods are useful for understanding the wear mechanisms and generating correlations. Erosive wear in industrial equipment is usually a combination of various wear mechanisms and a laboratory experiment is more significant if it either simulates only one wear mechanism or reproduces at the microscale the wear process from the particular situation of interest. A number of erosion simulation devices have been devised for use with slurries.

Impacting jets have been used in many erosion studies. They have the advantage that the variables of interest (including velocity and impact angle) can be determined accurately and varied widely. Unfortunately, most of the erosion results in the scientific literature refer to jets of particles in gases. Sargent et al. (1982) noted some significant differences between steel erosion results obtained with slurries and those previously reported for gas jets. The impingement angles which produced maximum erosion were significantly higher in water slurries.

The Miller test (ASTM G75–89, 1989; Miller, 1986) was originally developed to compare solid particles in terms of their effects in reciprocating pumps. It provides a measure of the effect of the particles on a standard surface. Because a normal force is applied to the particles when they are moved relative to the substrate, it would be difficult to use the results of such a test to predict pipe wall erosion quantitatively. However, the test does provide a qualitative indication of the relative abrasion tendency of a given slurry on a given wall.

A device which produces particle-wall abrasive wear without applying an external force is the Coriolis effect device shown schematically in Figure 8–8 (Tuzson, 1982). The slurry flows radially outward from the central chamber while the flow channel is rotated about the axis. The Coriolis effect forces the particles against the boundary of the flow channel, producing abrasive wear as the particles move outward. The force of the particles normal to the boundary is proportional to the product of the radial and tangential velocities.

In another method, Hoey and Bednar (1980) rotated cylindrical specimens in a baffled container of slurry. The slurry could be recirculated through the tank while being aerated or deaerated. The axis of rotation was parallel to the axes

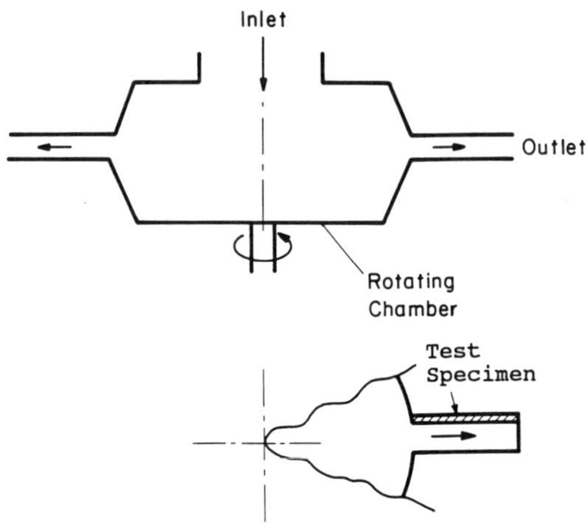

Figure 8–8. Test device for the Coriolis effect.

of the specimens so that impingement angles ranged between 0 and 90 degrees in a given experiment and the distribution of impingement angles would be velocity dependent. Considerably different corrosion–erosion patterns were evident on the upstream and downstream surfaces, demonstrating the importance of particle–surface interactions. Their device was similar hydrodynamically to the pot tester of Tsai (1981). Tests of this type are probably useful in screening materials for resistance to erosion or erosion–corrosion in a nonspecific sense, especially for situations where particle sizes or trajectories cannot be defined.

De Bree et al. (1982) used a device in which specimens were rotated in opposite directions, so that the concentration was uniform in the contact region. Impingement angles and velocities could be determined reliably for this device. The results agreed with those reported in the classical study of Bitter (1963a,b).

A flow-through rotating wear test apparatus was developed by Madsen (1988). The apparatus sketched in Figure 8–9 consists of a stator lined with about 16 plane specimens, a rotor with relatively small blades, and an open or closed circuit for the suspension. An analysis of the wear pattern on the probes indicated that directional impact was the predominant wear mechanism.

Two devices to simulate erosion wear by friction and by impingement in dense slurry flow are described as follows.

In the oscillatory apparatus for erosion by friction, the solid particles that cannot be maintained in suspension by hydrodynamic forces slide or rest on the solid boundary wall transmitting a normal stress (σ_C). Such a region can be observed, for instance, at the bottom of circular pipes, on the exterior wall of pipe elbows, in centrifuges, or in pump casings. To simulate the erosion by friction, an oscillatory device (Roco et al., 1987) is used in which the particles slide and rotate on the exposed surfaces. The moving frame has a flat oscillatory motion in such a way that the "bed of particles" slides on the bottom, where the specimens or

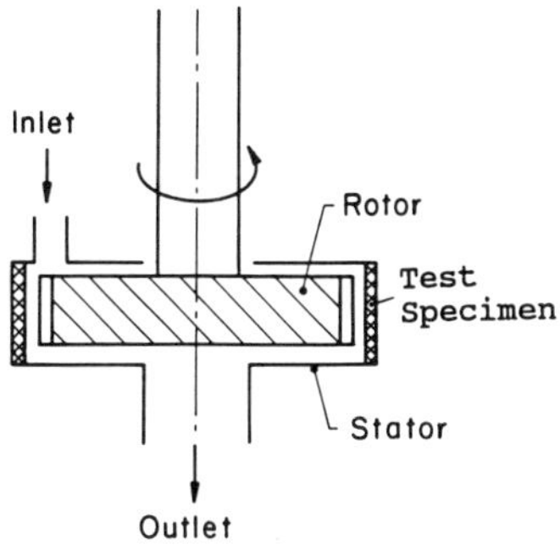

Figure 8–9. Flow-through rotating apparatus.

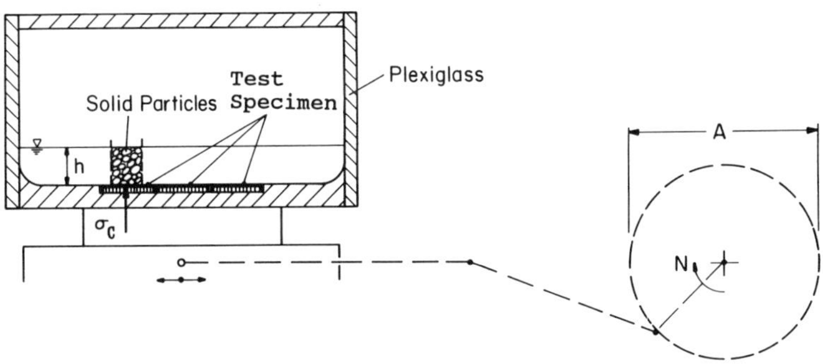

Figure 8–10. Test device for frictional wear pattern.

probes (exposed surfaces) are located (Figure 8–10). The reciprocating motion is in the horizontal plane. The stress due to the immersed weight of particles in fluid, transmitted downward, is $\sigma_C = (\rho_s - \rho_L)gCh$, where C is the solid concentration in the particulate bed, ρ_s is the solid density, ρ_L is the fluid density, and h is the bed height. The fluid may be any carrier liquid (usually water) or gas (usually air). The particulate material may be replaced periodically in a batch system, or continuously with a flow-through system. The amplitude (A) and frequency (N) of the oscillations may be varied. Various specimens of wearing materials may be tested in parallel for the same particulate system.

Results obtained with the frequency $N = 505$ rpm and amplitude $A = 50.8$ mm are illustrated in Figure 8–11. The first test (Figure 8–11a) was run with a mixture of water and sand particles $d_{50} = 0.27$ mm. The three plate samples were covered with a thin layer of polyamide resin ($s \simeq 30$ μm). The vertical stress is $\sigma_C = 135$ N/m^2. The reduction in thickness of the resin coating was measured in this case with a micrometer. For more complex geometries, another method had to be used. The curve given in Figure 8–11a averages the measurements made on three specimens. One observes the decrease of erosion rate in time because of the particle attrition and roughness reduction. $(\Delta \dot{s}_C) = \tan \theta_0$ is used for the

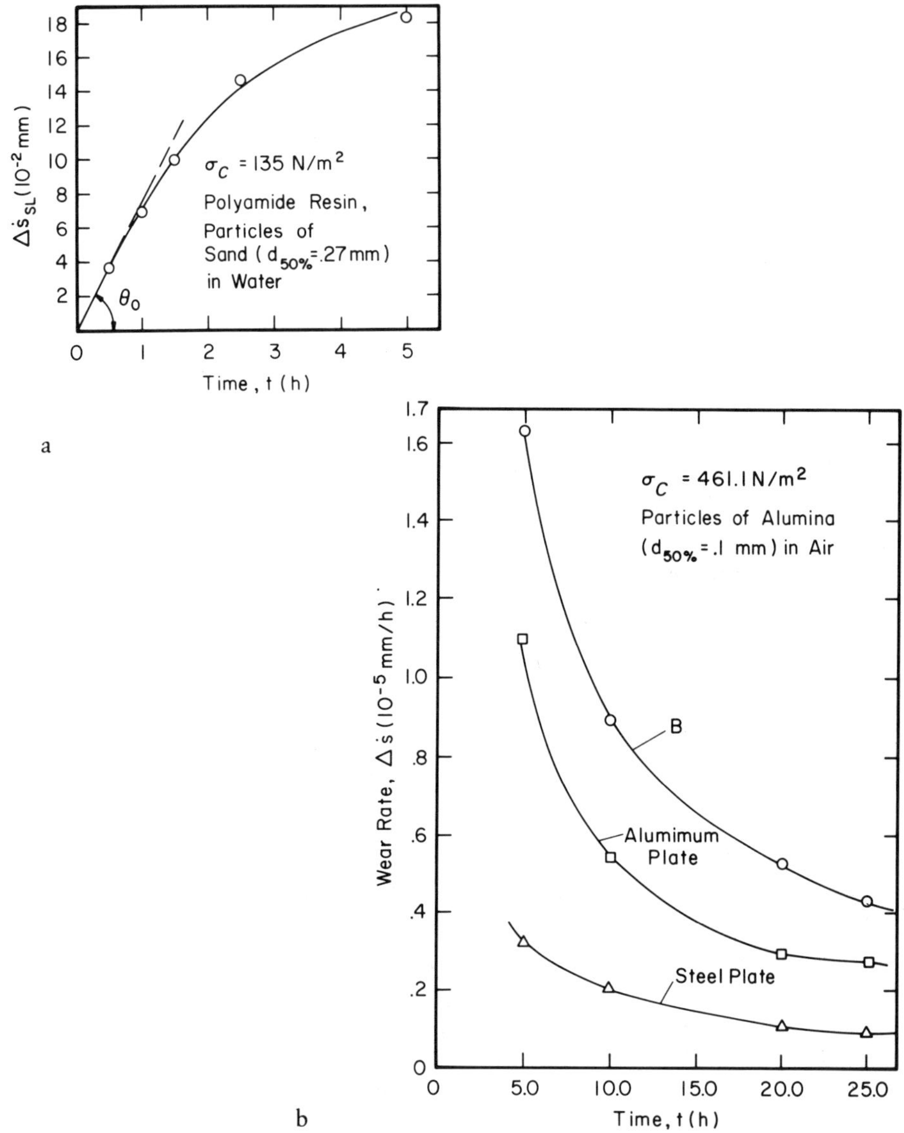

Figure 8–11. Experimental results for frictional wear pattern: (a) sand particles on polyamide resin and (b) alumina particles on plates of different materials.

calculations with new sand (θ_0 measures the inclination of the tangent to the curve at the origin). In Equation 8–3, $\dot{E}_i$ and $\dot{E}_{i0}$ are obtained from the product of the velocity of the particles and the appropriate wall shear stress. The shear stress τ_C is determined from the normal stress σ_C and a coefficient of dynamic friction $\tan \beta \approx 0.3$ ($\tau_C = \sigma_C \tan \beta$).

In Figure 8–11b we show some results obtained with alumina particles (d_{50} = 0.1 mm) in air. The exposed surfaces were of three different materials (brass, aluminum, steel). The decrease of the wear rate with time ($\Delta \dot{s}_C$ is proportional to $t^{-0.33}$) is due to the particle attrition.

In the inclined wall test for erosion wear by impingement, one simulates both directional and random impact with the same apparatus. Figure 8–12a shows a schematic view of the impact specimen, with an inclined surface at angle α to the mean velocity of the slurry stream. The specimen is placed in a confined stream of constant cross section so that a two-dimensional computational model can be used to predict concentration and velocities in this case. The two-dimensional assumption was verified from the erosion wear distributions on the specimen.

When the angle α is zero, only the random impact of particles in turbulent motion occurs. Erosion wear is important at high concentrations and high shear stresses. The stress in this case is the Bagnold dispersive stress σ_B at the wall.

The setup (two-dimensional inclined wall in confined stream of suspension) provides data for wear rates, which can be used for comparisons with numerical predictions. It has some advantages compared to the commonly used particle-laden free jet test. For any free jet angle, there is a spectrum of effective impact angles of individual particles, which is difficult to determine and to relate to the local wear distribution, particularly in inclined jets ($\alpha \neq 90$ degrees). Furthermore, the jet is not confined, and the mixing with the surrounding fluid causes supplementary computational problems.

The wear rate distribution $\Delta \dot{s}$ on the inclined wall at rows A, B, and C for plate inclination α is illustrated in Figure 8–12c. Figure 8–12d shows the average erosion rate on plates covered by polyamide resin, epoxy resin, and Ni-hard. Relative units are used for erosion rate $[\Delta \dot{s}(\alpha)/\Delta \dot{s}_{max}]$. The directional impact varies with the distance from the leading edge of the specimen (with l shown in Figure 8–12). The absolute value and the orientation of the impingement velocity vector change with l. This variation is reflected in the shape of the curves of erosion rate ($\Delta \dot{s}$) versus attack angle (α) for various rows of the measuring positions (Figure 8–12c).

In wear tests, the reduction of the wall thickness Δs_j over a relatively small interval of time Δt_j can be approximated by

$$\Delta s_j = \Delta \dot{s}_j \cdot \Delta t_j \tag{8–6}$$

where $\Delta \dot{s}_j$ is a constant wear rate. However, over a large interval, the wear can alter the wall geometry, wall roughness, and indirectly the flow indices. These changes may be important and require an iterative approach for wear prediction (Roco et al., 1987). The cumulative wear cannot be obtained by linear extrapolation.

A particulate wear process has a stochastic character because the factors determining wear are randomly distributed in time. The process is determined by the random particle dynamics in the vicinity of the bounding wall and by the probabilistic impingement event for a given pair of materials (particle-wall). The particle interactions with the wall can be described by probability density distribution

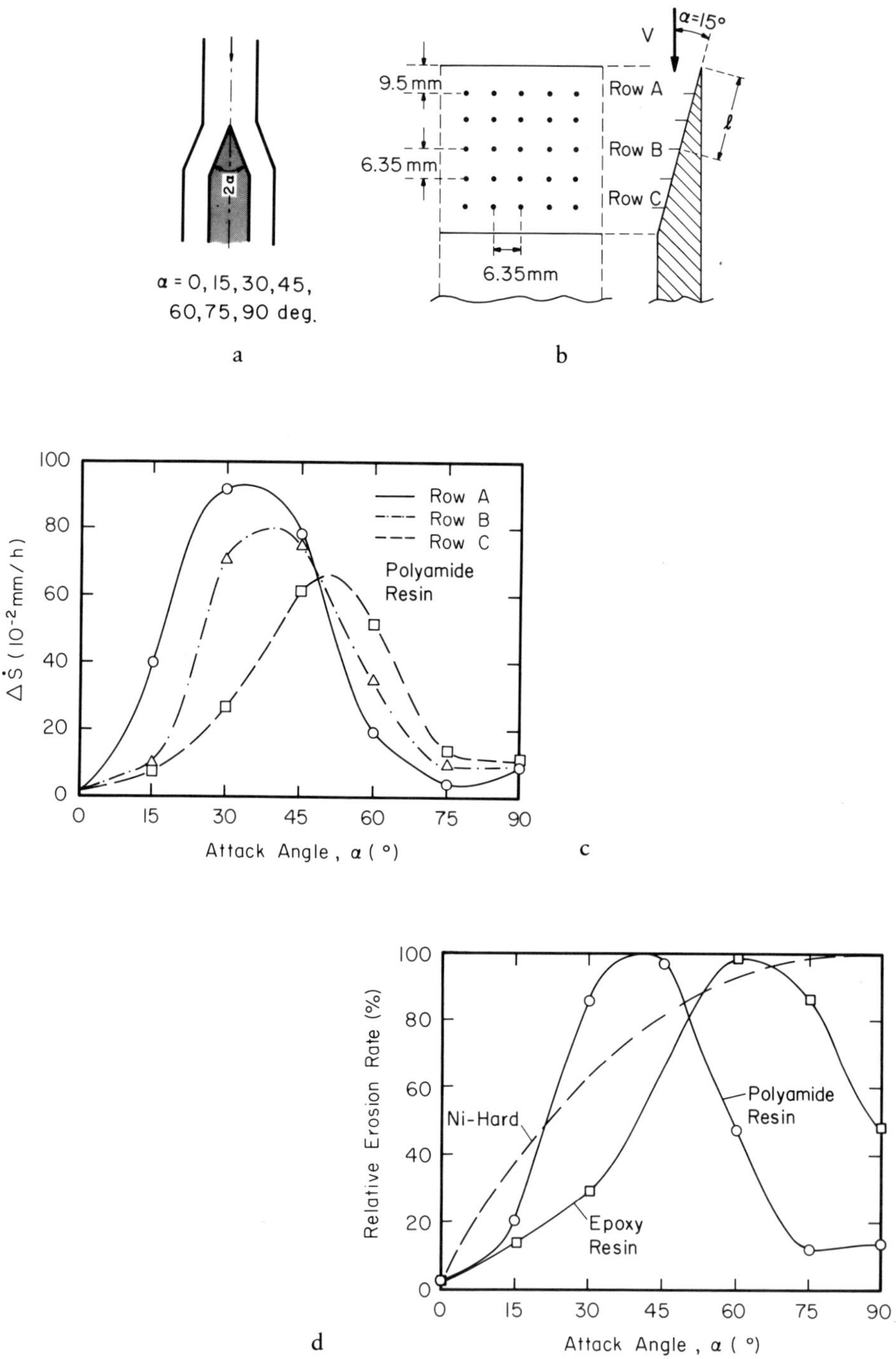

Figure 8–12. Inclined wall test for impingement wear: (a,b) schematic, (c) measured wear rates, averaged at different rows (different *l*), and (d) averages over the inclined probe.

functions: $P_v(V_s, \alpha_i)$ for the directional impact, $P_k(V_s', \alpha')$ for random impingement, and $P_{fr}(V_{slip}, V_{rol})$ for friction. These distribution functions and the mean values of the variables are to be obtained from experiments or two-phase flow modeling (V_{slip} is the particle sliding velocity, V_{rol} is the particle rolling velocity at the wall). For each interaction event, there is a probability that the total interaction energy will be dissipated in some proportion into the particles and wall. Only a part of the energy dissipated into the wall is used to create new surface and, therefore, produces wear.

Usually, the transported particles have a nonuniform size distribution. The diameter of uniform particles that would produce the same amount of wear as a given assembly of particles characterized by a broad size distribution, for the same flow conditions (average concentration and velocity, D, etc.), is defined as an equivalent wear diameter. Its value is different from the mean particle size because the interaction energy and the rate of material removal are nonlinear with particle size. Some experiments performed for materials with a broad quasi-logarithmic size distribution suggest that the equivalent wear diameter is about d_{15} for pumps [15% sieve diameter, larger than the average particle size (Roco and Minani, 1989)], and about d_{20} for slurry pipes (Roco and Cader, 1990).

The material removal can be characterized by a wear histogram. The effective wear rate varies in time and space, even if the mean flow is uniform in the vicinity of the wall. Consider, for example, uniform flow in a slurry pipeline. In an ideal situation, for which the instantaneous point flow parameters is constant, the pipe material homogeneous, the pipe alignment perfect, etc., the wear rate would be constant in time and along the pipe for the same position α. In a real case, the erosive process is stochastic because of the variation of the instantaneous solid velocities and other local parameters. Equations 8–1 through 8–6 predict the average wear rate for some given mean flow parameters in the pipe. However, the erosive process can deviate from its mean in time and along the pipe. The total wear (Δs) and the pipe thickness (s) can be characterized by probability distributions. The differences ($s - s_{av}$) are mainly caused by

1. the initial wall thickness probability distribution,
2. local flow disturbances,
3. the stochastic character of the particle impingement upon the wall,
4. nonuniformities of the pipe material and initial surface roughness.

Figure 8–13 illustrates the wear histogram. The probability that the pipe thickness would have a value (s) at a given operation time (t) is $(p_s)_t$. The initial probability distribution for the new pipe is defined by the curve t_{new} in Figure 8–13, the mean value of Δs at a time t is $(\Delta s_{av})_t$, and for the most exposed portion of the pipe the reduction in thickness is $(\Delta s_{max})_t$. The initial design and inspection forecasting during operation should take into account the safety factor $(\Delta s_{max})_t/(\Delta s_{av})_t$. Adjustable safety-risk levels can be determined after each pipe inspection during operation, in a manner similar to the known practice for corroded pipes (Buhrow, 1985).

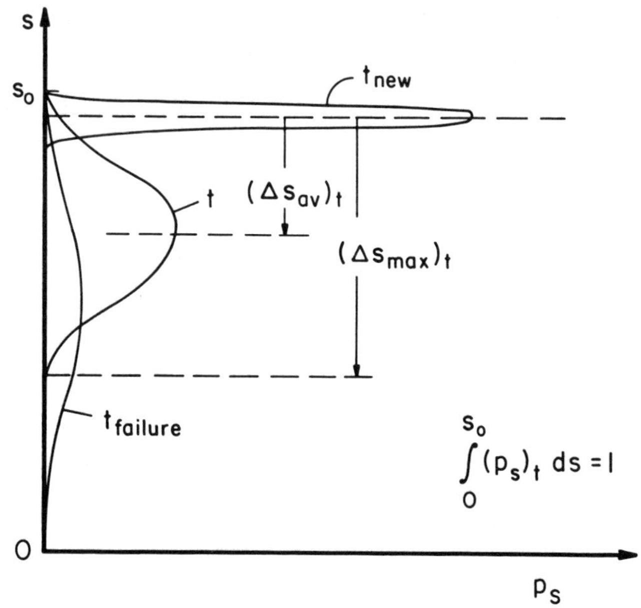

Figure 8–13. Wear histogram.

8.8 WEAR IN SLURRY EQUIPMENT

The velocities and trajectories of solid particles in a slurry pump are different from those of the carrier liquid. The relative velocity between the 1-mm glass beads and neutrally buoyant plastic particles in the vertical midplane of a rectangular venturi section is illustrated in Figure 8–14 (Ye et al., 1990). Compared to the plastic beads, the velocity of the glass beads is lower in the convergent section and in the venturi throat. After the throat, the velocity of the glass beads equals and exceeds that of the plastic beads (Figure 8–14c). Most of the particle-wall collisions occur downstream of the venturi throat at relative large impact angles (mostly between 5 and 30 degrees) for the plastic beads, and upstream of the throat at relatively small angles (several degrees) for the glass beads. Larger impact angles in the divergent section are caused by recirculations, which may explain larger wear rates generally observed after areas with flow obstructions. Figure 8–15a shows, as another example, the deviation of particle trajectories from the average motion of mixture in an impeller. At the impeller inlet, particles heavier than the carrier fluid have a tendency to migrate toward the leading face of the blade (Figure 8–15a) and to the back shroud (Figure 8–15b). At the outlet of a typical centrifugal pump impeller, under the effect of Coriolis and centrifugal forces, the finer particles (with trajectories A in Figure 8–15c) congregate close to the leading face of the blade and leave the impeller at angles of 5–15 degrees smaller than $\beta_{2,imp}$. Larger particles (B in Figure 8–15c) leave the impeller at angles larger than $\beta_{2,imp}$.

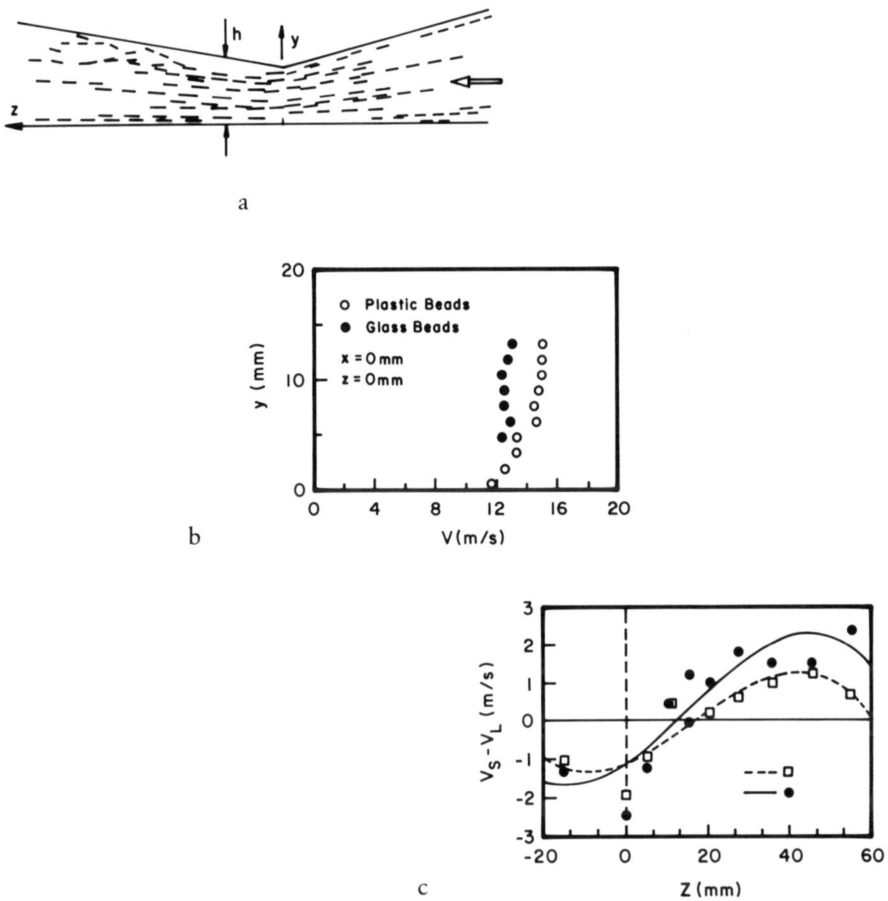

Figure 8–14. Velocity comparison between neutrally buoyant plastic beads and 1-mm glass beads in the vertical mid-plane of a venturi section: (a) schematic, (b) in the venturi throat at $Z = 0$, and (c) along the venturi section.

The areas of maximum wear may be explained by the shape of these trajectories. Typical areas with higher wear rates sketched in Figure 8–15 are the inlet section, leading face, and outlet section of the blades (zones I, II, and III, respectively) and the impeller back plate (zone IV). Larger particles would cause greater damage in zones I and IV, whereas finer particles would wear more in zone III. Generally, the wear is distributed nonuniformly on the impeller, casing, front and back plates, as well as in suction and discharge pump pipes. The wear pattern depends on factors such as the pump construction, solid particle size distribution, discharge flow rate, recirculations, mean concentration, and rotational speed. Some qualitative trends can be observed in experimental studies, but their indications cannot be taken out of the general flow environment (i.e., assembly of conditions) in which the experiments are conducted.

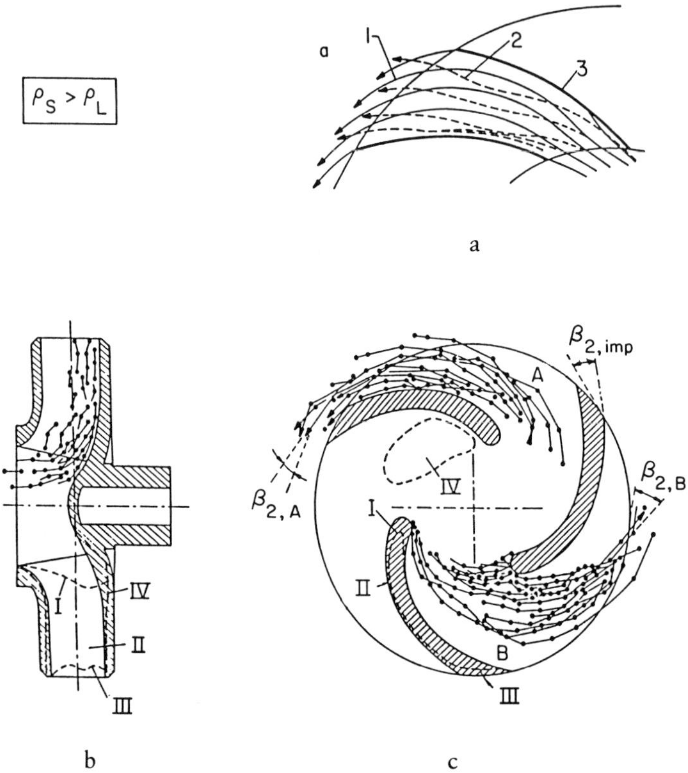

Figure 8–15. Solid particle trajectories and areas of maximum wear in impeller. 1—mixture streamlines; 2—solid particle trajectories; 3—blade; A—fine particles; B—coarse particles; I, II, III, IV—typical areas of maximum erosion wear.

One can determine the wear distribution on the wetted walls with the energy approach, which makes the connection between the two-phase flow and point wear rate. All four wear mechanisms coexist in dense slurry flow, but their relative importance varies in different zones of the pump.

1. Directional impingement is significant at the leading zone of the impeller blades and casing tongue, as well as in the zones with intense recirculation and rebounded particles. This wear mechanism becomes more important at off-duty flow rates.
2. Random impingement plays the major role at the pump best efficiency point (BEP) flow rate along the central zone of the impeller blades and casing wall.
3. Frictional wear is caused by the particles that cannot be maintained in suspension by turbulence and accumulate into sliding layers along the leading face of the blades and peripheral contour of the casing.

4. Grinding wear is encountered in the stuffing box and its seal, as well as in the gaps between the impeller shrouds and casing when large particles are handled.

The probabilistic character of the wear process must be reflected in the predictive approach in two ways: (1) the estimation of the minimum threshold solid velocity at which the wear process occurs and (2) calculation of the mean wear rate.

The impingement energy of a particle ($0.5 m_s U_s^2$ where U_s is the instantaneous solid velocity, $U_s = V_s + V_s'$, and V_s is the mean of U_s) has a probability distribution function (for instance, lognormal) characterized by a variance, which is about the same order of magnitude as the mean value in centrifugal pumps. Even if the mean impingement energy of solids is not sufficient to initiate the wear, some larger instantaneous values may well exceed that limit.

The wear rate increases with the variance of probability distribution function $P(\sigma_p)$ of the impingement stress $\sigma_p = 0.5 m_s U_s^2 \cdot U_s$. That is:

$$\frac{\int_0^\infty \sigma_p \cdot P(\sigma_p) \cdot d\sigma_p}{(\sigma_p)_{av}} > 1 \tag{8-7}$$

where $(\sigma_p)_{av}$ is the mean impingement stress. This ratio would equal unity only if $U_s = V_s$ and $\sigma_p = (\sigma_p)_{av}$. Generally, two-phase flow numerical models have used a deterministic approach because of its relative simplicity (see, for instance, Roco and Cader, 1988; Roco and Dehaven, 1989).

If the particle time scale is large compared to the flow time scale (i.e., the particle trajectory is not significantly affected by the boundary layer) and inertial effects dominate the flow, one might simplify the approach by using an inviscid flow simulation with slip boundary conditions for the momentum equation (Roco et al., 1984a,b; 1987). Figure 8–16 illustrates computational results on wear distribution in a centrifugal pump casing tested in the University of Kentucky Laboratory (Roco et al., 1989). The computational scheme was two dimensional (Figure 8–16a) and the wear rate was calculated as an average on the peripheral part of the casing (segment a–b in Figure 8–16a). The test was performed with an epoxy resin coating. Sand particles of $d = 0.27$ mm were pumped in a water slurry at concentrations varying between 5% and 30% by volume. Figure 8–16b–e, show parametric results for variable flow rate, concentration, particle size, and rotational speed.

Despite the large variety of wear patterns in different pumps, some trends can be observed. For example, in Figure 8–16 the high wear rates in the casing could be attributed to the sliding bed mechanism. In all tests the wear rate increased with the particle size and pump rotational speed. However, the behavior with flow rate and concentration variations was a function of the assembly of flow conditions. Other qualitative trends for the wear changes in the main pump components with the flow rate (Q/Q_{BEP}), concentration (C), particle size (d_s), rotational speed (N), and pump construction (geometrical dimensions, materials) have been previously reported by comparing the predictive approach to experiments (for example, Roco et al., 1984a,b; Roco and Cader, 1989).

a

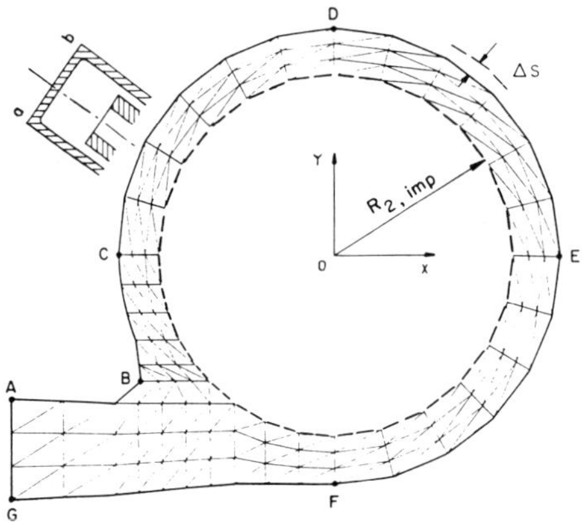
D
ΔS
a
b
Y
$R_{2,imp}$
C
O
X
E
B
A
G
F

b

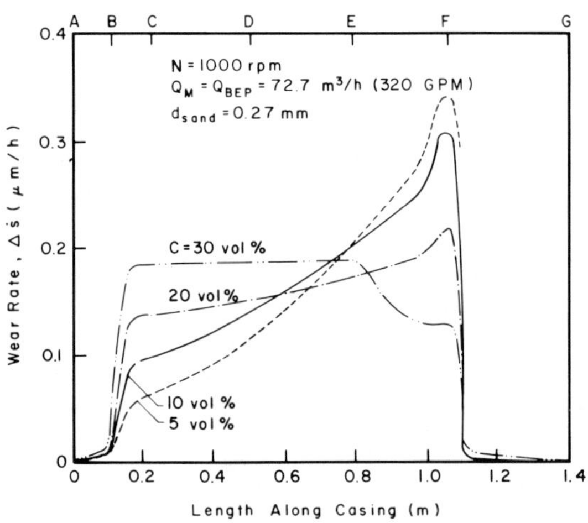
A B C D E F G
N = 1000 rpm
$Q_M = Q_{BEP}$ = 72.7 m³/h (320 GPM)
d_{sand} = 0.27 mm
C = 30 vol %
20 vol %
10 vol %
5 vol %
Wear Rate, $\Delta \dot{s}$ (μm/h)
Length Along Casing (m)
0
0.1
0.2
0.3
0.4
0.6
0.8
1.0
1.2
1.4

c

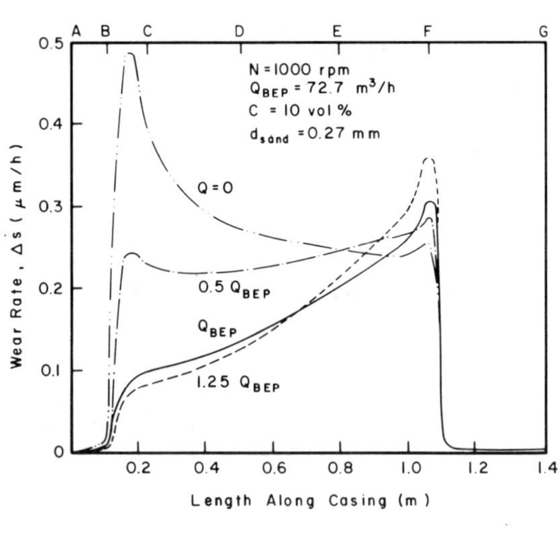
A B C D E F G
N = 1000 rpm
Q_{BEP} = 72.7 m³/h
C = 10 vol %
d_{sand} = 0.27 mm
Q = 0
0.5 Q_{BEP}
Q_{BEP}
1.25 Q_{BEP}
Wear Rate, $\Delta \dot{s}$ (μm/h)
Length Along Casing (m)
0
0.1
0.2
0.3
0.4
0.5
0.6
0.8
1.0
1.2
1.4

d

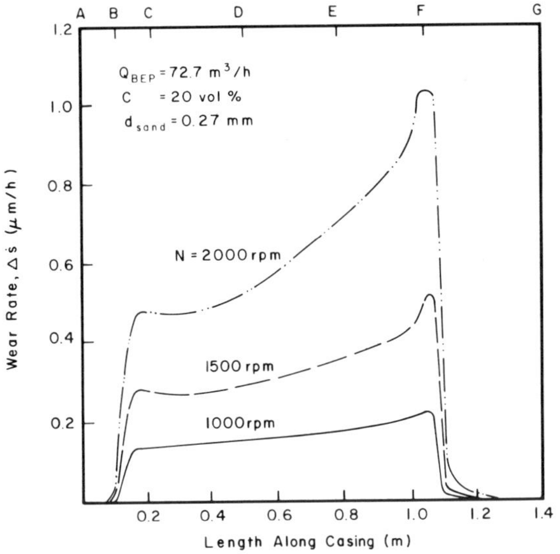

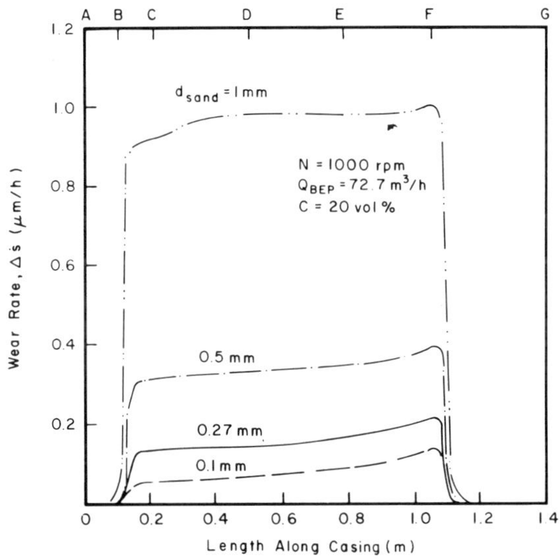

e

▲

◄ **Figure 8–16.** Sample of wear prediction in a centrifugal pump casing: (a) finite-element mesh, (b) effect of flow rate, Q, (c) effect of concentration, C, (d) effect of pump rotational speed, N, and (e) effect of particle size, d.

Two dimensionless numbers, the fluid–particle Froude Fr_* number and particle Reynolds number in pump, Re_p, offer an indication of the wear intensity in a pump channel. The first number is the ratio of the turbulent dispersion force to the inertial force acting on a particle

$$Fr_* = \frac{u_*^2}{ad(\rho_s - \rho_L)/\rho_m} \tag{8–8}$$

where a is acceleration (including gravity and centrifugal acceleration) and $u_* = (\rho_L/\rho_m)V/[5.98 + 5.75 \log_{10}(b/2k)]$ is the friction velocity. V is the mean velocity in the cross section of the pump channel, b is the cross-section width, and k is the surface roughness. The particle Reynolds number is

$$Re_p = \frac{|u_s - u_L|d}{\nu_L} \tag{8–9}$$

where $|u_s - u_L|$ is a function of acceleration at a representative location in the channel.

The average wear rate per channel increases with Re_p (this number affects the directional impingement wear component) and with $1/Fr_*$ (this number affects the frictional wear component). In the pump casing of Figure 8–16 the Re_p number is of order of 10 and Fr_* about 1. A Froude number Fr_* less than unity corresponds to a predominant frictional component.

For a long straight pipeline, abrasion is likely to be the dominant wear process. An idealized version of a particle-wall contact is shown in Figure 8–17. The penetration depth y represents plastic deformation so that the rate of abrasive wall wear per contact will be ryV_w m^3/s, where V_w is the velocity of the particles in the closest layer.

y is proportional to the normal force and inversely proportional to the hardness of the material. In fact, hardness measurements are obtained from an indenting particle–wall contact of this type. Evidently, abrasive wall erosion can be reduced by reducing y.

Since the properties of the particles must be regarded as independent variables, we can minimize wear by increasing wall hardness or by increasing the tendency to produce elastic (as opposed to plastic) deformation. The proof resilience is a measure of the ability of a material to absorb energy before producing irrecoverable plastic deformation. In terms of the yield stress σ_y and Young's modulus of elasticity E, the proof resilience is $0.5\sigma_y^2/E$. For metals, there is a fairly close correlation between proof resilience and hardness. However, for nonmetals, the relationship between these properties will be different so that it is possible to find soft abrasion-resistant materials. Rubber is the most obvious example of a material of this type.

The stress produced by the penetrating particle will depend on the applied force and will vary inversely as the area of the contact asperity. Evidently, large projections may not produce stresses high enough to exceed the yield stress of the wall material.

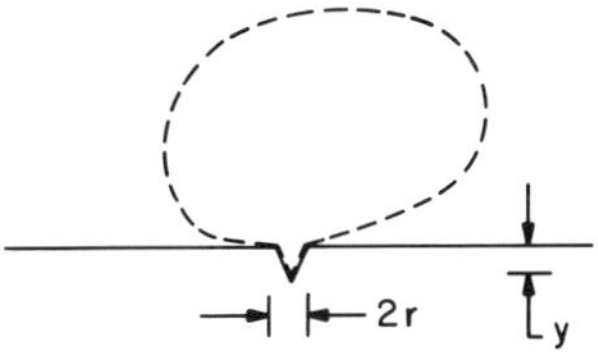

Figure 8–17. Idealized contact between a particle and a wall.

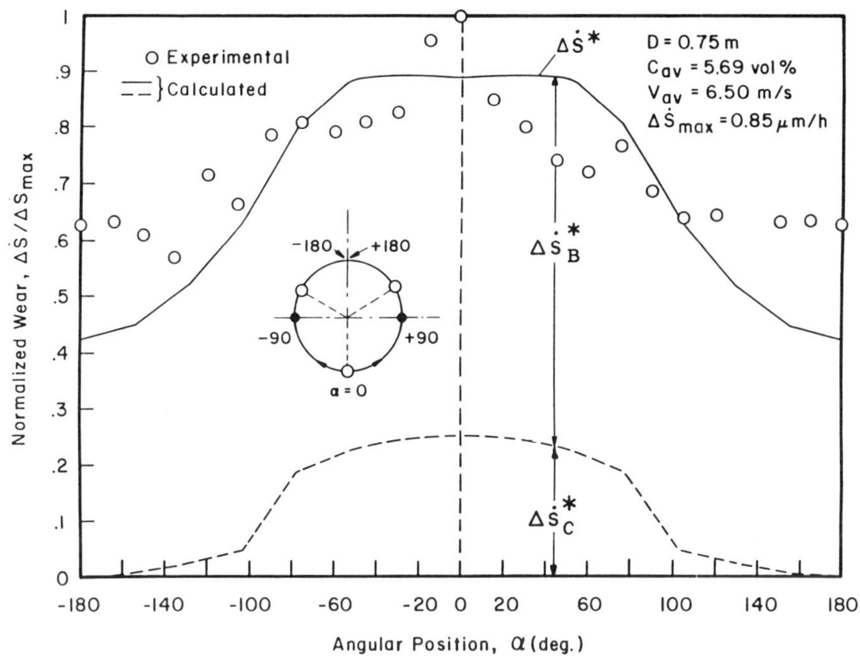

Figure 8–18. Sample of wear prediction in a dredging pipe.

The wear distribution in a cylindrical pipe is illustrated in Figure 8–18 (Okayama, 1986). For comparison, the wear prediction with an energy approach applied in conjunction with a one-equation kinetic energy turbulence two-phase flow model and a finite-volume method is given in the same figure. Other comparisons of numerical predictions to laboratory and industrial pipe wear measurements are presented by Roco and Cader (1988).

Typical wear distributions in horizontal slurry pipes are illustrated in Figure 8–19. The maximum wear rate may be located at the pipe bottom or higher up, depending on the solid-particle velocity and solid stresses close to the wall. The wear profile in a pipe cross section is important for estimation of pipe wear life or to find the time after which the pipe should be rotated. If the pipe is to be protected internally, the wear distribution can be used for the protection design.

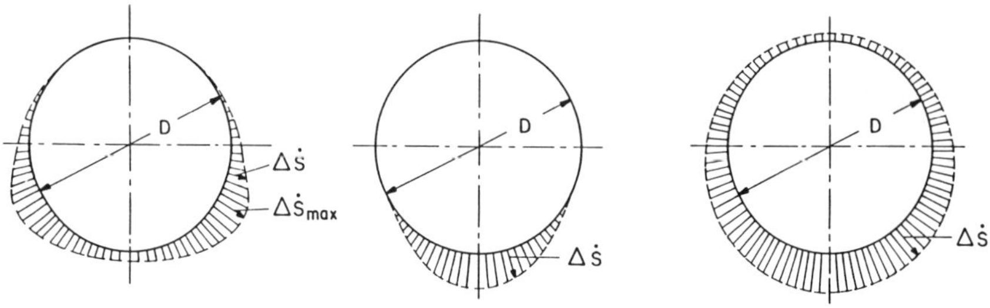

a b c

Figure 8–19. Typical wear distributions in slurry pipes: (a) near deposition velocity, (b) low concentration, above deposition velocity, (c) high velocity suspension flow. (Adapted from Roco and Cader, 1988.)

Methods to reduce wear must be evaluated in terms of their cost and effects on system operating conditions. Some general criteria to be followed in reducing wear are the following:

1. The flow should be smooth, without pronounced changes of direction or recirculations.
2. The maximum velocity should not exceed a limit which depends upon the pair of materials (transported particles and eroded wall).
3. The chemical environment should be suitable for the materials in contact with slurry.
4. If possible, one should reduce the particle size or surface roughness.
5. In some cases, higher solid concentrations are desirable so that the specific wear per transported material may decrease. This is particularly significant at concentrations of more than 10% to 15% by volume;
6. Materials should be chosen in accordance with the predominant wear mechanism; for instance, one should choose more resilient materials for large angles of impact. One may also employ composite materials which may satisfy the wear requirements for more than one wear mechanism.
7. The flow rate and concentration should be controlled within the design limits.
8. In an assembly, the worn pieces should be replaced in time to avoid the increase of wear damage of other components. Uniform wear may be achieved by applying protection on the most eroded areas (for instance, at junctions) or changing the design (for instance, using pocket elbows and special valves).
9. A wear-resistant surface layer (chemically reacted surface film) can be sometimes applied by using particular additives, for instance zinc dialkyldithiophosphate (Round, 1976; Spedding and Watkins, 1983).

Chapter 9
Pumps and Feeders

9.1 PUMPING EQUIPMENT FOR SLURRIES

A pumping system is required for any hydrotransport installation except for gravity-driven flows. The types of pumps for slurry transport are generally the same as those used for single-phase fluids, with some modifications. There are also specific devices, such as feeders, for the introduction of solid materials after single-phase pumps, in continuous or batch operation.

Centrifugal pumps are most widely used in solids transportation systems because of their adapability and economy for a large range of operational conditions. Positive displacement pumps are generally more expensive, except for very small flow rates (less than $Q = 10\ m^3/h$). However, they can provide performances that cannot be achieved with centrifugal pumps: producing a quasi-constant flow rate at different pumping pressures, handling slurries with relatively high concentrations, pumping at heads over 100 meters column of fluid, etc. The area of application of diaphragm pumps and feeder systems is for highly abrasive, coarse, and concentrated slurries, which may justify their higher cost than the standard positive displacement pumps. Jet pumps and gas-lift systems become economical alternatives only in specific situations, such as underground pumping, the oil industry (particularly gas-lift systems), or in mixing equipment.

The areas of applicability of the principal types of pumps as a function of particle size and discharge pressure is suggested by the diagram given in Figure 9–1 (after Gandhi et al., 1980). The diagram considers centrifugal pumps (rubber and metal lined), piston and plunger positive displacement pumps, and feeder systems such as the hydrohoist method. A simple guide for the preliminary selection of a slurry pump is given in Table 9–1 (after Odrowaz-Pieniazek, 1982).

A single-phase fluid pump is designed to reach high-energy efficiency at particular flow rate–pressure conditions. Slurry pumps are designed with supplementary requirements in mind:

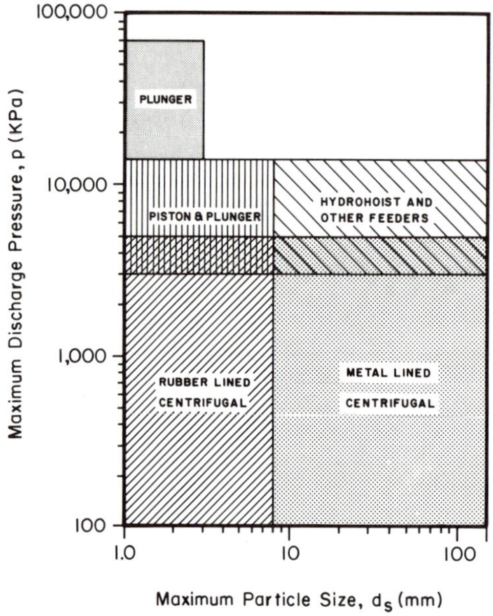

Figure 9–1. Areas of application of slurry pumps. (From Gandhi et al., 1980. *Proc. 5th Conf. on Slurry Transportation,* p. 268. Adapted with permission.)

1. To allow the passage of particulate or fibrous materials without blockage. If a slurry consists of particles with a narrow size distribution, the minimum opening of the pump channels should exceed the maximum particle size by a factor of 2 or 3 to preclude bridging. This factor is lower for particles with broad size distribution.
2. Good wear resistance to erosion, as well as to corrosion and cavitation which are more likely in the presence of solids.
3. Minimum particle attrition, which is essential in some applications (such as china clay pumping where the particle roughness is essential, or food product handling).
4. The ability to remove any blockage.
5. Ease of replacement of any worn or damaged components.
6. To deliver particular characteristic curves, which are recommended for the stability of operation of the pump-pipe system.

A pump-pipe system is more stable if the angle ϕ between the pipe curve and pump head–flow rate curve is sufficiently large (Figure 9–2). Positive displacement pumps are said to be unconditionally stable because the angle ϕ_v is positive and large. At low flow rates in the pipe, near the critical deposit velocity the angle ϕ_c for a centrifugal pump can become very small or even negative as a function of the pump and pipe curves. If the angle ϕ_c is negative the pipe could block because the resistive pipe head would exceed the available pump head for any small flow reduction. As a result, the fluid decelerates until blockage occurs. However, if the pump curve is sufficiently inclined to the horizontal so that $\phi_c > 0$, the pipe can operate safely even at flow rates below the critical deposit velocity.

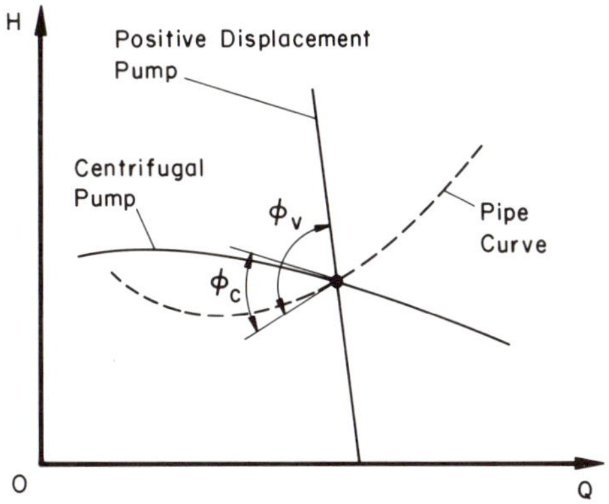

Figure 9–2. Stability in operation of a pump-pipe system.

9.2 CENTRIFUGAL SLURRY PUMPS

Centrifugal slurry pumps are designed considering four main aspects: hydraulic, mechanical, material selection, and wear resistance. The specific requirements for the slurry pumps originate from the effects of solids on the hydraulic performance, wear, cavitation, mechanical loads, and vibrations.

The construction of a slurry pump has some particularities, as compared to water pumps, to handle the particulate phase with reliability and efficiency. Impellers have larger cross-flow dimensions with greater distance between shrouds ($b_2 >$ twice the particle size, Figure 9–3a). There may be a reduced number of blades (for instance, two and three, Figure 9–3a,b), and no front shroud (open impeller) or a recessed impeller (the impeller blades only cover partially the radial cross section; $b_2' > 0$ in Figure 9–3c). If the impeller is completely recessed as in vortex pumps, the momentum is transmitted from the blades to the main flow *a* through an intermediate vortex *b* (Figure 9–4). The increased thickness of the impeller shrouds and blades for wear resistance has a negative effect on the hydraulic performance. Thicker blades reduce the effective flow cross section, and straighter shrouds and blades, as required for the manufacturing of highly resistant materials, will diminish the hydraulic efficiency. To obtain a falling head–flow rate characteristic curve, the angle of the blade outlet should be in the range of 20 degrees. The clearance recirculations, which may cause significant wear, are reduced by using expeller vanes (dorsal blades). The inlet diameter may also be changed to diminish the particle impact angle at the pump suction.

The construction of water pumps is a function of the pump specific speed $N_s = N(\text{rpm}) \cdot Q^{0.5}(\text{m}^3/\text{h})/H^{0.75}(\text{m})$. The dimensionless specific speed is $N_s^* = NQ^{0.5}/(gH)^{0.75}$. Some particularities can be defined for slurry pumps. As the pump specific speed increases, the casing shape generally changes from a quasi-spiral to a concentric configuration (Figure 9–5). The concentric configuration offers a flatter energy efficiency versus flow rate curve, a smaller radial force (F_r in Figure 9–5), and a more uniform frictional energy loss in the casing. This configuration is particularly recommended for pump specific speeds over $N_s = 150$.

Table 9–1. Pump Selection Guide

P_{max} (*MPa*)	*Slurry Abrasivity*	d_{max} (*mm*)[a]											
		0.5		*0.6*	*2*	*3*	*4*	*5*	*6*	*10*	*50*	*100*	*200*
0.2	S	MS	AD	CR	CR	CR	CR	CM	CM	CM	CD	CD	CD
	A					CR	CM						
	V					CM	CM						
0.45	S	MS	AD	CR	CR	CR	CR	CM	CM	CM	CD	CD	CD
	A					CR	CM						
	V					CM	CM						
0.6	S	MS	AD	CR	CR	CR	CR	CM	CM	CM	CD	CD	CD
	A					CR	CM						
	V					CR	CM						
1.0	S	MS		CR	CR	CR	CR	CM	CM	CM	CD	CD	CD
	A					CR	CM						
	V					CM	CM						
1.8	S	MS		CR	CR	CR	CR	CM	CM	CM	CD	CD	CD
	A					CR	CM						
	V					CM	CM						

4.0	S		WP	CR	CR	CM	CM	CM	CM	DS	DS	—
	A				CM							
	V				CM							
7.0	S	WP	DP	DP	DP	CM	CM	CM	DS	DS	DS	—
	A		WP	WP	WP							
	V		LD	LD	LD							
16.0	S	WP	DP	DP	DP	DS	DS	DS	DS	DS	DS	—
	A		WP	WP	DS						—	—
	V		LD	LD	DS						—	—
27.5	S	WP	WP	WP	WP	DS	DS	DS	DS	DS	DS	—
	A			WP	DS						—	—
	V			DS	DS						—	—

Source: Odrowaz-Pieniazek, 1982. *Chartered Mech. Eng.* 29(4):22. Reprinted by permission of the Council of the Institution of Mechanical Engineers from *Chartered Mechanical Engineers.*

[a] For $Q < 12\ m^3/h$; if $Q > 12\ m^3/h$ a correction coefficient is applied.

Pressures quoted for centrifugal pumps refer to pumps arranged in series where applicable.

Parameters: d_{max} = maximum particle size, P_{max} = maximum pressure, Q = flow rate.

Slurry abrasivity in terms of the Miller number: S = Slightly abrasive < 30 Miller, A = Abrasive 30–60 Miller, V = very abrasive > 60 Miller.

Pump types listed in this table:

AD	= Air diaphragm	CD	= Centrifugal dredge type
CM	= Centrifugal hard metal trim	CR	= Centrifugal rubber trim
CS	= Centrifugal self-priming	DA	= Displacement air type
DP	= Double acting piston	DS	= Displacement system
LD	= Liquid diaphragm	MS	= Mono type screw
SP	= Single acting piston	WP	= Washed plunger

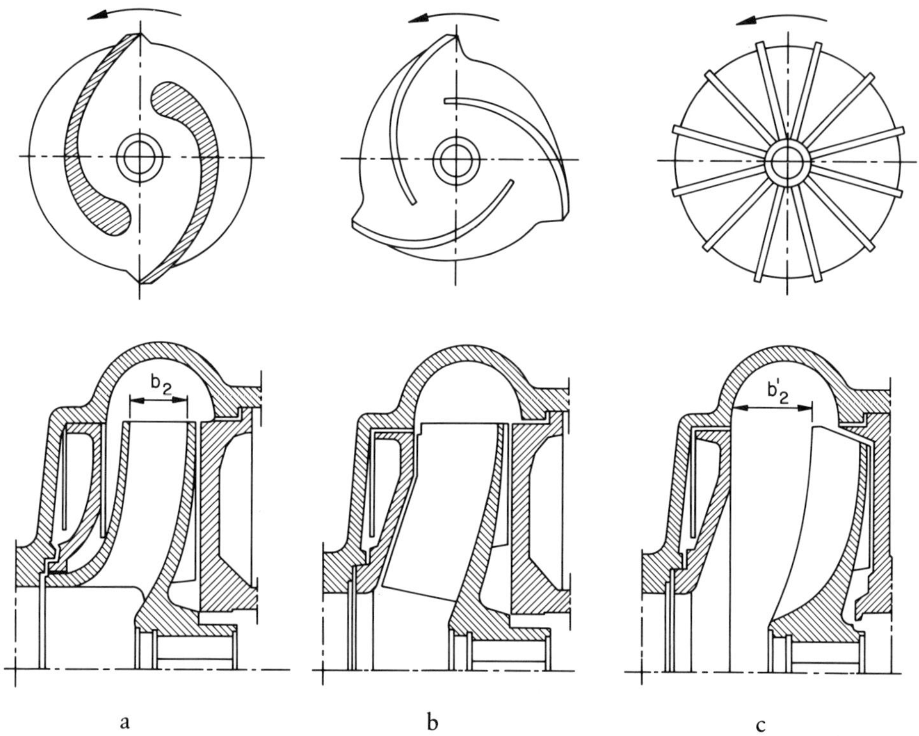

Figure 9–3. Impellers for centrifugal slurry pumps: (a) closed impeller, (b) open impeller, and (c) recessed impeller.

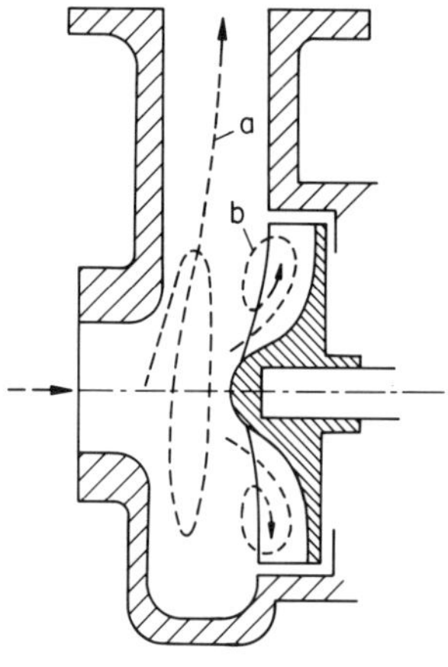

Figure 9–4. Schematic for the vortex pump: a—through flow; b—vortex flow.

The usual seal with a packed gland at the pump shaft often has to be serviced by a water flushing system at high pressure. A volumetric pump can deliver a quasi-uniform flow rate of this fluid. As an alternative, one can remove a small

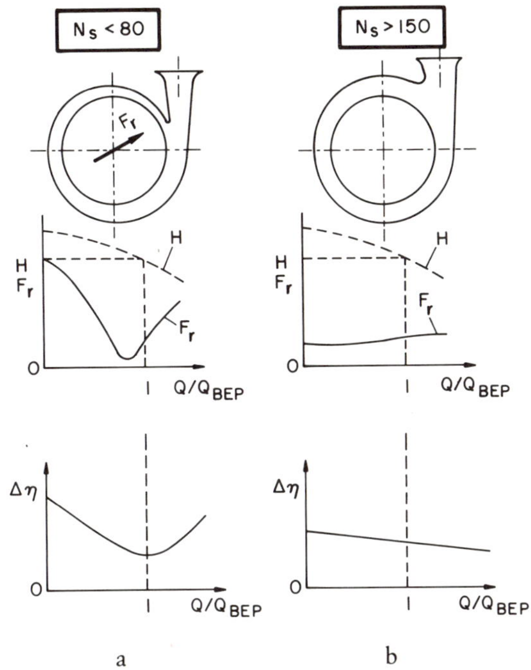

Figure 9–5. Slurry pump casings: (a) quasi-spiral casing and (b) concentric casing.

flow at the pump discharge and filter the stream before reaching the seal. A relatively more expensive solution is the use of mechanical seals, which eliminate or significantly reduce the additional flow through the seal. The seal is achieved between two smooth specially coated rings. A stream of water may still be necessary to cool the stationary ring of the seal.

Characteristic Curves

The operation of a centrifugal pump in a system can be displayed in four characteristic curves: head–flow rate $H(Q)$, power–flow rate $P(Q)$, hydraulic efficiency–flow rate $\eta(Q)$, and required net positive suction head $NPSH_r(Q)$. The head H is expressed as the height of a column of the mixture passing through the pump at a given moment. If not defined otherwise, P defines the power at the pump shaft ($P = T\omega = g\rho QH/\eta$, where T is the torque and ω is the angular rotational speed). Thus, the hydraulic efficiency η is the ratio of the hydraulic power ($g\rho QH$) to the shaft power.

The required net positive suction head is a pump characteristic which must be compared to the available net positive suction head $NPSH_a$ in the installation where the pump is mounted. The inlet pump design, and particularly the blade leading edge, determines the required suction head $NPSH_r$. When pumping

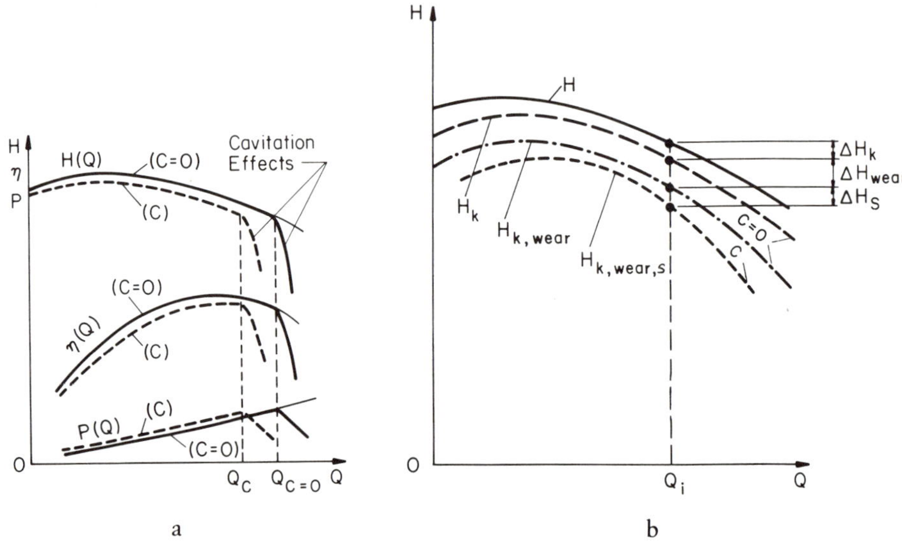

Figure 9–6. Effect of solids on the characteristic curves of a centrifugal pump: (a) head (H), energy efficiency (η), and power (P) versus flow rate (Q), and (b) schematic for the effect of roughness change (ΔH_k), dimension change by wear (ΔH_{wear}), and solids in suspension (ΔH_s) on the head–flow rate curves.

slurries, larger solids concentrations and particle sizes tend to decrease the available head and increase the required one. If $\text{NPSH}_a < \text{NPSH}_r$, then cavitation occurs and eventually pockets of air appear at the pump suction. The performance of the pump deteriorates substantially when this happens. The cavitation susceptibility increases when solids are added to the fluid because of air entrainment, air entrapped in the pores at the particle surface, and supplementary losses in the pump suction region.

The characteristic curves of slurry pumps (Figure 9–6) differ somewhat from water pumps because (a) the construction for improving the wear resistance and solids passage affects the hydraulic and energy performances, (b) deterioration of pump performance occurs during operation because of changes in pump dimensions by wear, and (c) solids affect the flow uniformity and mixture viscosity. The typical trends are a reduction of the pump head and efficiency and an increase of power and cavitation susceptibility (the difference $\text{NPSH}_a - \text{NPSH}_r$ decreases because of higher energy losses in the suction pipe and change of the flow at the pump inlet). However, the pump head may be increased by a roughness reduction in the first hours of operation, by the erosion of sharp edges, or by turbulence attenuation produced by fine particles in suspension.

The hydrodynamic effect of solids varies along the path of a fluid particle in the pump: suction (0–1 in Figure 9–7), impeller channels (1–2), transition between

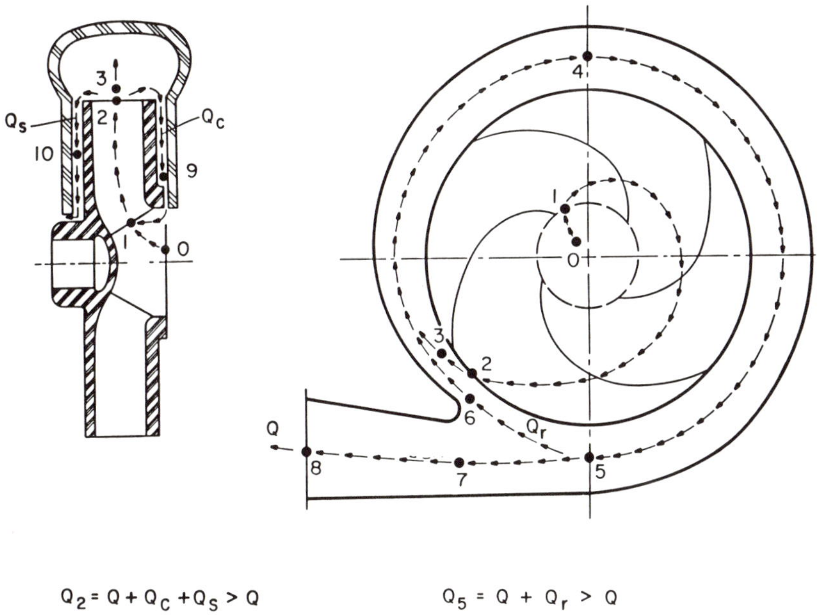

Figure 9–7. Streamline of a fluid particle in a centrifugal pump.

the impeller and casing (2–3), casing (3–8), and gap between the impeller and casing plates (9 to 10). The effective head (H) can be derived by subtracting from the theoretical head (H_{th}) the hydraulic resistances caused by friction (ΔH_{fr}), local losses (ΔH_{loc}, i.e., changes of cross section and of direction in pump), and secondary flows (ΔH_{sf}). A schematic is given in Figure 9–8. The theoretical head is a function of the velocity triangles (Figure 9–9) in the pump:

$$H_{th} = \frac{u_2 c_{u2} - u_1 c_{u1}}{g} \tag{9–1}$$

where u_1 and u_2 are the impeller blade velocities at the inlet and outlet diameters ($u_i = \pi D_i N/60$), c_{u1} and c_{u2} are the tangential projections of the absolute velocities at the inlet and outlet diameters, and g is the gravitational acceleration. For liquid–solid mixture flow, the velocities in Equation 9–1 are mass averaged:

$$u_i = \frac{u_{iL}(1 - C)\rho_L}{\rho_m} + \frac{u_{is} C \rho_s}{\rho_m}, \qquad i = 1,2 \tag{9–2}$$

The solid particle velocity at the impeller outlet can be obtained from experiments or numerical prediction of particle trajectory.

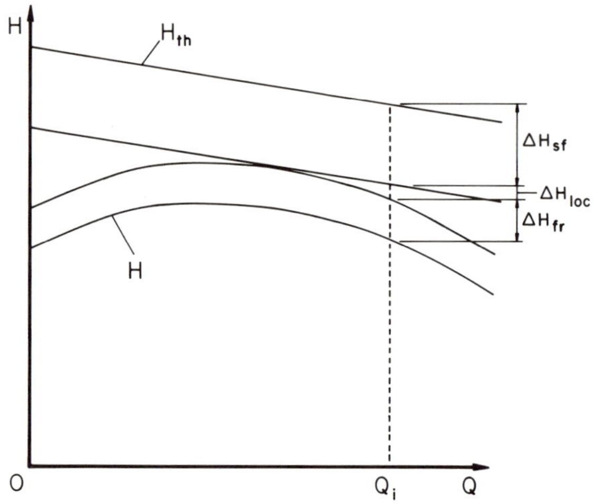

Figure 9–8. Headlosses in a centrifugal pump caused by secondary currents ΔH_{sf}, local ΔH_{loc}, and frictional ΔH_{fr} losses.

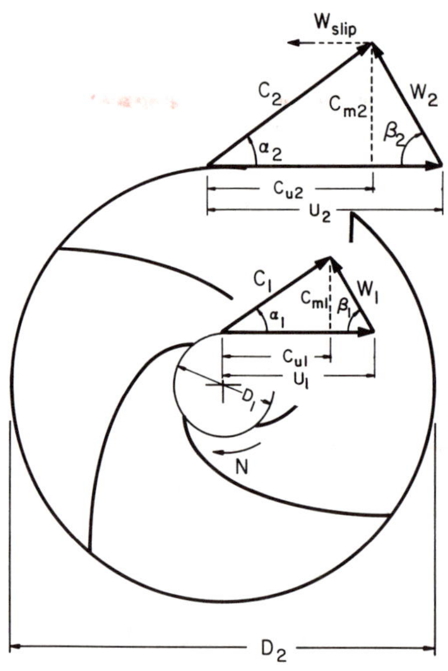

Figure 9–9. Velocity triangles in impeller: Subscript 1—denotes inlet, Subscript 2—denotes outlet.

The theoretical power is

$$P_{th} = T_{th}\omega \tag{9–3}$$

where the theoretical torque is $T_{th} = 0.5\rho Q(D_2 c_{u2} - D_1 c_{u1})$. For mixtures, the mass flow rate ($\rho Q = \rho_L Q_L + \rho_s Q_s$) and velocities ($c_{u1}$ and c_{u2}) are mass averaged.

The local losses ΔH_{loc} are generally larger at flow rates different from the best efficiency flow rate, Q_{BEP}. Figure 9–10 illustrates some typical recirculations caused by the wakes which develop at the impeller inlet, impeller outlet, and casing tongue when $Q \neq Q_{BEP}$. These areas are also subject to intense localized wear.

Some correlations have been proposed to evaluate the head reduction caused by solids. These are based on experiments (Wiedenroth, 1970; Vocadlo et al., 1974; Cave, 1976; Roco et al., 1986a,b; Sellgren and Vappling, 1986). The correlations were established for particular series of pumps and cannot be applied directly to another pump series. However, the dimensionless numbers used and

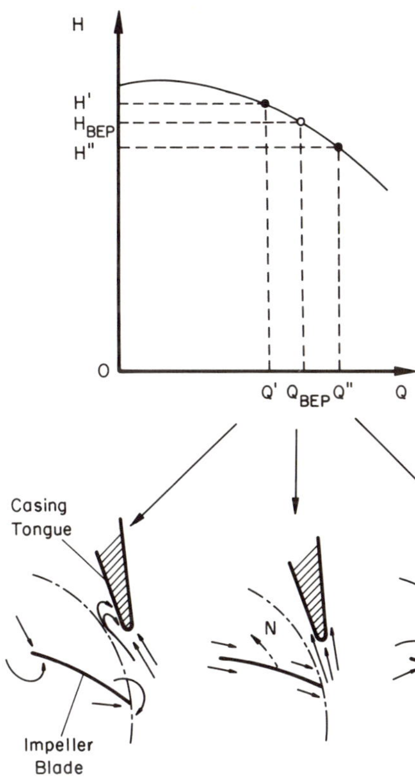

Figure 9–10. Typical wakes at the impeller blade inlet, impeller blade outlet, and casing tongue.

the trends they show can be employed to interpret other experiments. In these correlations, the head reduction is related to the density ratio S_s, particle size d, solids concentration C, particle drag coefficient C_D, pump specific speed N_s, and particle Reynolds number $\mathrm{Re}_{p\infty} = |V_\infty| d/\nu_L$ (Wiedenroth, 1970), particle–fluid Froude number Fr_{*2} and particle Reynolds number $\mathrm{Re}_{p2} = |u_s - u_L|_2 d/\nu_L$ (Roco et al., 1986a,b) at the impeller outlet, and pump Reynolds number $\mathrm{Re}_{\mathrm{pump}}$ (Sellgren and Vappling, 1986).

The experiments of Wiedenroth show that the solids effect on ΔH_s increases with $\mathrm{Re}_{p\infty}^{0.3}$ and diminish with $N_s^{2.46}$.

The particle–fluid Froude number is the ratio of the turbulent dispersion force to the inertial force acting on the particles, similar to Equation 8–9:

$$\mathrm{Fr}_{*2} = \frac{u_{*2}^2}{a_2 d(\rho_s - \rho_L)/\rho_m} \tag{9–4}$$

where u_{*2} is the friction velocity at the impeller outlet. The latter was selected at a characteristic flow cross section in the pump, namely, at the impeller outlet (see Roco et al., 1986a,b). The acceleration and friction velocity are defined in terms of c_{u2}

$$a_2 = \frac{c_{u2}^2}{0.5 D_2} \tag{9–5a}$$

$$u_{*2} = \left(\frac{\rho_L}{\rho_m}\right) \frac{c_{u2}}{5.98 + 5.75 \log_{10}(b_2/2k)} \tag{9–5b}$$

with

$$c_{u2} = u_2\left(1 - \frac{Q}{A_2(\tan \beta_2) u_2}\right) \tag{9–5c}$$

where D_2 is the impeller outlet diameter, a_2 is the centrifugal acceleration at D_2, b_2 is the width of the impeller channel at D_2, A_2 is the outlet cross-section area at D_2, k is the wall roughness, and $\tan \beta_2$ is the blade angle at D_2.

The particle Reynolds number at D_2 in the pump is

$$\mathrm{Re}_p = \frac{|u_s - u_L|_2 d}{\nu_L} \tag{9–6a}$$

with

$$|u_s - u_L|_2 = V_\infty \left(\frac{a_2}{g}\right)^{0.5} (1 - C)^E \tag{9–6b}$$

where V_∞ is the terminal settling velocity in the liquid due to gravitation alone, C is the average solids concentration by volume, and E is an exponent (Maude and Whitmore, 1958) as a function of particle Reynolds number $\mathrm{Re}_{p\infty} = V_\infty d/\nu_L$.

The experiments of Roco et al. show that the secondary flow losses ΔH_{sf} in the pump increase with $\mathrm{Re}_{p2}^{0.3}$ and N_s^{-1}, whereas the local losses ΔH_{loc} decrease with Fr_{*2}.

A pump Reynolds number is

$$\mathrm{Re}_{pump} = \frac{u_2 D_2}{\nu_m} \tag{9–7}$$

Experiments carried out for heterogeneous (Sellgren and Vappling, 1986) and homogeneous non-Newtonian (Walker and Goulas, 1984) mixtures show a decrease of the solids effect with increasing Re_{pump}. The influence of solid particles on the head reduction and efficiency reduction is usually negligible at $\mathrm{Re}_{pump} > 2 \times 10^6$.

An illustration of the characteristic curves measured in the laboratory is shown in Figure 9–11 (Roco et al., 1988). It is obvious that larger particles (sand 1.5 mm) have a greater effect on the head H and power P than the finer ones (sand 0.27 mm). The flow rate was varied by obstructing the discharge pipe. However, a slight variation of the rotational speed with the flow rate occurs as a function of the motor characteristics (see top of Figure 9–11a). The concentration in this laboratory experiment was constant only over a flow rate of about 300 m^3/h because of sedimentation in the pipes and mixing reservoirs. In large pumps the effect of solid particles is smaller because the ratio between the typical vortex size in the pump and particle size (i.e., Fr_{*2}) is larger. This effect can be observed by comparing Figures 9–11a and 9–11b for the same pump with two particle sizes (sand 0.27 mm and 1.5 mm), and Figures 9–11b and 9–12, where both tests were carried out with the same solid material (sand 1.5 mm).

The headlosses, characteristic curves, and wear intensity can be predicted using numerical simulations. Solutions of practical significance have been obtained with some simplifying assumptions such as steady-state flow. However, the actual flow is unsteady (because of the finite number of blades), nonuniform, turbulent, and with a second phase in suspension. Figure 9–13 illustrates laser doppler velocimeter measurements (Roco et al., 1990) performed at a point P, 8 mm in front of the impeller outlet in a radial cross section of a centrifugal pump casing. The pump is a laboratory model of a slurry pump (Worthington 311R pump), with a power of 15 kW, an outlet diameter of 75 mm, a best efficiency flow rate of 64 m^3/h and pumping head of about 10 m at a rotational speed of N = 1000 rpm. The specific speed is N_s = 1450 (N_s^* = 0.46). Eight transparent windows were mounted on the casing. A typical window and corresponding system of coordinates is presented in Figure 9–13a. After collecting a given sample size (1400) for each 5-degree interval of the angular position of the impeller (θ_i = 0–360 degrees), the frequency distribution function of the velocity can be analyzed statistically. Figure 9–13 (b, c, and d) illustrates the measurements as a function of the angular position of the impeller. One notes the change of the frequency distribution function of the circumferential velocity V_y as a function of the impeller angular position (θ_i = 0, 30, 60, 90, and 120 degrees) between the passage of two successive blades (360/3 blades = 120 degrees). The effect of θ_i

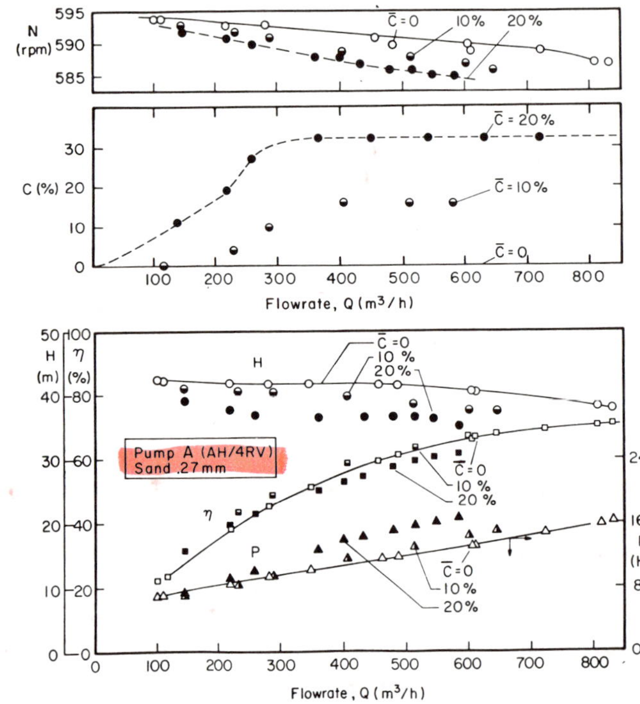
N (rpm)
C (%)
Flowrate, Q (m³/h)
H (m)
η (%)
Pump A (AH/4RV)
Sand .27mm
P (HP)
C̄ = 0
10 %
20 %
C̄ = 20 %
C̄ = 10 %

a

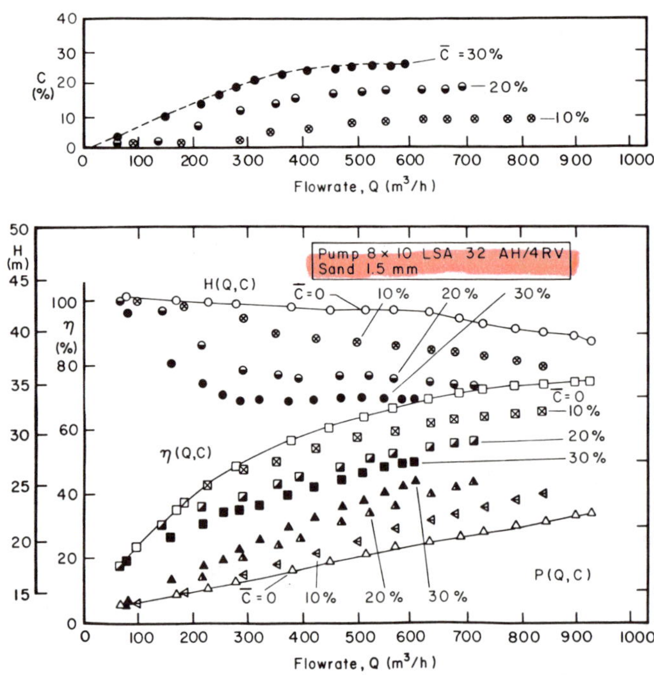
C (%)
C̄ = 30%
20%
10%
Flowrate, Q (m³/h)
H (m)
η (%)
Pump 8 × 10 LSA 32 AH/4RV
Sand 1.5 mm
H(Q,C)
η(Q,C)
P(Q,C)
P (HP)
C̄ = 0
10 %
20 %
30 %

b

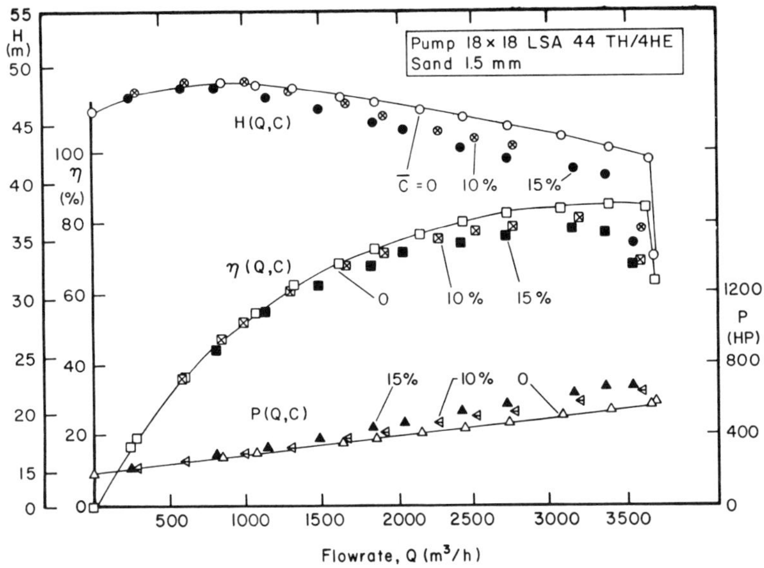

Figure 9–12. Effect of solids on a large pump $Q_{BEP} = 3200$ m^3/h, $H = 36$ m, $N = 450$ rpm, $D_2 = 1.13$ m: experimental result with sand $d_s = 1.5$ mm. (Adapted from Roco et al., 1986b.)

on the mean point velocity, its standard deviation, flow angle, and Reynolds stress is exemplified in Figure 9–13c. The periodicity of 120 degrees induced by the impeller blades is clearly evident in these graphs. The curves do not repeat identically after each 120 degrees because of small differences in the blade sizes. The velocity variation ($\overline{V_y}$) between the passage of two successive blades is about 3 m/s at the chosen point. The shape of the velocity profile suggests the development of wakes after the passage of each blade. Figure 9–13d illustrates the slip velocity of solids $V_{\theta s} - V_{\theta L}$ at the impeller outlet as a function of the impeller position. The circumferential slip velocity takes both positive and negative instantaneous values during a full impeller rotation. The slip velocity averaged over the impeller circumference has specific trends (Cader et al., 1991). The radial component of solids velocity is larger than the fluid velocity, presumably due to the centrifugal and Coriolis acceleration. The circumferential component is smaller than that of the carrier fluid, i.e., the particles are dragging behind the fluid in the tangential direction. These measurements also have shown that the presence of solid particles at low concentrations reduces the turbulent velocity fluctuations and Reynolds stresses.

◄ **Figure 9–11.** Effect of solids on the characteristic curves of a medium size pump $Q_{BEP} = 950$ m^3/h, $H = 37$ m, $N = 590$ rpm, $D_2 = 0.81$ m: experimental results with two sand particle sizes (a) $d = 0.27$ mm and (b) $d = 1.5$ mm (Adapted from Roco et al., 1986b).

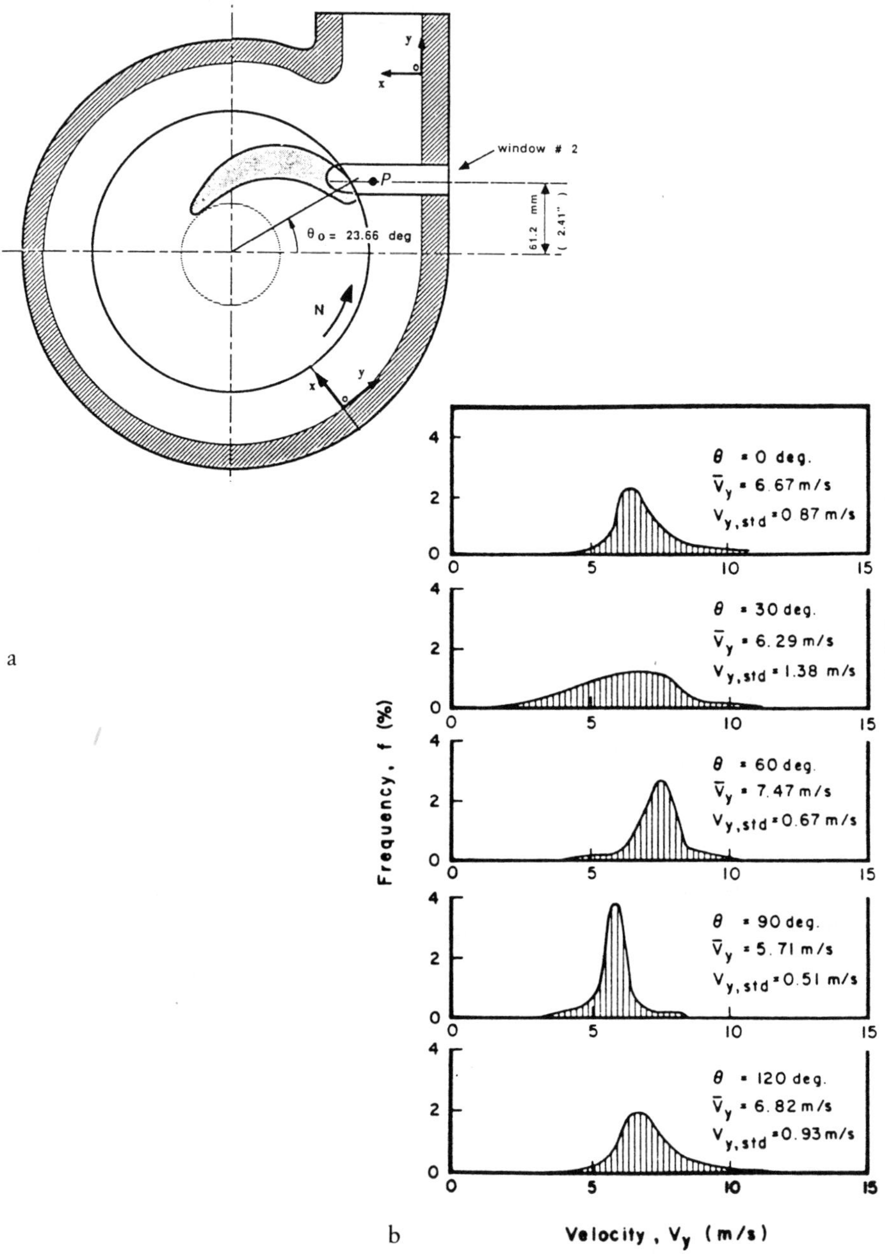

Figure 9–13. Velocity measurements at one point: (a) pump casing, (b) frequency distribution function of the velocity component V_y at five impeller positions (θ_i = 0–120 degrees, where 120 degrees = 360/3 blades), (c) effect of the encoder position on V_y, its standard deviation, angle of the velocity vector to the x-direction α_{xy}, and shear stress $-\tau_{xy}/\rho$, and (d) mean slip velocity of a glass sphere d_s = 0.8 mm in the circumferential direction (Adapted from Roco et al., 1990, and Cader et al., 1991).

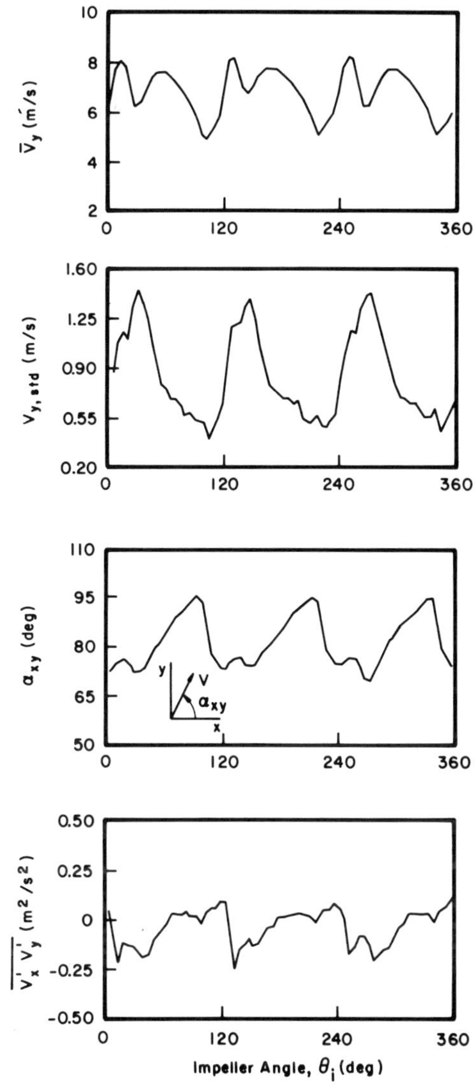

$\bar{V}_y$ (m/s)
$V_{y,std}$ (m/s)
α_{xy} (deg)
$\overline{V'_x V'_y}$ (m²/s²)
Impeller Angle, θ_i (deg)

c

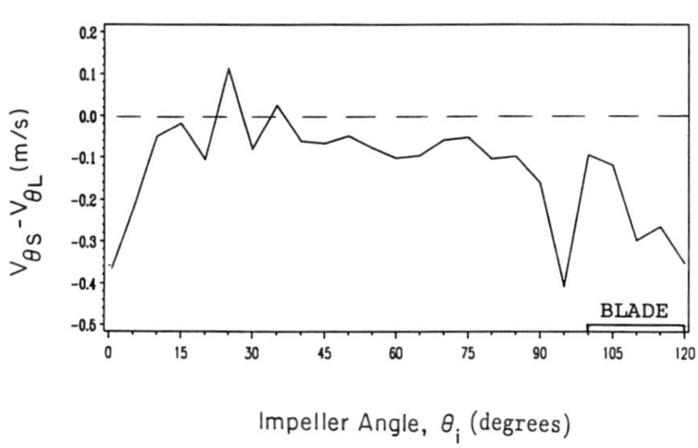

$V_{\theta S} - V_{\theta L}$ (m/s)
BLADE
Impeller Angle, θ_i (degrees)

d

Figures 9–14a and 9–14b show the measured circumferential velocity $V_{\theta L}$ and secondary currents in a radial cross section for the impeller angle $\theta_i = 0$. Figure 9–14c shows the mean velocities in the casing cross sections for two pump flow rates—the best efficiency flow rate and half of it. One observes that the mean velocity for $\theta = 60$ to 360 degrees in the casing is higher for the lower pump flow rate. These measurements demonstrate the complexity of the flow, where the recirculation between the casing and the impeller plays an important role.

Figures 9–15 and 9–16 illustrate the velocity and concentration computed in a pump impeller and casing using time-averaged methods. The basic microscopic equations used in both situations are Equations 7–8 and 7–11, with two particularities. The gravitational acceleration is replaced by an acceleration field determined by the centrifugal force and Coriolis force. Second, the diffusion coefficient is computed as a function of the specific geometry of the flow domain (Roco and Reinhard, 1980; Roco and Dehaven, 1989).

Centrifugal Pump Selection

Pump selection is a function of the mean and extreme operating conditions: flow rate, required head, particle size, concentration, and available suction head. These must all be considered in the context of the pump and system operation. However, some general selection criteria for centrifugal slurry pumps can be formulated:

1. It is preferable to operate the impeller at relatively lower speeds with larger diameters to keep the wear rate low even if the investment is increased.

2. Since the operation of slurry pumps is usually at variable flow rates, it is desirable to have a flatter curve of the energy efficiency versus flow rate than that required for water pumps.

3. The mean operating conditions of the pump should be close to the best efficiency point. Overdesign is damaging for energy and water consumption, and increases wear rates and cavitation susceptibility. If the same solids flow rate is transported with an overdesigned pump, then a larger water flow rate has to be pumped with a lower solids concentration. Alternatively, the pump has to operate at flow rates less than the Q_{BEP}. In this case, a recirculation flow Q_{rc} develops in the casing, as shown in Figure 9–17. As a result, the velocity in the casing cross section increases, together with the wear rate and energy dissipation, even if the discharge flow rate diminishes (see measurements from Figure 9–14c). This means that a fraction of fluid and solid particles recirculate back in the casing in the tongue area. Besides the loss of efficiency, at $Q < Q_{BEP}$ the wear per mass of solids transported increases significantly (see measurements by Roco et al., 1984b, Table 1).

If the solids flow rate is variable, one must consider measures to limit the fluctuations of the pump operation. These measures include increasing the storage capacity upstream of the pump, using variable speed motors (Q in the pump characteristic curve is quasi-proportional to the square of the pump rotational

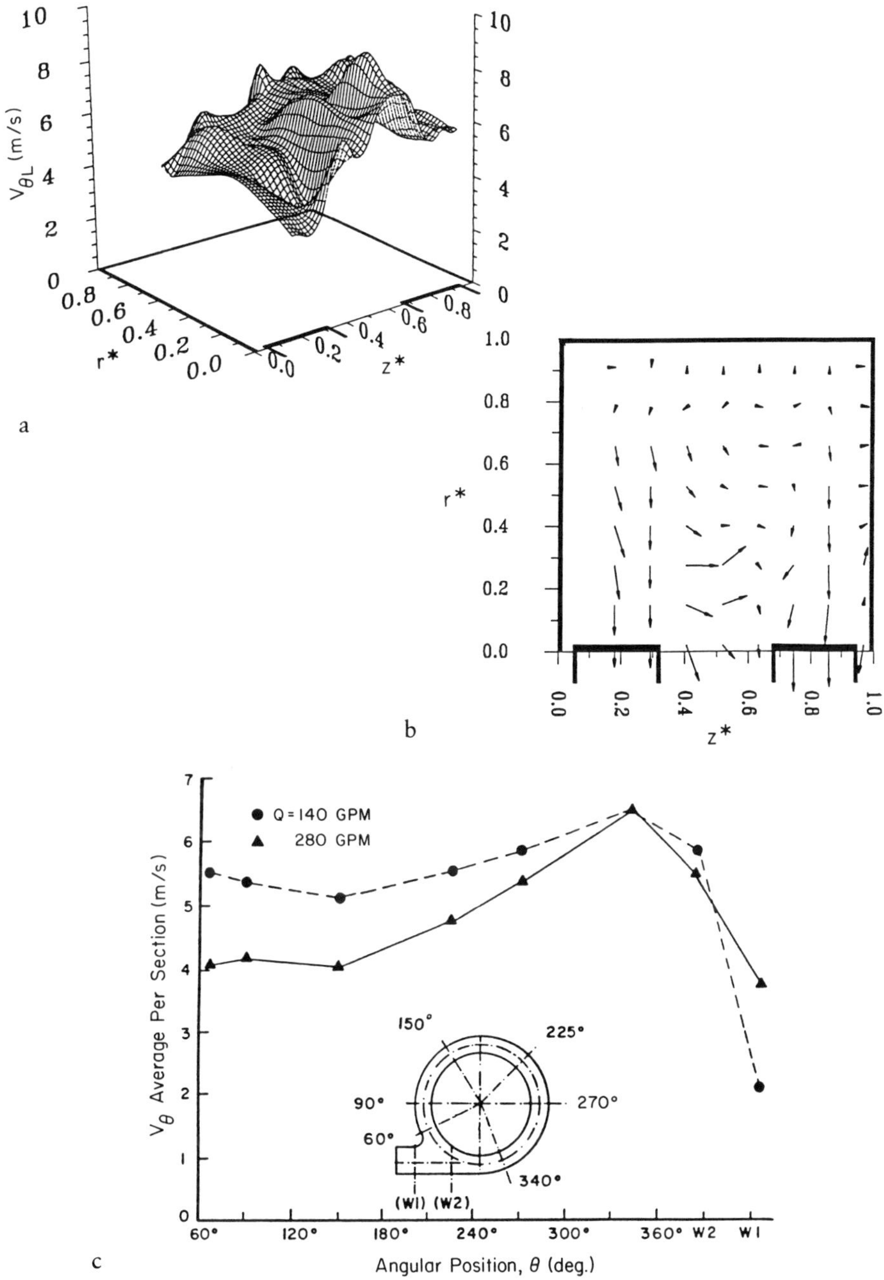

Figure 9–14. Measured velocity distribution in a pump casing: (a) circumferential velocity in a cross section, at window #2 (see Figure 9–13a), (b) secondary currents in cross section, (c) mean velocity in the casing cross sections (for Q_{BEP} and $0.5Q_{BEP}$. (Adapted from Roco et al., 1990.)

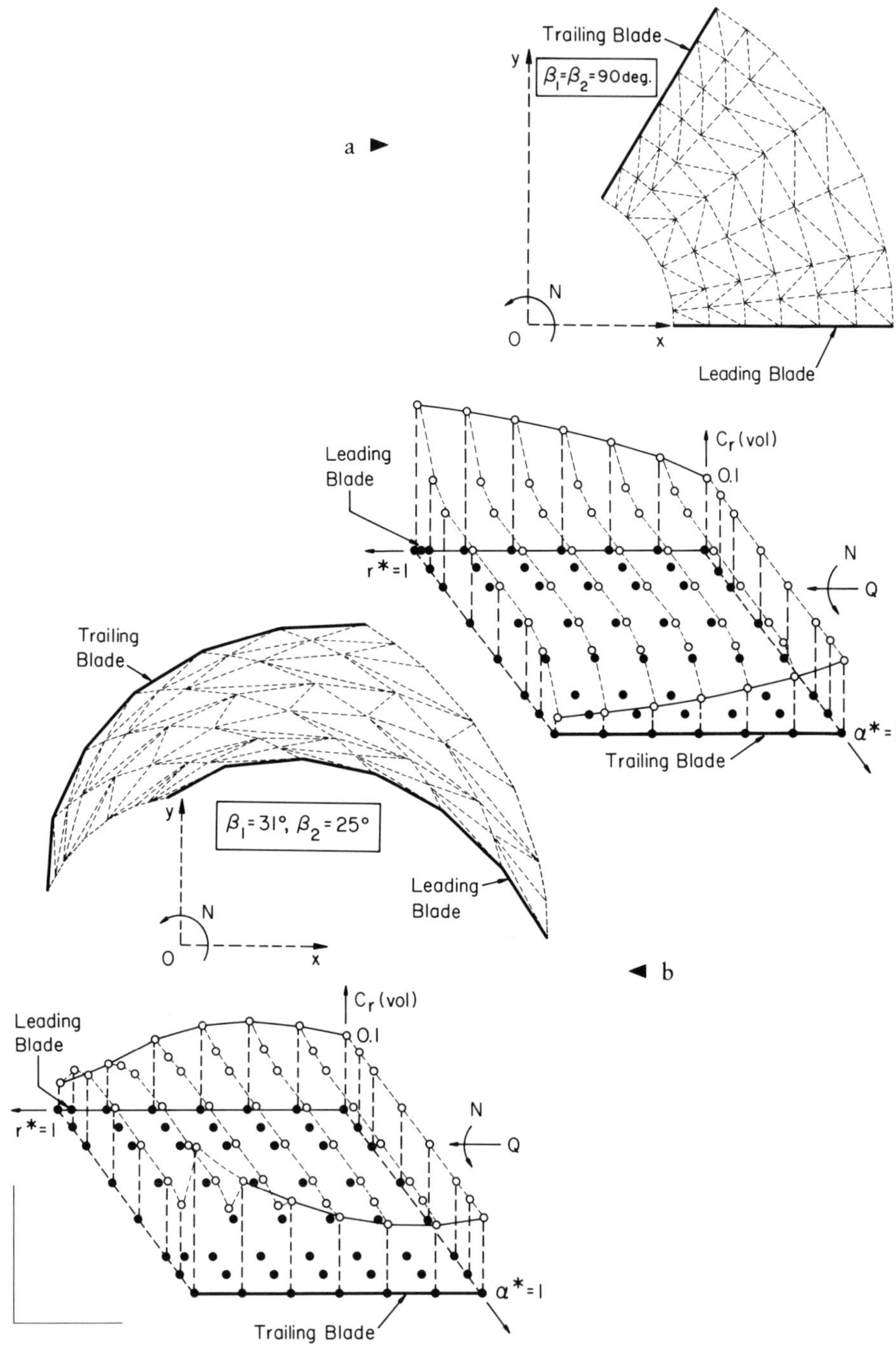

Figure 9–15. Concentration distribution between two impeller blades: (a) radial blades, (b) blades with single curvature. (From Roco and Reinhart, 1980. *Proc. Hydrotransport 7 Conf.* pp. 372–373. Reprinted with permission.)

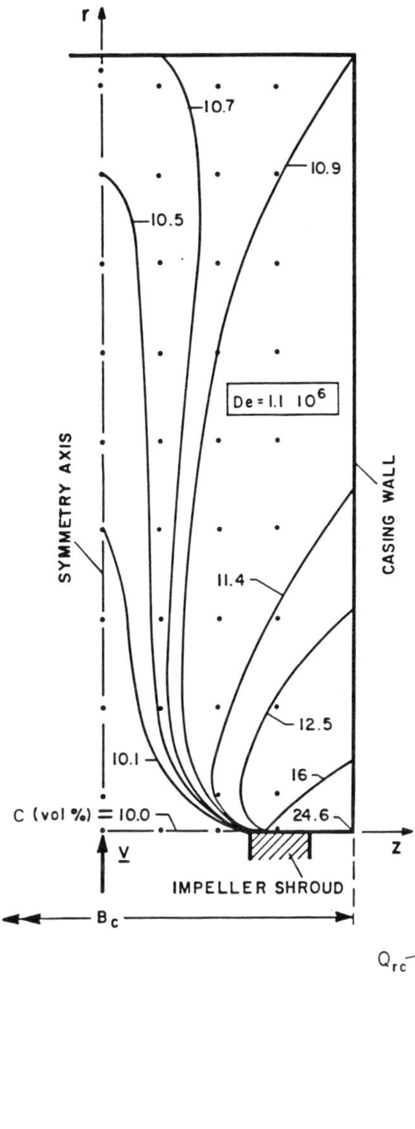

Figure 9–16. Concentration distribution in a radial cross section of a casing. (Adapted from Roco and Dehaven, 1989.)

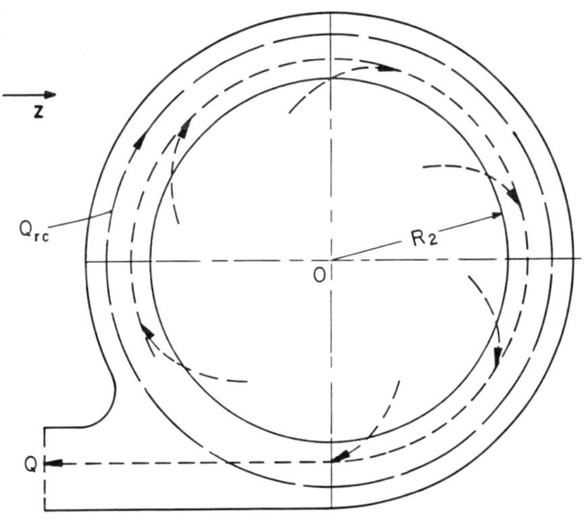

Figure 9–17. Recirculation flow Q_{rc} in casing.

speed, N^2), suction reservoirs with adjustable suction head (Q may be increased if the available suction head $H_{suction}$ is larger), pumps in parallel (that are connected to the main pipe as a function of the required Q at a given moment in operation), or even throttling the discharge pipe (which can reduce Q as a function of the head–flow rate pipe curve).

4. Material selection is determined by the particle size, shape, and expected impingement velocity. The peripheral impeller velocity ($u_2 = 0.5 D_2 \omega$) is used as a rough indication for the impingement velocity. Soft liners (natural rubber, neoprene, polyurethane) are used for flows where $u_2 < 20$–30 m/s, with irregular particles smaller than 2–3 mm and round particles smaller than 5–6 mm. Hard metal impellers are used for $u_2 < 40$–50 m/s, and steel impellers are used up to $u_2 = 50$–60 m/s. Generally, soft materials are adopted for high directional wear (caused by the impingement of particles with large impact angles, see $\Delta \dot{s}_v$ in Chapter 8) and hard metals for friction or abrasive wear (see $\Delta \dot{s}_c$ in Chapter 8).

5. The rotational speed of centrifugal slurry pumps is selected as a function of the pump characteristics, with the material limitations discussed above. If a slurry pipeline has booster pumps in series, at least one should have variable speed to adapt to changes of concentration, particle size, and density. The location of the pump(s) with adjustable speed is a function of the operation conditions to avoid NPSH problems and optimize the system reliability during transitory phenomena.

9.3 PISTON AND PLUNGER PUMPS

Besides water and oil pumping, piston pumps have been used for more than 60 years for mud handling in oil drilling. For this reason, piston pumps have been developed in connection with the oil field equipment companies (Ingersoll-Rand, Wilson-Snyder, Wirth, etc.). Piston pumps can be single or double acting, with one or more cylinders (duplex, triplex, etc.), placed in horizontal, vertical, or V positions. The most common are the duplex and triplex double-acting pumps. A schematic of the double-acting piston pump is given in Figure 9–18.

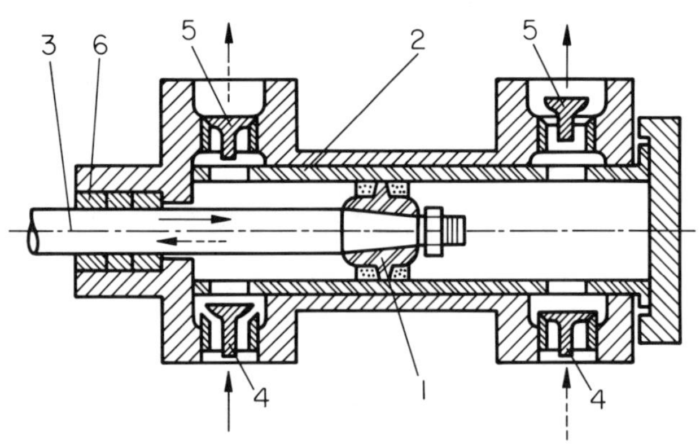

Figure 9–18. Schematic for double-acting piston pump: 1—piston, 2—liner, 3—rod, 4—inlet valve, 5—outlet valve, 6—seal.

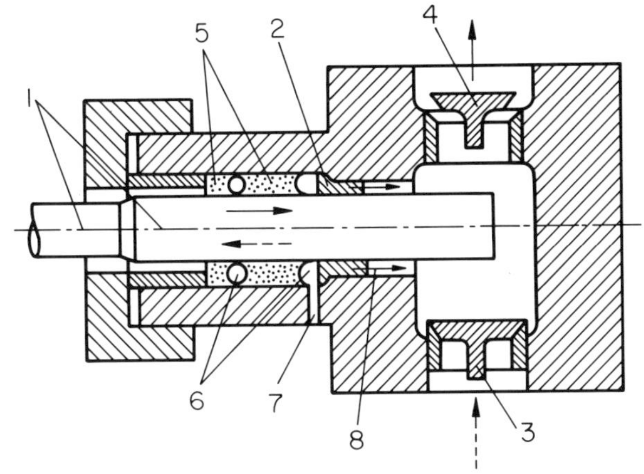

Figure 9–19. Schematic for a plunger pump: 1—plunger, 2—throat bushing, 3—inlet valve, 4—outlet valve, 5—packing, 6—rings, 7—flushing channel, 8—flush liquid.

The first significant large-scale application of piston pumps was the coal slurry pipeline of the Consolidation Coal Company in Ohio, IL in 1957, where duplex double action Wilson-Snyder pumps were used. The use of piston pumps for discharge pressures between 4 and 40 MPa is now common for fine suspension flow. Slurries with particles up to 3 mm can be pumped at pressures in the range of 15 MPa. The maximum particle size (of about 6 mm) is determined by the valve construction. The relative stroke (stroke length divided by the piston diameter) is between 0.75 and 2.5. Small piston diameters are adopted for very dense slurries (concrete, pastes, etc.).

Plunger pumps are constructed to reduce wear rates, particularly at the contact between the moving part and cylinder. The piston is replaced by a plunger (Figure 9–19), whose contact on the cylinder is flushed continuously. The plungers are best operated in the vertical position, with two or three in parallel.

The pressure fluctuations generated by reciprocating pumps can cause performance, reliability, and safety problems. Material fatigue arises in high-pressure systems. A numerical method to evaluate the pressure pulsations in discharge lines of reciprocating pumps, in the presence or absence of pulsation dampers, was presented by Vetter and Schweinfurter (1987).

The characteristic curve $H(Q)$ is represented by a quasi-linear curve (Figure 9–20). Its deviation from the vertical direction (corresponding to Q = constant) is due to the change of the volumetric efficiency as a function of the discharge pressure. If a pump has I cylinders, each with K action sides ($K = 1$ or 2), then the flow rate in m^3/s is

$$Q = \eta_v Q_t = \eta_v \left(\frac{\pi D^2}{4}\right) s \left(\frac{N}{60}\right) IK \tag{9–8}$$

where D is the piston or plunger diameter, s is the length of its stroke, and N is

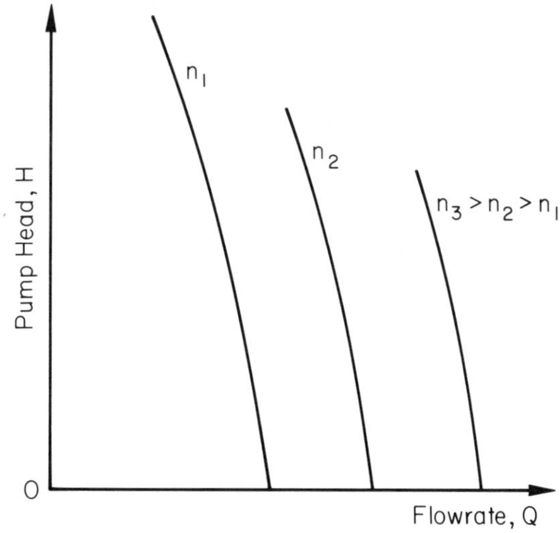

Figure 9–20. Characteristic curves $H(Q)$ at different speeds n for a positive displacement pump.

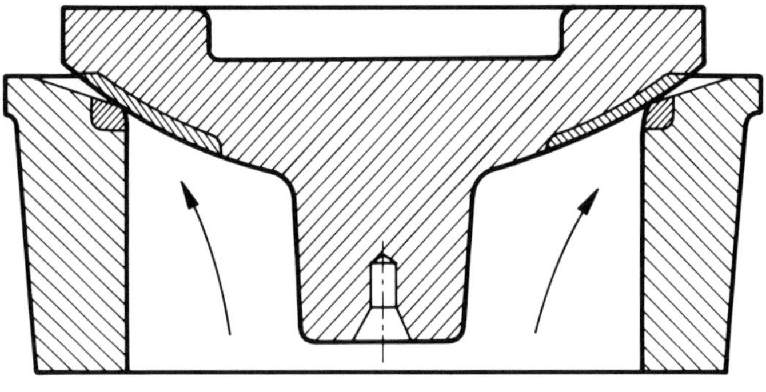

Figure 9–21. Spherical valve.

the rotational speed in rpm. The volumetric efficiency is determined by backflow during valve closing. For single-phase fluids, η_v is about 95%. Because the solid particles hinder the valve (Figure 9–21) closing, the volumetric efficiency of slurry pumps is between 90% and 95% for discharge pressures less than 10 MPa, and between 85% and 90% for higher pressures. The pump operational speed usually is in the range of 100 to 140 strokes per minute, with extremes between 40 and 300 strokes per minute.

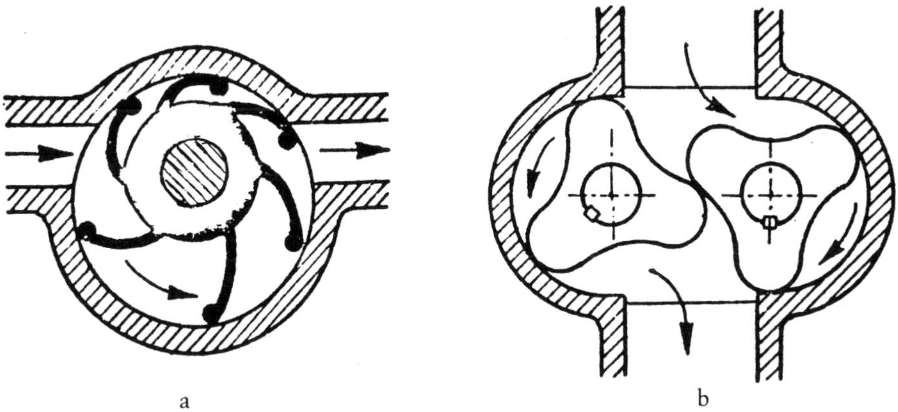

Figure 9–22. Rotary displacement pumps: (a) with elastic walls, (b) three-lobe pump ($p < 1$ MPa, $Q < 400$ m^3/h).

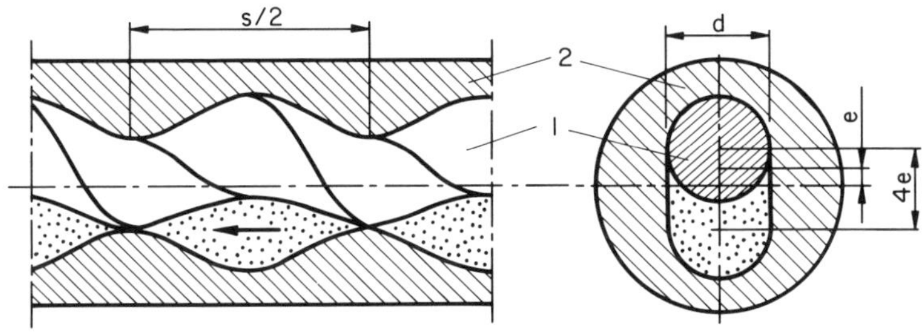

Figure 9–23. Progressive cavity screw pump ($p < 450$ kPa per stage with one rotor, $Q < 200$ m^3/h).

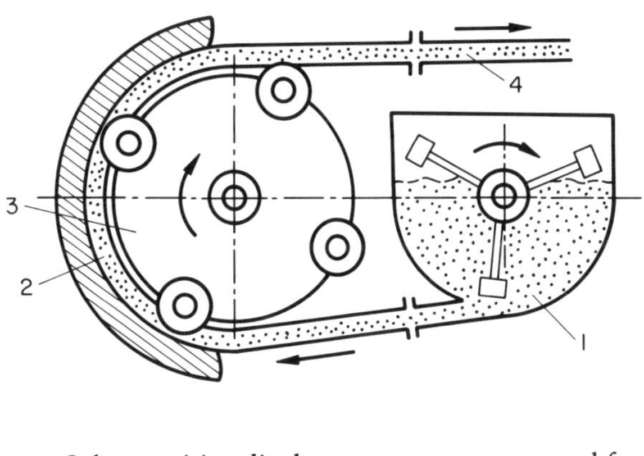

Figure 9–24. Progressing cavity peristaltic pump ($p < 60$ kPa, $Q < 1500$ m^3/h).

Other positive displacement pumps are used for relatively lower flow rates and pumping heads. Some of the more frequently used pumps include the rotative pumps with elastic walls and two or three lobes (schematic in Figure 9–22), progressing cavity screw, and peristaltic pumps (Figures 9–23 and 9–24, respectively).

9.4 DIAPHRAGM PUMPS

The obvious advantage of a diaphragm or membrane pump is the separation of the pump liquid end from the driving mechanism (moving parts). Between the motor piston and membrane there is an intermediate fluid, typically water or oil. Initially, membrane pumps were developed for metering systems or handling radioactive and chemically aggressive fluids without leakage (Bristol, 1981; Vetter and Hering, 1980). More recently, diaphragm pumps have been applied for highly abrasive slurries, with an abrasivity over Miller number 50 (Miller, 1986, 1987).

The mechanical properties of the membrane limit the pump performance. The maximum membrane amplitude (elasticity) limits the flow rate, and the membrane resistance limits the discharge pressure. The membranes may be planar, cylindrical, or of other shapes Figure 9–25 gives a schematic of the piston double-flux membrane pump (GEHO system) built by the Holthuis company (Holthuis and Simons, 1981).

Diaphragm pumps for slurry service have powers up to 500 kW and discharge pressures to 10 MPa. The volumetric efficiency of the membrane pumps is about 90% to 95%. The operating life of the elastic membranes used with abrasive slurries is in the order of 1000 hours. While the head–flow rate characteristic curves are similar to those of piston pumps, the energy efficiency of membrane pumps is lower because of the intermediate fluid motion. Diaphragm pump performances are functions of their construction and slurry characteristics (Eng. and Mining J., 1987).

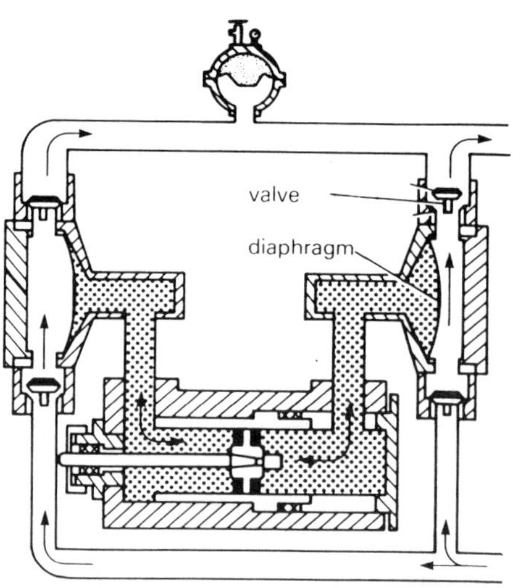

Figure 9–25. Double-flux GEHO pumping system.

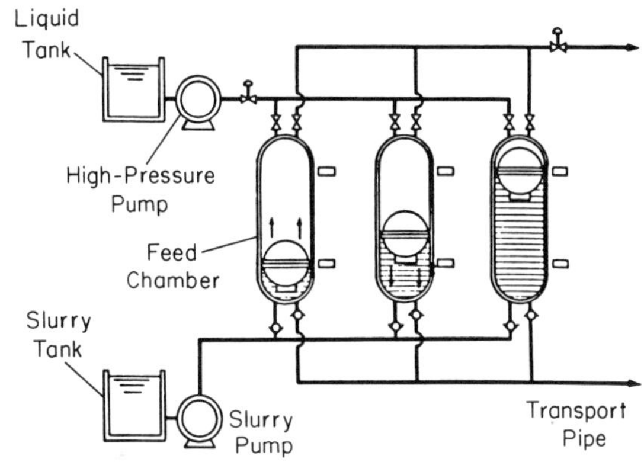

Figure 9–26. Hydrohoist pumping system.

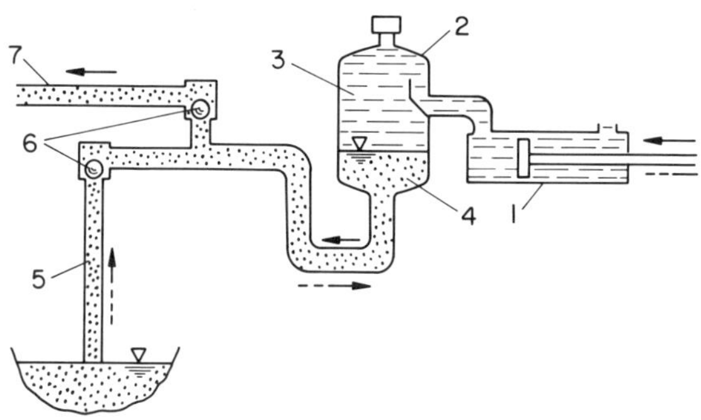

Figure 9–27. Mars pump: 1—piston, 2—reservoir, 3—oil, 4—suspension, 5—suction pipe, 6—valve, 7—discharge pipe.

9.5 FEEDERS

Feeder systems are used for pumping abrasive slurries of high concentrations and/or coarse particles. Their role is to eliminate or reduce the direct contact between suspensions and the active components (rotating impeller, piston, plunger, etc.) of the main pump. The following types of feeders have found acceptance:

1. Hydrohoist (Figure 9–26), initially developed in Japan and applied for underground mining (Sakamoto et al., 1983).
2. Three-pipe feeder system, developed in West Germany and Hungary (Kocsanyi and Maurer, 1972). This system was tested in coal mining, ash handling from electric power stations, and coal transportation.

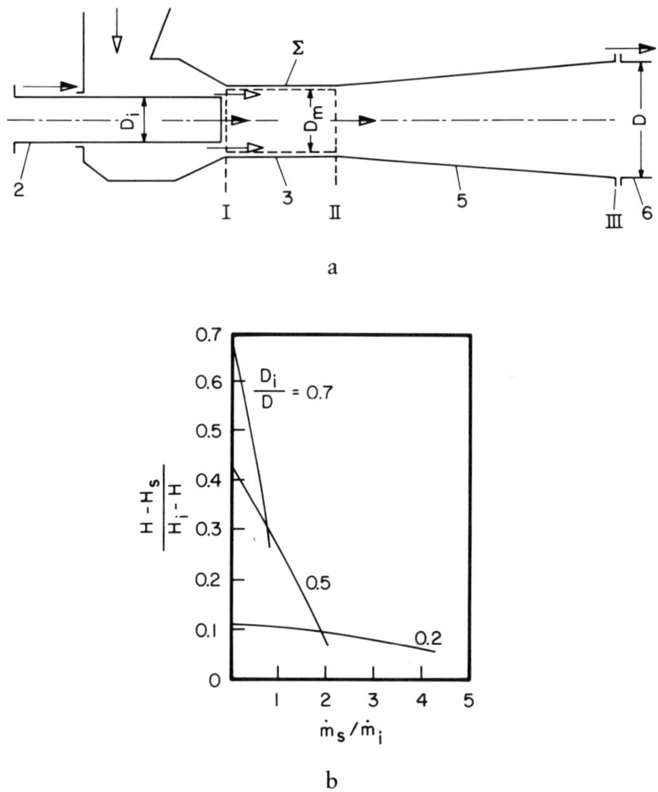

Figure 9–28. Jet pump: (a) construction, (b) characteristic curve ($\dot{m}_i$ and $\dot{m}_s$ are the mass rates of the injection and suction flow, respectively; H_i, H_s, and H are the injection, suction, and discharge pump head, respectively).

3. Helical feeder, proposed by Harvey (1982) and investigated by the Department of Energy in the United States.
4. Two-chamber batch system, developed in England.
5. Rotary system, developed and tested in Australia and Sweden (Bhattacharayya and Imrie, 1986).
6. Batch systems with intermediate fluids and membranes. In this group may be included the Mars pump (see schematic in Figure 9–27). These pumps have been applied to fine iron and copper concentrate long distance transportation.
7. Jet pump feeders (Figure 9–28), largely applied in chemical processes, mining, and dredging. Their use for long slurry pipelines is limited by their lower efficiency than centrifugal pumps (by 15–30%) and the limited domain of variation for the discharge pressure in stable conditions as compared to other pumping systems. The main parameter for optimization is the diameter of the mixing chamber (D_m in Figure 9–28) as a function of the injection diameter (D_i). The momentum equation written for the fluid within the mixing chamber

(control volume Σ) provides the basic equation. Its formulation for single-phase flow was discussed by Mueller (1964). Weber (1972) introduced the effect of the relative velocity between solids and fluid in the calculation of the discharge pressure and pump efficiency. Weindner (1955) determined the minimum length of the mixing chamber for satisfactory particle acceleration and mixing. The advantage of a jet pump compared to other feeders is the lack of valves, which create the major difficulty in the operation of other feeders. Recent experiments were performed on a special type of jet pump with an annular jet (Engerlin et al., 1988) and a flat jet (Khuntia and Murty, 1988).

Chapter 10

Instrumentation

10.1 MEASUREMENTS AND MATERIAL BALANCES

For the plant engineer, it is the concentrations and velocities used in material balances which are of greatest interest. We saw in Chapter 2 that volumetric flow rates of solids and fluid could be expressed in terms of the area-averaged mean velocity, V, and the delivered mean volumetric concentration of solids, C_v. Although mass fractions are slightly more convenient for process calculations, volume fractions are better indices of slurry concentration from a physical standpoint.

In this chapter we will describe some of the devices which can be used to determine slurry concentrations and velocities. A recurring problem with any device concerns the relationship between the measured quantity and the mean value which is required for process calculations. Some of this difference arises from the fact that concentration and velocity vary over the pipe cross section. As we have seen in Chapter 7, our understanding of these variations is still incomplete.

10.2 WALL SAMPLING

The simplest way to determine the concentration of a flowing slurry is to remove a sample through an opening in the pipe wall. However, even if the slurry concentration is uniform over the pipe cross section at the sampling point, a systematic error must be expected to occur since the particles and the fluid will not have identical velocity vectors as they flow into the sample tube. This can be seen if we consider the momentum equation for the solids. Neglecting concentration variations, Equation A2–8b becomes

$$\rho_s \frac{D\mathbf{v}_s}{Dt} = -\nabla P + \rho_s \mathbf{g} + \mathbf{f}_{sL} + \mathbf{f}_{ss} + \mathbf{f}_{sw}$$

If we neglect interparticle forces produced by the wall or interparticle contact and adopt a Lagrangian frame of reference, so that the coordinate system moves with a particle, the particle equation of motion can be written in dimensionless form for the i-direction as

$$\frac{dv_{si}^*}{d\tau} = -\left(\frac{1}{S_s}\right)\nabla^* P^* + \frac{(C_{Ds}\,\mathrm{Re}_0/24K)(v_L^* - v_s^*)|v_L^* - v_s^*|}{(1 - c)^m} + \mathrm{Fr} \tag{10–1}$$

where

$$v_{si}^* = \frac{v_{si}}{U}, \quad v_L^* = \frac{v_{Li}}{U}, \quad \tau = \frac{tV}{d_t}, \quad \nabla^* P^* = \left(\frac{d_t}{\rho_L U^2}\right)\nabla P$$

$$S_s = \frac{\rho_s}{\rho_L}, \quad \mathrm{Re}_0 = \frac{d_t \rho_L U}{\mu_L}, \quad K = \frac{\rho_s d^2 U}{18\,\mu_L d_t}$$

$$\mathrm{Fr} = \frac{g_i d_t}{U^2}, \quad x^* = \frac{x}{d_t}, \quad y^* = \frac{y}{d_t}$$

In these definitions, d_t is the sampler diameter and U is the sampling velocity. The inertial parameter K is a measure of the tendency of the particles to be carried past the sampler aperture. Figure 10–1 shows this segregation (Nasr-El-Din et al., 1985) in the form of particle trajectories (dots) for sampling in an idealized situation. The flow in this simulation is two dimensional, with no velocity components in the z-direction perpendicular to the page. In these calculations, fluid velocities were estimated using the ideal irrotational flow approximation for the mixture. This gives a velocity profile which is flat upstream of the sampler.

The example shows how the solids, with their greater inertia resulting from their higher density, are not collected at the same volumetric rate as the fluid. To minimize the tendency of the sampled concentration to be lower than the true values, we should use a high sampling velocity. The wall aperture should also be as large as possible. The method will be most reliable for small particles, high fluid viscosities, and high solids concentrations since all of these factors tend to increase the interfacial drag force f_{sL} per unit volume.

Figure 10–2 shows some experimental results, obtained with a vertical pipe, which illustrate the effects of particle size and sampling velocity for sand–water slurries. The surface mean particle diameters in these tests were 0.72 (coarse), 0.33 (medium), and 0.17 (fine) mm.

The results in Figures 10–1 and 10–2 were obtained in vertical flows for which the variation of solids concentration over the cross section is small. They show that it is probably impossible to eliminate error completely from wall sampling measurements although it may be feasible to reduce it to an acceptable level. Ultimately, it should be possible to predict the error for an actual (three-dimensional) flow to correct the measured values. In addition to the effect on the

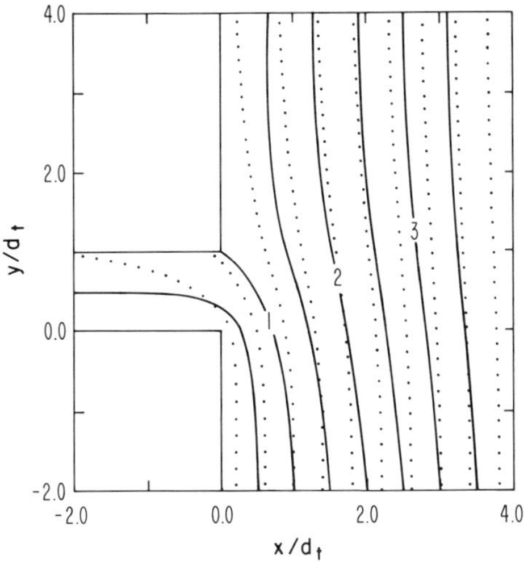

Figure 10–1. Particle segregation in wall sampling for a slurry with C_v = 0.25 and particle inertial parameter K = 25. (From Nasr-El-Din et al., 1985. *Can. J. Chem. Eng.* 63:748. Reprinted with permission.)

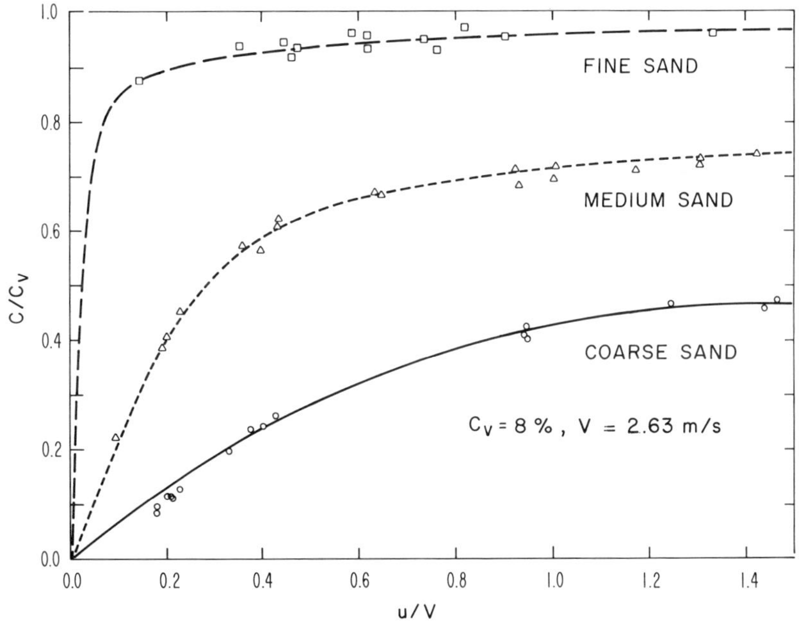

Figure 10–2. Effect of wall sampling velocity for a wall aperture 8 mm in diameter. (From Nasr-El-Din et al., 1985. *Can. J. Chem. Eng.* 63:749. Reprinted with permission.)

measured concentration, the particle size distribution of the sample will tend to be somewhat finer than that of the whole mixture because the finer particles are collected more efficiently.

For horizontal flows, errors which arise from inertial effects in wall sampling are complicated by the variation of solids concentration and velocity over the cross section. Results obtained with horizontal pipes have been reported by Nasr-El-Din et al. (1989). An important observation in this later study is that the mean particle diameter in the sample obtained from the bottom of the pipe was close to the mean for the pipe as a whole.

10.3 SAMPLING: L PROBES

An L-shaped probe, projecting into the flow, eliminates some of these disadvantages. If the sampling velocity is the same as the upstream mixture velocity and if the probe has an infinitesimal wall thickness, the flow is not disturbed and the sample should have the same concentration as the flow upstream. If the sampling velocity is too low, particle inertia will produce high sample concentrations. The opposite effect will occur if the velocity is too high. Assuming the flow into the probe is axisymmetric, a theoretical prediction of the effect of sampling velocity on measured concentration can be made (Nasr-El-Din et al., 1984). Figure 10–3 compares theoretical predictions and experimental measurements for a sand of surface mean diameter 0.33 mm. C_0 and U_0 in this figure are the upstream concentration and velocity.

The agreement shown in Figure 10–3 is indirect evidence of the utility of the equations of motion given in Chapter 2. The small systematic deviation is probably due to the fact that steady flow drag coefficients (as functions of Re_s) were used to evaluate f_{sL}.

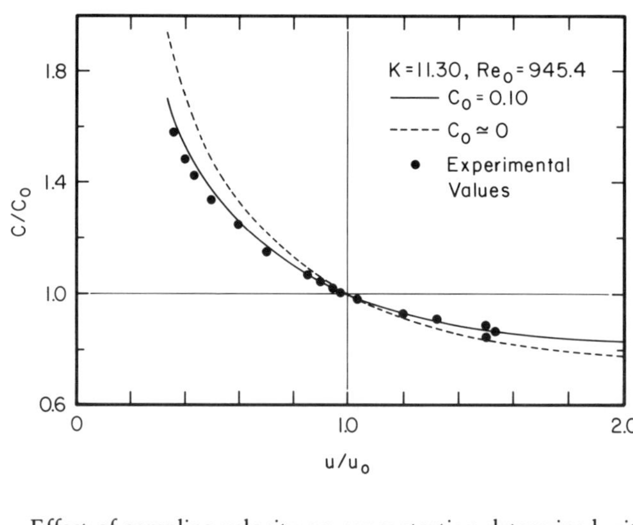

Figure 10–3. Effect of sampling velocity on concentration determined with an L-probe. The experimental values were determined for $C_0 = 0.10$. (From Nasr-El-Din et al., 1984. *Can. J. Chem. Eng.* 62:183. Reprinted with permission.)

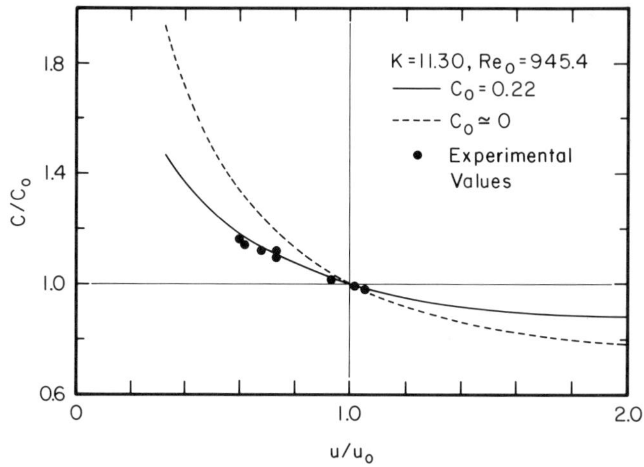

Figure 10–4. Effect of sampling velocity on concentration for $C_0 = 0.22$. (From Nasr-El-Din et al., 1984. *Can. J. Chem. Eng.* 62:184. Reprinted with permission.)

As in the case of wall sampling, the error decreases as the concentration increases because of the increased drag force. Experimental demonstration of this effect is given by the higher concentration results in Figure 10–4. Again, we note the good agreement between predicted and measured values.

Particle and fluid properties have long been known to influence the errors which result from anisokinetic probe sampling. Isokinetic sampling is used widely with gas–solid flows and the errors resulting from inertia effects can be very large because the density ratio is very high. However, the principles are the same in liquid–solid mixtures. Figure 10–5 shows the inertial effect at infinite dilution in terms of the particle inertial parameter K.

The experimental measurements in Figures 10–3 and 10–4 were obtained with thin-walled, tapered tubes. With blunt probes of significant wall thickness, a systematic error is observed (Nasr-El-Din et al., 1984). This is shown in Figure 10–6. The error is attributed to particles bouncing from the sampler wall, and being collected by the sampler as a result of their reduced velocities.

Figures 10–3 to 10–6 show that satisfactory values of local solids concentration can be obtained from an L-shaped probe if the upstream velocity is known. Although this velocity can be measured, it will generally be necessary to estimate the local velocity from the mean value V and this introduces some additional error. A probe which combines the velocity and concentration measurement functions is described by Cooke and Lazarus (1988).

The concentration which is determined by the isokinetic sampling process is the local value of the delivered concentration. If the distribution of this concentration is flat, as it is for slurries of fine particles at the inlet to a pipe, a single L-probe

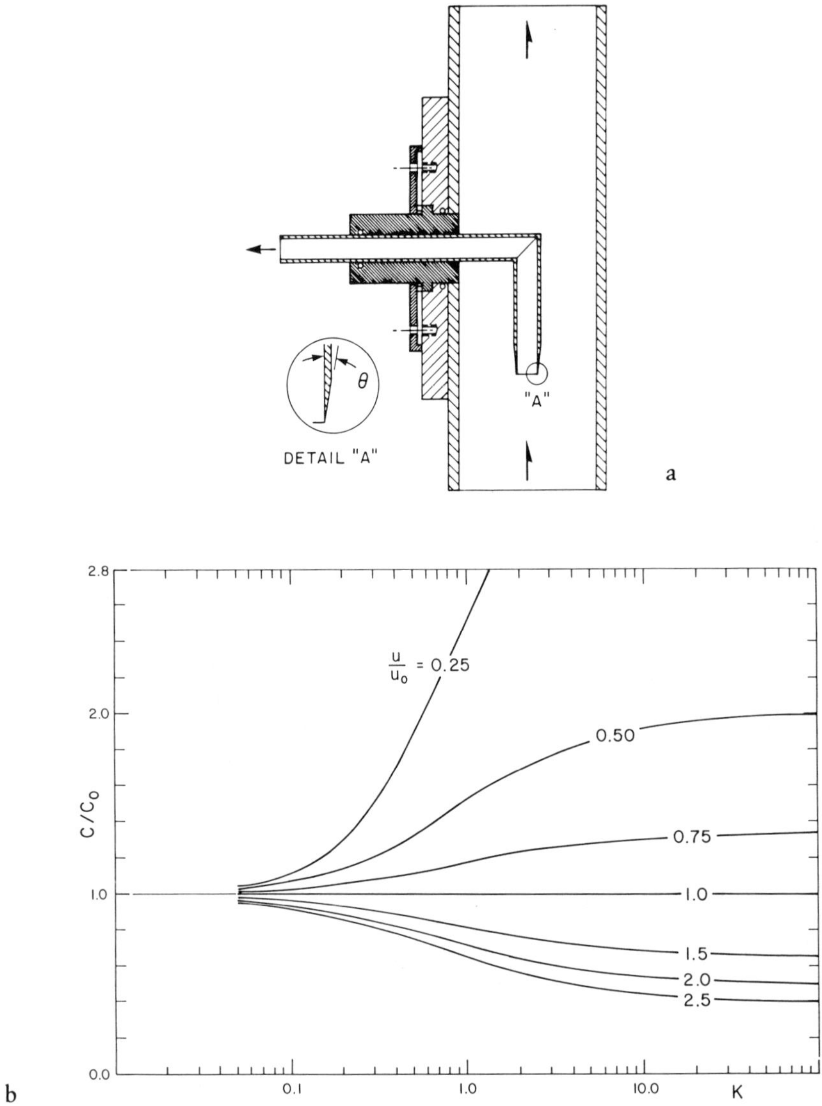

Figure 10–5. Effect of inertial parameter *K* on sample concentration. (10-5a from Nasr-el-Din, 1984. *Can. J. Chem. Eng.* 62:181. Reprinted with permission. 10-5b from Nasr-el-Din. Ph.D. thesis.)

measurement gives a reliable value of C_v. Otherwise, multiple measurements and integration will be necessary. These integration processes are so cumbersome that alternative methods, which attempt to give spatially averaged concentrations and which do not require manual operation, are clearly of interest.

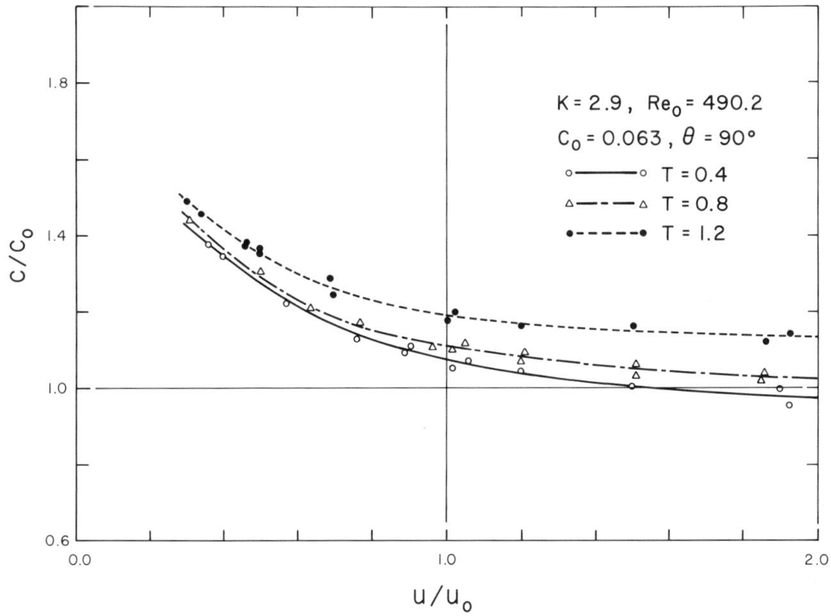

Figure 10–6. Effect of probe wall thickness on sample concentration. (From Nasr-El-Din et al., 1984. *Can. J. Chem. Eng.* 62:182. Reprinted with permission.)

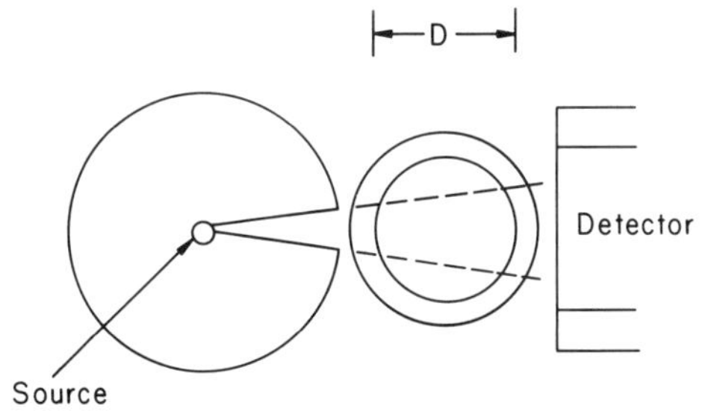

Figure 10–7. Schematic radiation densimeter.

10.4 RADIATION ABSORPTION

Radioactive sources have been used for concentration measurements for many years. A schematic representation of the device is shown in Figure 10–7.

The absorption of a monochromatic beam of radiation of intensity N in medium j obeys the Beer–Lambert law

$$\frac{dN}{dx} = -\mu_j N \tag{10–2}$$

where μ_j is the absorption coefficient. μ_j is a function of the photon energy (wavelength) of the radiation and the composition of the absorber. For gamma rays from a Cs-137 source (0.66 MeV) the absorption coefficient is nearly proportional to the density of the material.

The intensity of the beam, after passing through a number of absorbers in series, is obtained by integrating Equation 10–2 for each absorber and combining the results:

$$\ln\left(\frac{N}{N_1}\right) = -\sum \mu_j x_j = -\mu_w x_w - \mu_L x_L - \mu_s x_s \tag{10–3}$$

In this expression w denotes the pipe wall and N_1 is the unattenuated beam intensity. For a beam of small cross-sectional area, the path length $(x_s + x_L)$ is equal to the diameter of the pipe. The mean in situ concentration along the beam is $x_s/(x_L + x_s)$, which we denote as C_t. The beam intensity passing through the slurry-filled pipe may be related to C_t and the intensity N_L observed when the pipe is full of water:

$$\ln\left(\frac{N}{N_L}\right) = -C_t(\mu_s - \mu_L)D \tag{10–4}$$

Alternatively, in terms of the intensity N_0 observed with the empty pipe

$$\ln\left(\frac{N}{N_0}\right) = -[\mu_L + (\mu_s - \mu_L)C_t]D \tag{10–5}$$

Since the μ values are nearly proportional to their densities, Equation 10–5 shows that the attenuation of the beam is very nearly a function of the density of the mixture in the path. Equation 10–5 also shows that if concentration C_t is to be determined accurately, there must be a significant difference between the absorption coefficients for the fluid and the solids.

The precision of the device is directly dependent on the precision of the counting process. Since radioactive decay obeys a Poisson distribution, the standard deviation of a measurement which detects N disintegrations in a given time is essentially $\sqrt{N}$. (Actually the expression should be $\sqrt{N - 1}$, but N is very large.) This means that the fractional error of an N measurement varies as $1/\sqrt{N}$. To improve precision, one can increase N by increasing the source strength, the cross-sectional area of the beam, or the length of time for which N is measured. The absorption coefficients should be measured for the two components of the mixture.

If the flow is vertical and axisymmetric, the value of C_t is identical to the mean concentration C_r which is usually close to the delivered concentration C_v (Section 5.7). If the flow is horizontal or inclined, the beam should span a vertical diameter of the pipe and C_t will be a satisfactory approximation to C_r in most circumstances. As we saw in Section 2.1, the difference between C_v and C_r is frequently significant for horizontal flows. Although a beam of large cross-sectional area has the advantage of averaging over a larger region, the path length deviates from D as this occurs. A more important practical advantage of a large beam is that N is maximized for a given source strength.

If a third component is present (such as air), concentrations that are calculated from the intensity N can be seriously in error. A solution to this problem is to combine a gamma-ray gauge with other sensors (Fanger et al., 1978, 1986; Verbinski et al., 1979). For example, neutrons are strongly absorbed by hydrogen-containing species such as water and this can give a second radiometric determination. Alternatively, one can use the fact that the absorption of low-energy gamma rays is more strongly dependent on atomic number than that of Cesium-137. Americium-241 (0.06 MeV) has been used as a second source to take advantage of this effect.

Considering Figure 10–7, we can understand the constraints imposed by geometry. Suppose we have a cesium-137 source in a lead shield of effective radius 4 cm. In addition, suppose the shield is machined to allow a conical beam of diameter 1 cm to exit from it. The source emission is absorbed by the shield, except for the conical beam. The fraction of the source strength which leaves the shield is the ratio of the beam area to the area of a sphere of radius 4 cm. This fraction is $[\pi(1^2)/4]/4\pi\,(4)^2$ or 1/256.

This is the fraction of the source emission which impinges upon the pipe. If the pipe has an internal diameter of 10 cm and a wall thickness of 5 mm, the gap between the source and the detector must be at least 11 cm to allow the pipe to be placed in the beam path. As it exits from the pipe, the beam diameter will increase to (15/4) cm. The diameter of the detector must be greater than this if counting is to be conducted at high efficiency.

To illustrate the absorption process, consider a source with an activity of 1 millicurie. The source emits 3.7×10^7 disintegrations per second so that in air (negligible absorption) the counting rate at the detector would be 1/256 of this or $1.45 \times 10^5\ s^{-1}$.

For a steel pipe the wall absorption coefficient is $0.62\ cm^{-1}$. The beam must pass through both walls so, neglecting wall curvature effects on the path length through the pipe wall, the value of N_0 is $\exp[-(0.62 \times 2 \times 0.5)]$ times the impingement rate at the detector or $10.8 \times 10^4\ s^{-1}$.

If the slurry contains sand ($\mu_s = 0.20\ cm^{-1}$) at a concentration of 20% by volume in water ($\mu_L = 0.08\ cm^{-1}$), the value of N for this situation is further reduced according to Equation 10–5. In fact, the efficiency of the counting device will not be 100% so that the source activity would have to exceed 1 millicurie to deliver this quantity of radiation in the 0.66-MeV peak of the spectrum.

If measured absorption coefficients are used, and if they are standardized or recalibrated periodically, these devices are very reliable.

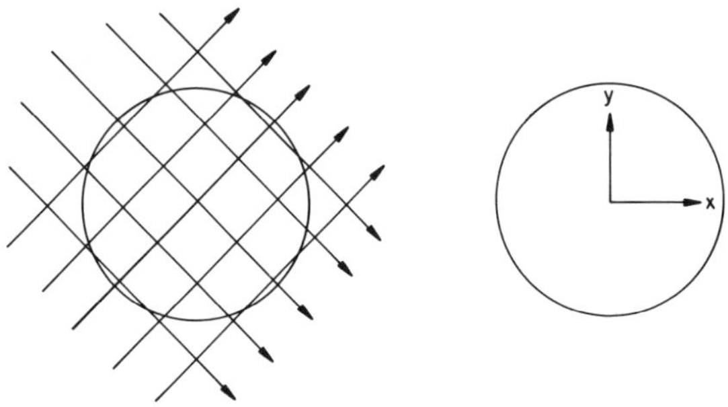

Figure 10–8. Grid for nonintrusive determination of $c(x,y)$.

10.5 RADIOMETRIC CONCENTRATION DISTRIBUTIONS

A horizontal beam which can be traversed vertically will give the mean in situ concentration when the C_t values are weighted by the appropriate path length and integrated over the cross section. For fine particles at moderate concentrations, flowing in large pipes, the concentration varies principally in the vertical direction.

If the beam can be rotated and traversed in other directions, the magnitude of the lateral variation can be determined. The method of polynomial approximation (Michalik et al., 1968) has been shown to give satisfactory results. Consider the concentration to be a function of x and y defined in Figure 10–8. The linear absorption coefficient of the mixture μ (see Equation 10–4) is

$$\mu = c(\mu_s - \mu_L)$$

so that the absorption coefficient can be written as a double power series in x and y:

$$\mu(C) = \mu(x,y) = \sum_{i,j=0}^{n} a_{ij}x^{n-i}y^j$$

For the chord defined by coefficients α_k and β_k

$$y = \alpha_k x + \beta_k$$

intersecting the pipe wall at x_1 and x_2, the total absorption will be

$$\ln\left(\frac{N}{N_L}\right) = \int_{x1}^{x2} \sum_{i,j=0}^{n} a_{ij}x^{n-1}(\alpha_k x + \beta_k)^j dx$$

The left-hand side of this equation is a measured quantity and the right-hand side is a function of the known coefficients α_k and β_k defining the chord. It is also a function of the known limits of integration x_2 and x_1 and the $(n \times n)$ coefficients a_{ij} that defined the mixture absorption coefficient. If we measure the total absorption for p chords, where $p \geq n^2$, we can use the least-squares method to determine these unknown coefficients. A suitable grid for these measurements in a horizontal pipe uses α_k values of -1, 0, and $+1$ (the beam horizontal and inclined at ± 45 degrees). The beams must be located very carefully with this method because random errors in location will produce larger negative errors than positive ones. Since chord length errors will be largest near the pipe wall, the method will tend to indicate low concentrations in this region, in the presence of random experimental errors.

10.6 WEIGHING

In principle, the simplest method for determining the mean solids concentration is to weigh a section of the pipe. The technique has been used in a number of laboratory pipelines of small diameter and a commercial device based on this principle has been tested (Gillies et al., 1984). Careful alignment of the weighed section was necessary to eliminate systematic errors proportional to the square of the velocity. Weighing provides the mean in situ concentration C_r and it has the advantage of being suitable for horizontal pipelines.

10.7 CONCENTRATION FROM PRESSURE DROP

Another simple technique is to measure the pressure drop for a vertical (or even inclined) flow. If the pipeline has a fairly large diameter, the frictional component of the total pressure drop will be small and can be estimated satisfactorily using a single-phase model. The pressure drop is then given by the mechanical energy balance

$$-\frac{dP}{dx} = \rho_m g \frac{dh}{dx} + \frac{2fV^2\rho_m}{D} \tag{10-6}$$

ρ_m is the slurry in situ density and f can be estimated from the viscosity and density of the carrier fluid.

If a vertical U-bend is used, the pressure drops for the upward and downward flowing sections can be combined on the assumption that the frictional effects are the same in the two limbs of the U (Figure 10–9). This yields

$$\rho_m = \frac{(P_1 - P_2) + (P_4 - P_3)}{2gL} \tag{10-7}$$

ρ_m is related to the mean in situ concentration C_r by Equation 2–5.

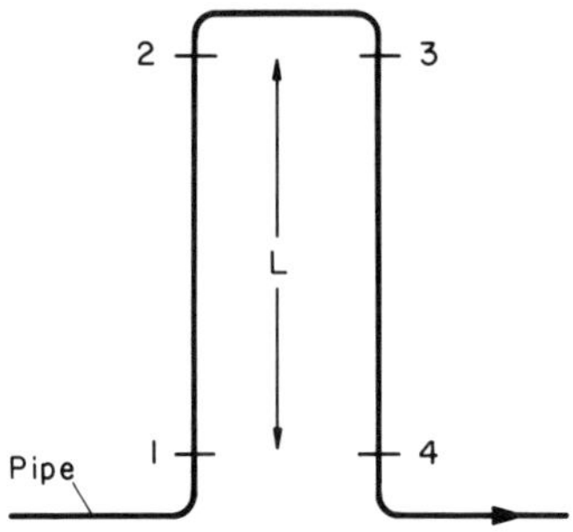

Figure 10–9. Vertical U-bend used for concentration measurement.

Equation 10–7 neglects the effect of slip between the particles and the fluid and this slip will be different in upflow and downflow. Although corrections for this effect have been suggested, we saw in Section 5.7 that the effect is small, except for large particles at low concentrations. Since this method is not very suitable for low concentrations, the slip correction is probably no more important than the effect on the pressure drops produced by the distorted velocity and concentration distributions arising from the bends. These distortions are described in Section 11.2.

10.8 ELECTRICAL METHODS FOR CONCENTRATION

Electrical properties of slurries have been used for many years in laboratory sensors and commercial devices based upon these principles are now available (Auburn). The capacitance of a mixture is directly proportional to the dielectric constant which, according to Keska (1978), is a linear function of the volumetric concentration. If a high-frequency AC source is applied between electrodes mounted on the pipe wall, the impedance will depend on the dielectric constant in the region between the electrodes. A device to be used with cellulose suspensions of low solids content was reported by Gaigolas and Whetstone (1986).

When mounted in a vertical pipe, for which the concentration distribution is uniform over the cross section, there is no difficulty interpreting the readings of such a device. The fact that vertical installation is usually impractical was recognized by Keska (1978). He examined the spatial averaging which occurred when a capacitance device was used in coarse particle flows with strong segregation. The difference between capacitances sensed by electrode pairs on vertical and horizontal diameters is a measure of the inhomogeneity of the concentration profile.

The electrical resistance of a mixture depends upon the resistance of the solids and that of the carrier fluid. A relationship between resistance and concentration was derived for a rather idealized geometry by Maxwell (1892). If the solids are perfect insulators and R_0 is the fluid resistance

$$\frac{R}{R_0} = \frac{2 + C}{2 - 2C} \tag{10–8}$$

More recent experiments suggest that this expression can be applied to aqueous slurries. If the current and the potential difference between two electrodes immersed in the flow are measured, one can find the slurry resistance and thereby the concentration. Surface polarization resistance in such a device can be minimized with high-frequency AC. In the Auburn device, pairs of electrodes are selected to span the pipe cross section sequentially and provide an area-averaged in-situ mean concentration.

Unfortunately, polarization is difficult to eliminate completely and moreover, it is velocity dependent. Also, if concentration is to be determined in this way the carrier fluid resistance must also be measured. The deposition velocity sensor of Ercolani et al. (1979) responds to resistance changes produced by concentration fluctuations which become pronounced at velocities near deposition. Because the fluid resistance R_0 is not measured, the relationship between the R fluctuation and the local C value is essentially a qualitative one in their device.

A method for minimizing the polarization and velocity-dependence effects has been developed (Nasr-El-Din et al., 1987). This allows a local concentration measurement to be made with a probe inserted into the flow. This is shown schematically in Figure 10–10.

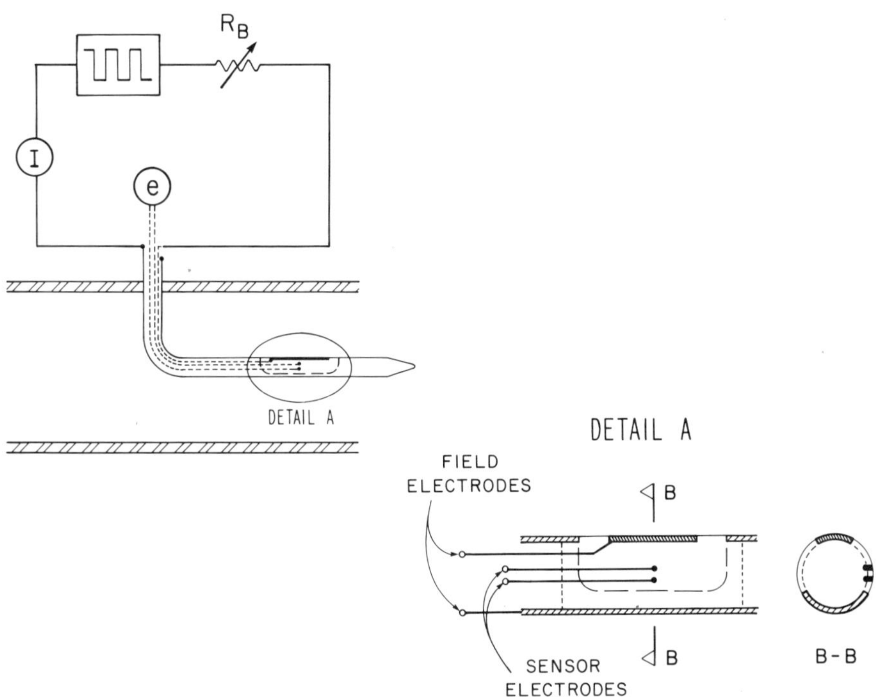

Figure 10–10. Probe for electrical determination of local concentration. (Reprinted with permission from *Int. J. Multiphase Flow* 13:368, Nasr-El-Din, H., et al., A conductivity probe for local concentration measurement in slurry flows, Copyright 1987, Pergamon Press PLC.)

The applied potential is fixed and the ballast resistance R_B is varied to maintain a constant current flowing between the field electrodes as the probe is moved from point to point in the flow. The potential between the two sensor electrodes is measured and since the impedance of the measuring system can be very high, virtually no current flows in the sensor potential circuit. Thus, polarization on the sensor electrodes is negligible. This eliminates the velocity dependence and gives the device a rapid response. The latter feature is exploited in a velocity-measuring mode which will be described later.

The potential between the sensor electrodes is directly proportional to the resistance in the region between them since the current flow is constant. If AC excitation is used, as shown in the diagram, the frequency should be moderate (say 1000 Hz with slurries of conducting fluids such as tap water). This should minimize the capacitive reactance effect and allow the concentration to be obtained from the resistance increase and Maxwell's equation.

To eliminate a pipe wall effect, the field electrode closest to the wall should have a large area, as shown in the diagram. The pipe wall effect diminishes in importance as the pipe size increases and is lower for metal than for insulating pipes. Satisfactory performance of this device has been found in a large (500 mm I.D.) pipe. Since it provides a local measurement, its output requires integration to obtain the mean concentration. It is likely that a device of this type would be useful primarily in establishing the validity of devices which provide some spatial averaging.

10.9 VENTURI METERS

A constriction of a flow produces a pressure drop in the region where the velocity increase occurs, as shown schematically in Figure 10–11. For homogeneous flows,

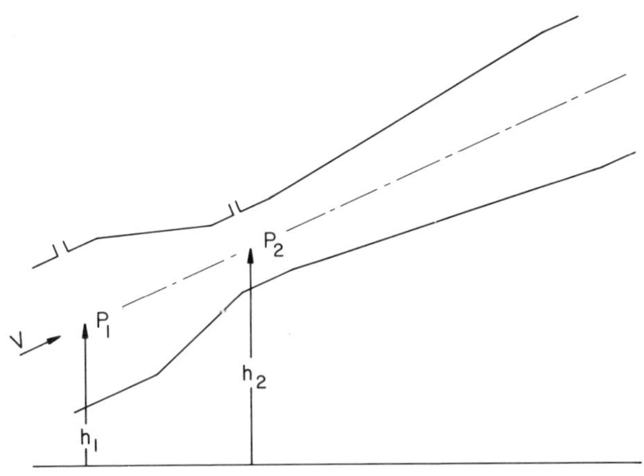

Figure 10–11. Schematic Venturi flowmeter.

Equation 1–4 can be used to relate the mean velocity to the measured pressure difference

$$V = C_{Dv}\beta^2\left(\frac{2(P_1 - P_2) + 2\rho g(h_1 - h_2)}{\rho(1 - \beta^4)}\right)^{1/2} \tag{10–9}$$

In comparison with other constriction devices, the advantages of the venturi are that the coefficient C_{Dv} is close to unity and the pressure recovery in the diverging section is high. For C_{Dv} to be near unity, the flow should be turbulent in the upstream pipe and wall friction must be a small fraction of the pressure drop. Equation 1–3 shows that this requires $4\tau_w/D$ to be small compared to $\rho V\, \partial V/\partial x$.

For slurries, we would expect the device to be suitable without any complications as long as the flow could be considered homogeneous. The constriction produces considerable acceleration over a short distance so that inertial differences between fluid and particles may be more pronounced in a venturi tube than in macroscopically steady pipe flow. Slip between the phases would increase with the density difference between them.

The common situation, with the particles denser than the fluid, was considered by Shook and Masliyah (1974). Using the equations of motion from Chapter 2 and neglecting wall friction, the pressure change $(P_1 - P_2)$ can be calculated for a two-phase mixture. If this is inserted in Equation 10–9 and if ρ is considered to be the approach mixture in-situ density, then a value of C_{Dv} can be predicted. For coarse (2mm) dense ($S_s = 12$) particles, C_{Dv} was predicted to be greater than unity. This interesting effect was corroborated by experimental measurements.

Similar C_{Dv} values (greater than unity) were observed for a horizontal meter. In a later study using very coarse (13mm) gravel and a horizontal meter, coefficients higher than the clear water values were again observed. These could be explained with the two-layer model (Shook, 1982).

If *fluid* values of C_{Dv} are used in Equation 10–9, then the density ρ can be obtained from the delivered concentration C_v as $\rho_L[1 + C_v(S_s - 1)]$. Evidence of this for venturi tubes in horizontal pipeline flows has been given by Gillies et al. (1988). This means that in combination with a velocity-measuring device, the venturi meter is a useful method for determining C_v.

A recent solution for two-phase flow in a venturi has been presented by Sharma and Crowe (1989).

If wear is significant in normal pipe flow, it would be expected to be more severe at the constriction of these meters and this would cause the coefficient to change with time.

10.10 PIPE ELBOW METER

Because pipelines nearly always contain bends, the pressure drops associated with these fittings have some attraction for velocity determinations. However, the location of the tappings and the orientation of the bend (whether in a horizontal or vertical plane) affect the pressure difference measured in the direction of flow. As described in Section 11.1, for bends in a vertical plane, the direction of the flow also matters. To this list of variables affecting the pressure drop in the direction of flow, we must add the geometry of the bend, the properties of the fluid, and the properties of the particles because these affect pressure drops in straight pipes.

The radial pressure difference between the inside and outside surfaces of a pipe bend in the horizontal plane can also be measured. For a homogeneous mixture, the momentum equation A1–4 shows that the radial momentum equation is

$$\frac{\rho_m v_\theta^2}{r} = \frac{\partial P}{\partial r}$$

if gravity and wall friction effects are negligible.

If we approximate the left-hand side of this equation using V and the radius of curvature R_B, we can find the pressure difference between the inside and outside surfaces of the pipe bend:

$$(P_0 - P_i) = \frac{D\rho_m V^2}{R_B}$$

Interpreted in terms of a discharge coefficient defined by

$$V = C_{De}\sqrt{2\Delta P/\rho_m} \qquad (10\text{–}10)$$

this suggests that the discharge coefficient for a horizontal bend should approach $\sqrt{R_B/2D}$. This relationship applies to single phase flows when the elbow has a suitable approach length.

Colwell (1987) placed a standard 90 degree pipe elbow at the discharge from a centrifugal pump and found C_{De} to be the same for water and sand slurries with d_{50} as high as 0.9 mm. ρ_m was calculated from the mean in situ concentration C_r in his experiments and V was always high enough so that deposition did not occur. The actual value of C_{De} was different from the theoretical value, probably because the inside surface of the bend had the abrupt transitions associated with a threaded elbow.

10.11 MAGNETIC FLUX FLOWMETERS

This device is shown schematically in Figure 10–12. The conducting mixture flows through the magnetic field and generates a transverse potential difference in accordance with Faraday's law of electromagnetic induction.

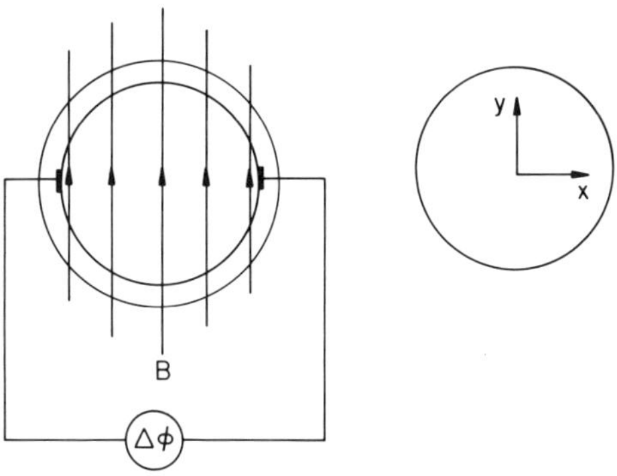

Figure 10–12. Schematic magnetic flux flowmeter.

Maxwell's equations for electric and magnetic fields are

$$\nabla \times \mathbf{E} - \frac{\partial \mathbf{B}}{\partial t} = 0$$

$$\nabla \times \mathbf{H} - \frac{\partial \mathbf{D}}{\partial t} = \mathbf{J}$$

where **D** is the electrical displacement and **J** is the current density.

We use the steady-state forms of these relationships so that a potential Φ can be defined such that $\mathbf{E} = -\nabla\Phi$. For isotropic, homogeneous media, the field intensity **H** is related to the magnetic induction **B**, the permeability of free space μ_0 and the relative permeability κ_m:

$$\mathbf{B} = \mu_0 \kappa_m \mathbf{H}$$

Ohm's law for a moving medium is expressed in terms of the conductivity σ and the velocity **V**:

$$\mathbf{J} = \sigma(\mathbf{E} + \mathbf{V} \times \mathbf{B})$$

Using the identity $\nabla \cdot (\nabla \times \mathbf{B}) = 0$ and assuming the material properties constant gives

$$\nabla \cdot (-\nabla\Phi + \mathbf{V} \times \mathbf{B}) = 0 \tag{10–11}$$

For the meter shown in Figure 10–12, the magnetic field has only the component B_y. We can understand some features of the meter by neglecting $\partial^2\Phi/\partial y^2$ in Equation 10–11 and any variation of axial (z-wise) velocity. The potential difference is measured at $y = 0$ for which

$$\frac{d}{dx}(d\Phi/dx + B_y v_z) = 0$$

or

$$\Delta\Phi = \int_{-D/2}^{D/2} (C_1 - B_y v_z)\, dx$$

The meter output is adjusted to give zero potential at zero flow or, on integration,

$$\Delta\Phi = KB_y DV \tag{10–12}$$

Shercliff (1962) solved Equation 10–11 in terms of a weight function $w(x,y)$ such that in Equation 10–12

$$K = \iint v_z(x,y) w(x,y)\, dx\, dy - \iint v_z(x,y)\, dx\, dy$$

The weight function was evaluated for uniform magnetic and velocity fields. It was shown to be a maximum near the electrodes and a minimum at the top and the bottom of the pipe in Figure 10–12. With $a = 0.5D$,

$$w = \frac{a^4 + a^2(y^2 - x^2)}{a^4 + 2a^2(y^2 - x^2) + (x^2 + y^2)^2} \tag{10–13}$$

This variation of the weight function might lead to some doubt about the area-averaging properties of these meters in flows with considerably distorted velocity distributions. Distorted distributions would occur in horizontal flows with slurries of particles with high settling rates. In terms of the concepts introduced in Chapter 6, such slurries would have substantial contact loads. This problem should not be exaggerated because there are a number of other possible complications and yet the net effect of all these has been found to be minimal in calibration studies, as long as the flow is deposit-free and free from transverse velocity components.

With vertical flow the area-averaging uncertainty with slurries is reduced because the velocity profiles are at least as flat as those in the clear carrier fluid. Moreover, wear by particle-wall abrasion should be uniform and low in vertical flow.

However, vertical flow is often impractical in plant operations. Furthermore, an elbow a short distance upstream of this type of meter has been found to produce systematic errors. This may be due to the tangential velocity component (swirl) produced by the elbow. Pairs of 90 degree bends at 90 degrees produce particularly strong swirls, according to Ito (1987). If a straight approach calming section is provided, so that flow disturbances are damped out, horizontal installation of magnetic flux flowmeters has been shown to be satisfactory.

Fluctuations in $\Delta\Phi$ for slurries were reported by Kazanski and Bruhl (1972) and interpreted as indications of "macroturbulent" velocity components in the flow. Their frequencies corresponded to motion of disturbances with wavelengths a few multiples of the pipe diameter (Peters and Shook, 1981). The amplitudes of these fluctuations were found to be fairly strongly dependent on mean concentration, particle size, and density. This suggests that they originate in concentration variations caused by velocity fluctuations. A time-averaging technique is normally employed with these meters to eliminate such fluctuations.

Equation 10–12 shows that the $\Delta\Phi$ versus V or Q relationship is fixed at the time the device is manufactured and should not change with time unless B_y or D change. Fouling of the electrodes with a nonconducting contaminant such as a hydrocarbon could cause failure of the meter, however. Ultrasonic cleaning methods have been employed to remove such deposits. A useful summary of user experience with these (and other) meters is given by Heywood and Mehta (1988).

10.12 ULTRASONIC FLOWMETERS

These devices represent the ultimate in convenience. The sensor elements are small, portable, nonintrusive, and easily installed. Various types of meter are available and two of these are shown schematically in Figures 10–13 and 10–14. A useful review of both types has been given by Sanderson and Hemp (1981).

In the Doppler flowmeter, the signal is reflected from the suspended matter in the flow. The difference between the transmitted and reflected signal frequencies is a measure of the velocity of the medium through which the signal moves.

In Figure 10–13 the beam is propagated into the mixture with a period $t_0 = 1/f_0$, where f_0 is the entering frequency. The wave travels with an effective velocity $(a - U\cos\Theta)$, where U is the velocity of the flowing medium. Neglecting any slip between the medium and the reflecting species, the wavelength at the reflector is $(a - U\cos\Theta)t_0$, because of the Doppler shift, and the frequency is f_0.

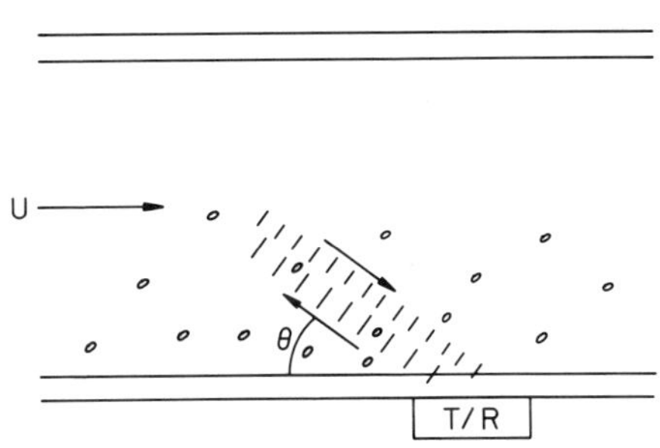

Figure 10–13. Ultrasonic Doppler flowmeter.

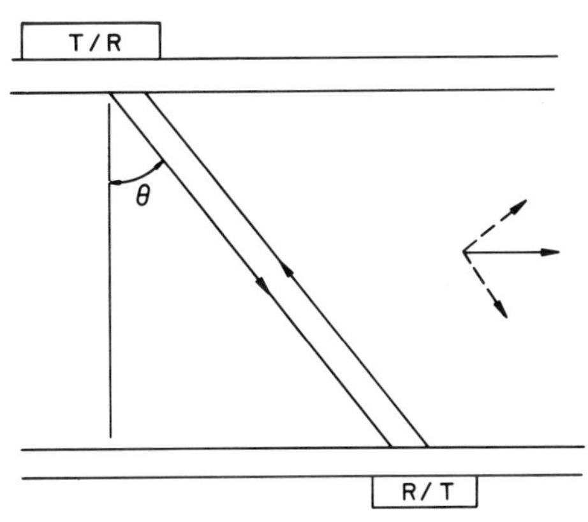

Figure 10–14. Ultrasonic transmission flowmeter.

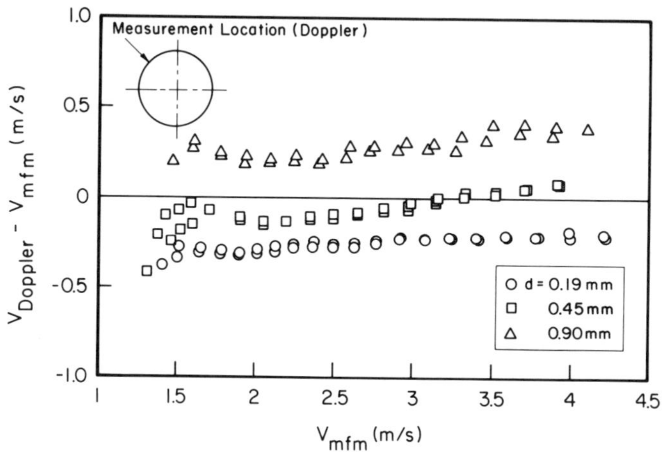

Figure 10–15. Difference between Doppler flowmeter velocities and magnetic flux flowmeter velocities for slurries in a 52-mm pipeline ($C_r = 0.10$). (From Colwell et al., 1989. *J. Pipelines* 7:133. Reprinted with permission.)

The reflected beam wavelength is also $(a - U \cos \Theta)t_0$ but the velocity is now $(a + U \cos \Theta)$. Since $U \langle\langle a$ the frequency of the reflected beam is $f_0[1 + (2U/a) \cos \Theta]$ (Ensminger, 1973). The frequency shift is, therefore, $2f_0U \cos \Theta/a$. The ratio $(\cos \Theta/a)$ can be expressed in terms of the properties of the transducer using Snell's law for refraction of the beam at an interface.

The device does not require calibration but the area-averaging performance must be established experimentally. This depends on the location of the meter on the pipe wall, the concentration and velocity distributions in the flowing mixture, and the beam attenuation properties of the particles. The normal location for a Doppler meter would be at or above the midpoint of the pipe, for which the first studies (Faddick et al., 1979) showed satisfactory performance.

Figure 10–15 (Colwell et al., 1989) shows the variation of measured velocity, compared to that from a magnetic flux flowmeter, in a 52-mm-I.D. pipeline. The meter location was 45 degrees above the midplane. Deposition occurred near 1.5 m/s for all three slurries and we observe that the error is relatively insensitive to mean velocity.

It is known that as the particle size increases, velocity distributions become more distorted, with higher velocities in the upper half of a horizontal pipe. Figure 7–9 showed the velocity distribution for a sand similar in size to the coarse sand used by Colwell et al. (1989). It seems that the ultrasonic flowmeter responds preferentially to the closest particles, when the particle diameter is large.

From the discussion in Section 2.12, we note that to a first approximation it is the velocity of the "medium" to which the instrument responds and not just that of the particles. However, the penetration of the wave or beam depends upon scattering by the particles and this, in turn, determines the area-averaging ability of the device. Scattering of the beam increases with solids concentration and is

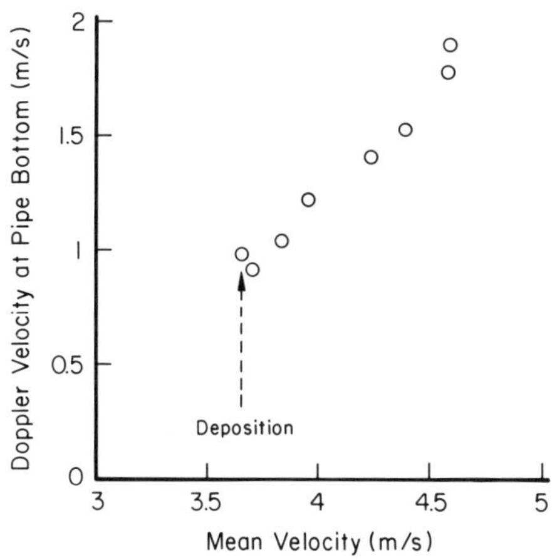

Figure 10–16. Doppler flowmeter velocities measured at the bottom of a pipe, as a function of mean flow velocity. (From Colwell et al., 1989. *J. Pipelines* 7:139. Reprinted with permission.)

very sensitive to the ratio of particle diameter to the wavelength of the ultrasound. Other mechanisms contribute to beam attenuation so that the phenomena of wave transmission and reflection in a slurry are fairly complex (Ensminger, 1973). This means that for slurries with broad particle size distributions, laboratory verification of the performance of an ultrasonic meter is desirable.

Figure 10–16 shows velocities measured with a Doppler flowmeter located at the bottom of a 495-mm pipe (Colwell et al., 1989), compared to the mean velocity. As the mean velocity falls, the Doppler meter reading also falls. However, a stationary deposit on the bottom of the pipe does not necessarily produce a zero Doppler meter reading. Even with a fairly thick (~50 mm) stationary deposit, the meter reading was higher than the velocity in the vicinity of the transducer. It appears that the meter responds to conditions at points on the pipe wall at some distance from the transducer.

In an ultrasonic transmission flowmeter, if the beam is planar the path length depends on the pipe diameter and the displacement angle Θ. Neglecting any variation in the flow velocity U, the transit time in the downstream direction (Figure 10–14) is

$$t_d = \frac{D}{(a + U \sin \Theta) \cos \Theta}$$

For the upstream direction the corresponding expression is

$$t_u = \frac{D}{(a - U \sin \Theta) \cos \Theta}$$

The velocity of transmission of the pressure wave can be obtained from a no-flow measurement:

$$t_0 = \frac{D}{a \cos \Theta}$$

Using these expressions the velocity is obtained as

$$U = \frac{D(t_u - t_d)}{t_0 \cos \Theta \sin \Theta \, (t_u + t_d)} \tag{10–14}$$

We see that U in Equation 10–14 is the chord-average velocity which differs somewhat from the area-averaged velocity V. Furthermore, since the measurement relies upon the velocity of the pressure wave in the fluid, factors such as temperature or the presence of gas bubbles will affect the readings.

To function satisfactorily, the transmission flowmeter requires the beam to be transmitted across the pipe and back. For this reason the susceptibility to attenuation in a particular slurry would have to be ascertained before it could be installed with confidence.

To determine local velocities, a pulsed beam can be directed into the mixture. The signal returned after a time interval Δt will be associated with a particular region in space, at a distance $L = 0.5a\Delta t$ measured along the beam. The velocity of the particles at this position determines the frequency shift of the reflected energy (Hirsimaki, 1978; Scrivener, et al., 1986; Takeda, 1986). A deposition sensor probe using the ultrasonic principle is described by Lazarus and Lazarus (1988).

A concentration sensor using the effect of solids on wave velocity, described in Section 2.12, has been presented by Williams (1991).

10.13 TRANSIT TIME METHODS

If we can measure the time required for a tracer to travel a known distance, the velocity can be obtained. Since fluid conductivity or temperature can be measured with devices with short response times, tracers may be injected to move with the fluid. The transit time between two sensors may be obtained from the maximum cross-correlation of the signals S_u and S_d from the upstream and downstream sensors:

$$F(\tau) = \int_{\lambda=0}^{\tau} S_d(t) S_u(t - \lambda) \, d\lambda \tag{10–15}$$

where λ is the time delay. The velocity is then obtained from the spacing between the sensors and τ_{max}, the value of τ which maximizes F: $v = x/\tau_{max}$. Again, the problem with the method is the interpretation of the measurement in terms of the mean velocity V. If the flow is turbulent and axisymmetric, as in the case of homogeneous slurries, radial mixing will be rapid. The tracer concentration will then be nearly constant over the cross section so that the location of the sensor is not critical. Since slip between the particles and the fluid is negligible for these slurries, the time to the peak of the crosscorrelation will correspond to the time required to traverse the distance between sensors at the mean fluid velocity V_L. This approaches V very nearly for homogeneous slurries.

If a pronounced solids concentration variation over the cross section occurs (as with flows with high contact loads), a substantial velocity variation is produced. The mean velocity of the fluid V_L will then differ from V, even when the local slip is negligible, as shown in Section 2.1. The velocity variation over the cross section tends to increase the spread of the tracer with time or distance of downstream travel. Radial diffusion opposes this spread, tending to keep the tracer together, but the peak in the tracer concentration versus time relationship broadens in these slurries.

A tracer can be generated in a flowing fluid or slurry by neutron bursts from sources of the type developed originally for oil well exploration. If the mixture contains an appropriate element (O, Na, Si, F), the neutron irradiation will produce a short-lived gamma emitter which can be detected at a counter downstream (Porges et al., 1988). The mean flow velocity is determined from the travel time. Using rather elaborate numerical processing methods, the "signature" of the pulse at the detector can be interpreted in terms of the rheological parameters of the slurry if the flow is laminar. Although the technique is expensive, its nonintrusive nature is an important advantage.

In Section 10.11, we noted that concentration variations are observed in most slurry flows. These can be used (Beck et al., 1970) as naturally occurring tracers. In this case it is the particle velocity that is obtained from the transit time. Again, only in the case of axisymmetric flows with negligible slip can we escape from the area-averaging difficulty of interpretation. Unfortunately, concentration fluctuations diminish as the particles become finer so that the method is difficult to use with truly homogeneous slurries. Despite these practical problems, the concept was an invaluable innovation.

The area-averaging difficulty can be overcome by placing the upstream and downstream sensors on an L-shaped probe. If slurry resistivity is the property to be sensed, the double electrode configuration described in Section 10.8 can be used. The high impedance of the sensor electrode circuit eliminates polarization and gives the device a very rapid response. This response allows the axial displacement of the sensor electrodes to be of the order of 1 cm. The device is shown schematically in Figure 10–11. In this case the field electrodes are shown on the pipe wall. This is neither necessary nor particularly desirable since, as Figure 10–10 shows, these can be located on the probe itself.

Comparing Figures 10–17 and 10–10, we see that either pair of sensor electrodes could be used to measure the solids concentration, if two circuits of the type Figure 10–10 were used. The velocity probe must be carefully aligned so that the sensors are not in the wake of the upstream portion of the probe. Of course, there must be some retardation of the particles by the probe surface so that this device would be expected to read low. Calibration tests indicated that this effect was of the order of 2–3% (Brown et al., 1983) for turbulent water slurries of particles coarser than 180 μm. With higher viscosities, higher errors can be expected unless a significant proportion of coarse particles is present.

Because its surface would be subject to abrasion in long-term use, a device of this type is likely to be useful only for verifying instruments whose area-averaging

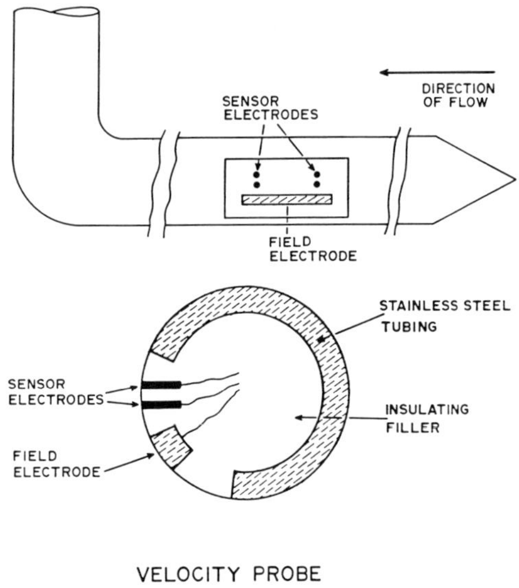

Figure 10–17. Probe for determining local particle velocity.

performance may be unknown. The device is a useful complement to the Pitot tube which can be used with slurries of very fine particles.

10.14 CORIOLIS METER

Figure 10–18 illustrates the principle of the Coriolis meter which can be used to obtain the total mass flow rate. The flow tube consists of a rigid pipe loop which is vibrated magnetically (motion shown as A–A) about an axis (in this case, a horizontal one). This produces an oscillatory gyroscopic torque about the axis B–B. The magnitude of this torque is proportional to the product of mass flow rate through the pipe and the velocity of the vibration (Heywood and Mehta, 1988).

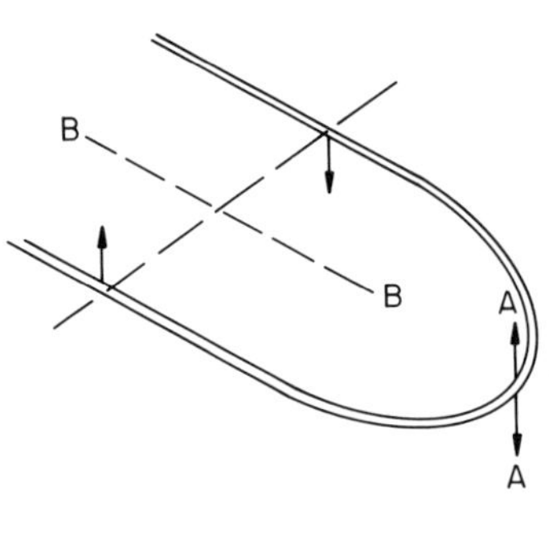

Figure 10–18. Schematic Coriolis meter.

Most commercial flowmeters split the flow and vibrate the two tubes with respect to each other. The flows are combined before leaving the meter. The difference between the motion of the tubes is proportional to the mass flow rate. Mathur and Ekmann (1982) observed satisfactory performance of a meter of this type when used with a coal–water fuel; the meter performance had been previously verified using water. The pressure drop and wear performance of this type of meter with coarse particle slurries would be of interest.

10.15 VISUALIZATION AND IMAGE ANALYSIS

Visualization of particle motion is an important source of information about slurry flow. Such information can be used to describe the flow patterns, to explain energy losses of containing walls, as well as to provide data for micromechanical and microscopic modeling. Flow visualization was used to measure particle rotation shown in Figure 1–13. The capabilities of the flow visualization techniques have been enhanced by recent developments in image analysis and computer technology which make it possible to measure point velocities simultaneously in a flow field (Adrian, 1986), to track an assembly of particles in a control volume (Majumdar et al., 1987), and to determine local concentration and particle size. A variety of methods have arisen as extensions of previous work in single phase flow visualization (Buchlin, 1984; J.S.M.E., 1988; Yang, 1989).

In general terms, image analysis provides a way to quantify a particulate flow microstructure and to determine statistical data on the flow parameters within a control volume. In dilute slurries the particles and the carrier fluid may have different refractive indexes. In more concentrated slurries, particles are chosen to have the same refractive indexes as the carrier fluid and only a small fraction of them are used as tracers (Zisselmar and Molerus, 1979; Edwards and Dybbs, 1984).

Figure 10–19 illustrates a typical result obtained by visualization with high-speed cameras and image analysis (Majumdar et al., 1987). The left-hand side of this figure shows the measuring control volume and the right-hand side presents the velocity fluctuations for a tracked particle. Two cameras at right angles are used to follow the three-dimensional motion of particles. The recorded data provide the information needed to determine trajectories, velocities, accelerations, inertial stresses, and kinetic energy. The relative positions and relative motion of neighboring particles can also be measured.

Other measuring principles for particle velocity have emerged in recent years:

- particle image displacement velocimetry (PIDV) (Adrian, 1986; 1991),
- laser speckle velocimetry (LSV) (Barker and Fourney, 1977),
- nuclear magnetic resonance imaging (NMRI) (Cho et al., 1982; Majors et al., 1989; Caprican and Fukushima, 1990),
- particle holography (Thompson et al., 1967),
- X-ray computer tomography (Miller et al., 1989),
- radioactive particle tracking (Devanthan et al., 1990),
- acoustic methods (Atkinson and Kytomaa, 1991).

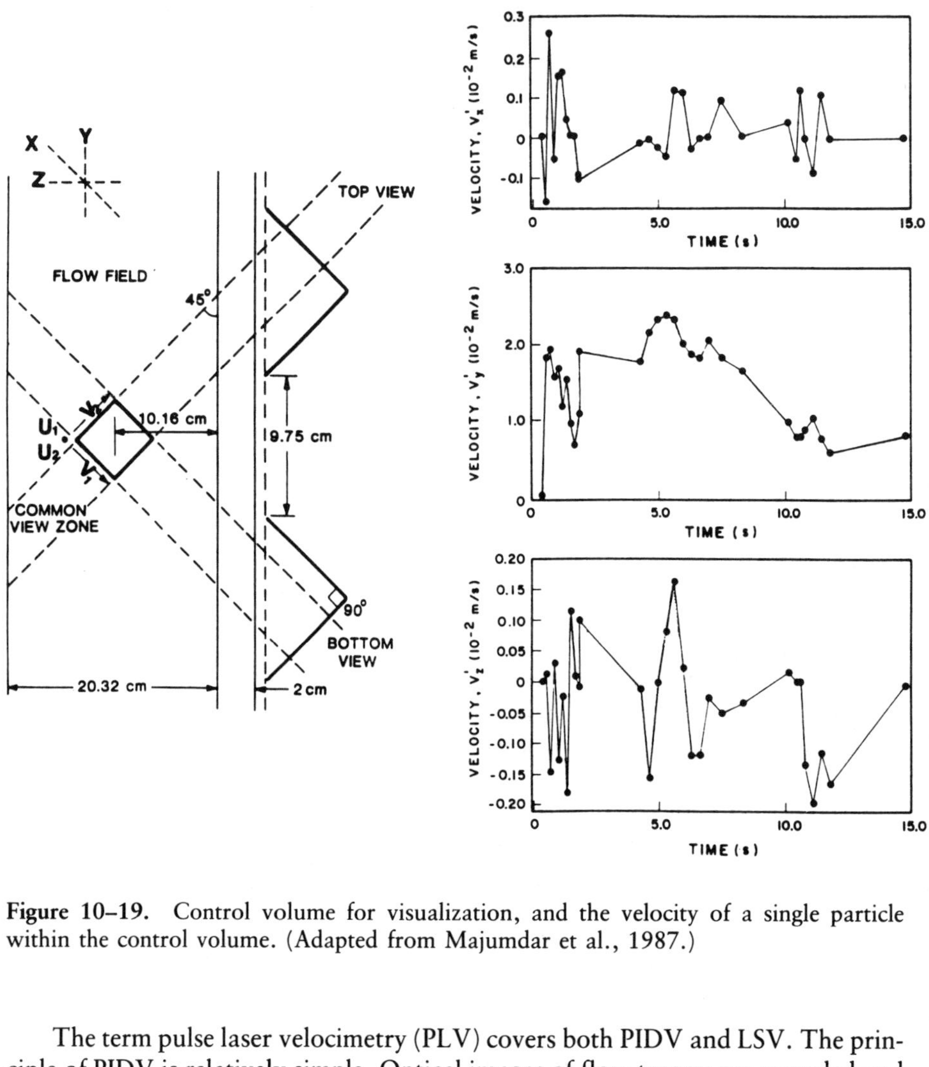

Figure 10–19. Control volume for visualization, and the velocity of a single particle within the control volume. (Adapted from Majumdar et al., 1987.)

The term pulse laser velocimetry (PLV) covers both PIDV and LSV. The principle of PIDV is relatively simple. Optical images of flow tracers are recorded and velocities are deduced from displacements during short time intervals. Recent advances in image processing make it possible to measure velocities of thousands of particles in two- or three-dimensional domains.

10.16 LASER DOPPLER VELOCIMETRY

There is a variety of laser-based techniques to measure particle velocity, particle size, and concentration. The most widely used of these is laser Doppler velocimetry for fine (micron size) particles. Using a two-beam laser system to form a measuring control volume at their intersection, light scattered or reflected from the fine

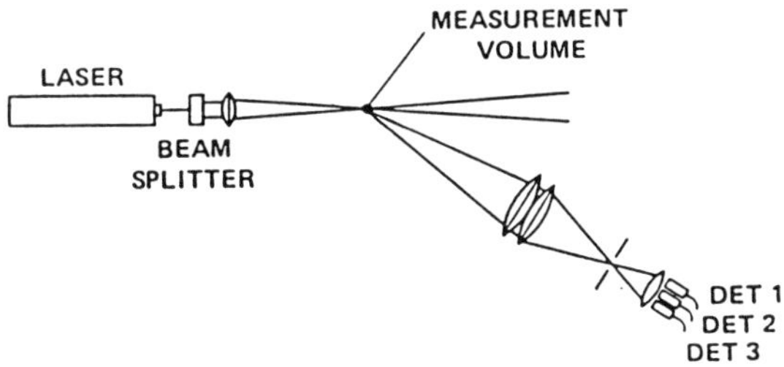

Figure 10–20. Schematic of Doppler particle analyzer. (From Bachalo, 1980. *Applied Optics* 19(3):365. Adapted with permission.) Det 1, 2, and 3 denote detectors.

particles penetrating the control volume can be related to particle velocity (Durst, 1973, 1982). Here, "fine particles" means those whose size is much smaller than the measuring control volume.

For larger particles, a number of techniques based on amplitude, frequency, or phase discrimination have been developed. The most straightforward technique, which is an extension of the LDV method for fine particles, uses amplitude discrimination. The amplitude of the light scattered or reflected by a larger particle is higher because the radius of curvature of the surface is larger. The LDV equipment is basically the same as for fine particles. The amplitude discrimination technique was used for the velocity measurements shown in Figure 9–13d.

A special place among other laser-based methods for individual particles should be given to the Phase Doppler Particle Analyzer (PDPA) which allows simultaneous measurement of particle size and velocity (Bachalo, 1980). The method, illustrated in Figure 10–20, consists of an optical system which is approximately the same as an LDV except that three detectors (instead of one) are located at selected positions behind a single receiver aperture. The spatial frequency of the interference fringe pattern produced by a particle passing through the measuring volume is inversely proportional to the particle size. This technique was developed initially for spherical particles and research is in progress to extend it to nonspherical particles.

Recent techniques to measure the average particle velocity and size in a control volume were reviewed in the Symposium on Particle Sizing (Hirleman, 1990).

New developments in laser technology make it possible to bring the laser beam(s) near the measuring locations with fiber optics and to monitor the parameters (velocity, particle size, concentration) of a flow process continuously by "on line" laser systems.

Chapter 11

Design and Operation Considerations

11.1 LOSSES IN FLOW THROUGH FITTINGS

These losses are important because fittings produce a substantial portion of the total flow resistance in most plant piping systems. Since the influence of a flow disturbance extends downstream, determining the friction losses for bends and fittings requires a number of pressure measurements to be made. By extrapolating the ultimate frictional pressure gradient to the bend (Figure 11–1, after Toda et al., 1972), the pressure drop attributable to the bend is determined as BB′. A part of this pressure drop is associated with the physical length of the fitting, which replaces an equal length of straight pipe.

For single-phase fluids, these losses are frequently expressed as K values where

$$\Delta P = 0.5 K \rho V^2 \tag{11–1}$$

We have seen that the frictional resistance mechanisms are significantly different for homogeneous and nonhomogeneous suspensions in pipe flow and these differences extend to flow through fittings.

For shear thinning homogeneous slurries Ma (1987) determined losses for 45 and 90 degree elbows, 90 degree bends (bend radius R_B/pipe diameter D = 4.5, 9, and 12.5), contractions, expansions, gate valves, and globe valves. For the elbows, bends, contractions, expansions, and the gate valve at high velocities, the losses could be expressed in the form of K values (Equation 11–1) that were the same as those obtained with the clear carrier fluid. At low velocities, where the flow in the fitting was likely to be laminar, the K value varied inversely as the Metzner–Reed Reynolds number $D^{n'} V^{2-n'} \rho_m/(K' 8^{n'-1})$. K' is defined in Equation 4–6.

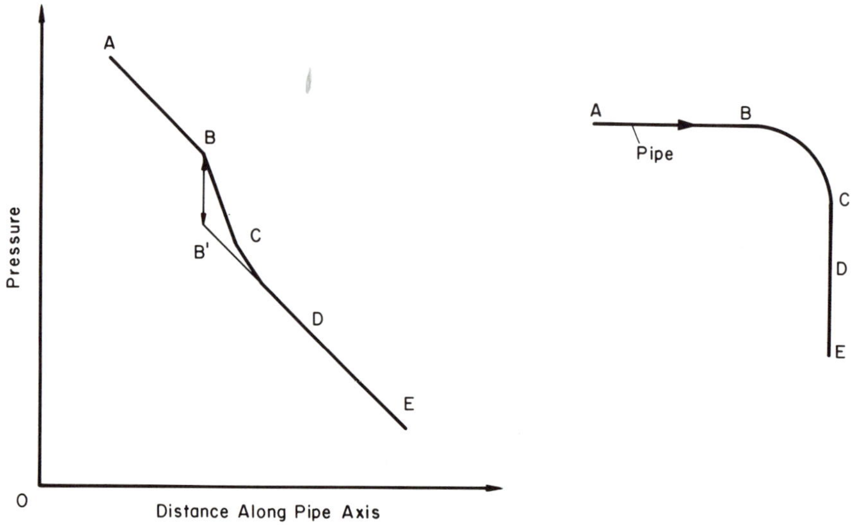

Figure 11–1. Pressure variation for flow through a bend.

The globe valves did not display this increased resistance in the laminar flow regime, presumably because the shear rates in these valves were always very high.

For settling slurries flowing through bends, Ayukawa (1973) gave the first quantitative interpretation of the resistance in terms of concentrations and velocities averaged over the pipe cross section. In so doing he pointed out the distinction between the two types of bends in a vertical plane: horizontal to vertical (H→V↑) and vertical to horizontal (V↑H→).

Pressure drops for settling slurries flowing through bends should exceed those for the carrier fluid. The latter can be evaluated from Ito's (1959, 1960) expression for homogeneous fluids and 90 degree bends. This is equivalent to the Fanning friction factor

$$f = \frac{0.00725 + 0.076\ \mathrm{De}^{-0.25}}{(2R_B/D)^{0.5}} \tag{11-2}$$

for values of the Dean number $(DV\rho/\mu)(2D/R_B)^2$ in the range of $300 > \mathrm{De} > 0.304$. R_B is the radius of curvature of the bend. In the absence of experimental information about bend pressure drops for the slurry, the pseudo-homogeneous approximation would have to be used. Additional allowances for contact load effects, augmented by inertia, would also be required.

Using glass spheres of $d = 1.89$ mm and 0.99 mm in a pipeline 30.2 mm in diameter, Toda et al. (1973) found significantly different pressure drops for water and for slurries in horizontal to vertical (H→V↑) bends. For bends in a horizontal pipeline, a difference between water and slurry flow pressure drops could only be detected at R_B/D (bend radius/pipe diameter) values greater than 8. When plotted against velocity, the pressure drops at constant solids concentration gave curves parallel to those obtained with the pure fluid.

Similar results were obtained by Kalyanaraman et al. (1973) using 2 mm sand particles. This suggests Coulombic friction as the source of the incremental effect due to solids. However, a simple interpretaton of the increment is not possible. We recall that for horizontal pipe friction, when Coulombic friction is dominant, Newitt's equation shows that the dimensionless group: $Fr = gD(S_s - 1)/V^2$ governs the dimensionless incremental headloss $\Phi = (i - i_L)/C_v i_L$. If centrifugal force (inertial) effects cause the increased friction, g would be expected to be replaced by (V^2/R_B) in Fr. However, the results of Toda and coworkers (1972) suggested that the dimensionless incremental pressure drop for bends, analogous to Φ, varied with $g\ R_B(S_s - 1)/V^2$.

Wildman et al. (1984) and Ekmann et al. (1985) found that the pressure drop for a coal slurry flowing through an H→V↑ bend exceeded that for a V↑H→ bend significantly at high slurry concentrations. The reverse was true for water. For the H→V↑ bend, the gravity and inertial effects in a slurry act in the same direction, whereas for the V↑H→ one they oppose each other. It was interesting that these friction effects were detected with slurries of fine coal, for which inertial effects might be expected to be small. Little effect of velocity was evident in their results, however. As the concentration increased there was an increased tendency toward homogeneous behavior.

11.2 FLOW PATTERNS AND SEGREGATION IN FITTINGS

For clear fluid flows, two vortices have often been reported for flow around bends (Figure 11–2a,b). The right-hand side of Figure 11–2b is at the inside of the bend. If we apply the momentum equation (Equation A1–4) for the r-direction, where v_θ is approximately V, $\partial P/\partial r$ is positive. Thus, the pressure is lowest at the inner ("suction") wall. We also see that in the central portion of the pipe, where velocities in the θ-direction are high, an (outward) velocity in the r-direction arises because of the centrifugal effect. This flow is balanced by an inward flow near the top and bottom walls where the velocities in the θ-direction are low and the pressure gradient forces fluid inward.

For slurry flows, Wildman et al. (1984) used a flow visualization technique and found complicated secondary flow patterns at the outer wall. Somewhat different behavior was observed for slurries of sand particles in water (Nasr-El-Din and Shook, 1987) at high Reynolds numbers. Particle and fluid motions in these flows represent the combined effect of inertia, drag, and fluid pressure effects in an H→V↑ elbow. Coarse particles were found to move outward under the influence of inertial forces, but fine particles actually concentrated at the suction wall.

With a standard ($R_B = 0.9D$) elbow, separated flow was observed at the inner wall, together with swirl at the exit plane. In this case, one of the vortices

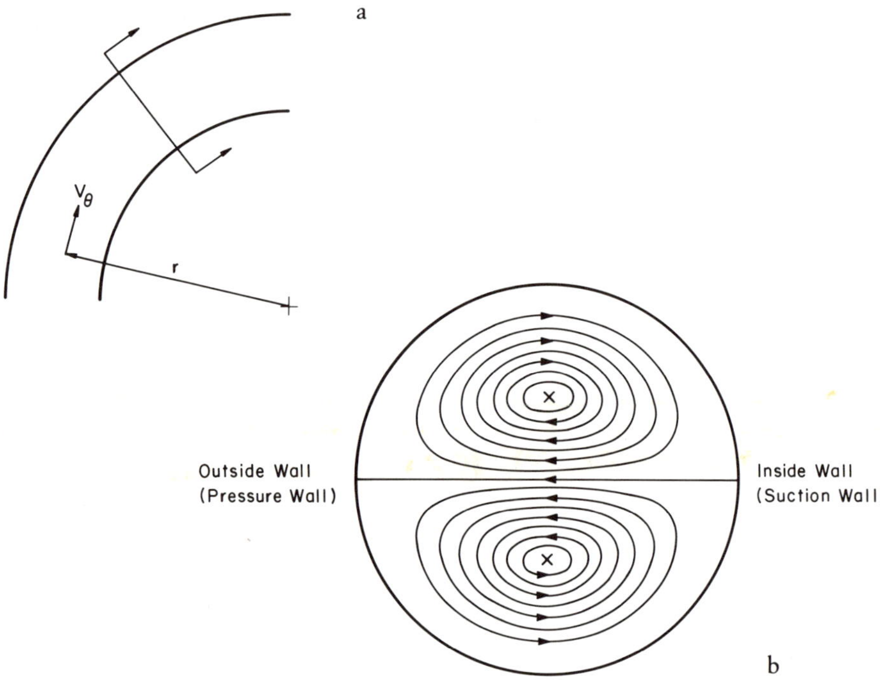

Figure 11–2. Idealized secondary flow resulting from a bend. (From Nasr-El-Din and Shook, 1987. *J. Pipelines* 6:241. Reprinted with permission.)

seemed to become dominant to produce the swirl. The effect of the rotational secondary flow was evident in solids concentration measurements made 22*D* downstream of the short radius bend. These are shown in Figure 11–3 as values of local concentration *C*, normalized with the mean delivered concentration, C_v. The concentration is a minimum at the center of the pipe and increases significantly as the walls are approached. There is evidently a centrifugal separation produced by the tangential component of the flow. The mean velocity was 2.6 m/s.

A longer radius bend showed less variation of concentration but gave evidence of the double helical flow shown in Figure 11–2. The axial velocity distribution was also different for short radius and long radius bends. Figure 11–4 shows that the longer ($R_B = 6D$) radius bend causes a drift of the maximum velocity toward the outer (pressure) wall. In the short radius bend the helical motion is evidently closer to axisymmetric.

For particles with a high settling tendency, if the exit pipe leading from a bend is horizontal, the effect of the bend-induced swirl will die out quickly. For the vertically exiting flows of Figures 11–3 and 11–4, gravity does not act to restore the solids distribution. Eventually diffusion, drag, and interparticle forces cause equilibrium to be established but this takes a greater distance in a vertical flow than in a horizontal one.

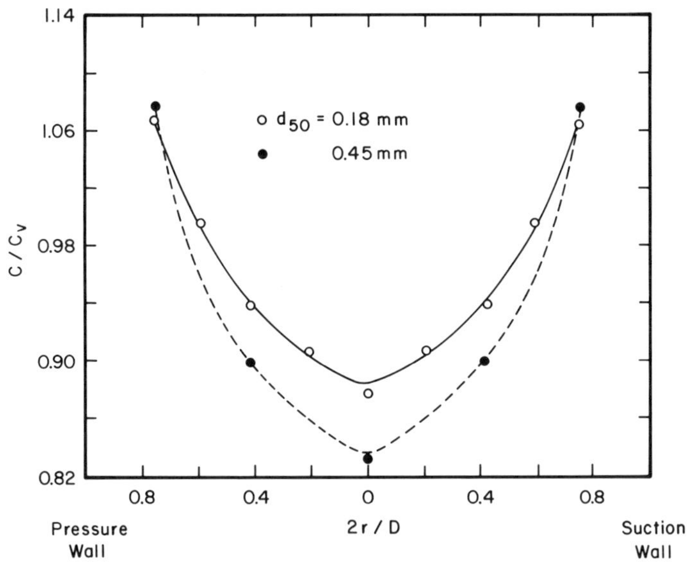

Figure 11–3. Concentration distribution downstream of a short radius H→V↑ 90 degree bend. (From Nasr-El-Din and Shook, 1987. *J. Pipelines* 6:245. Reprinted with permission.)

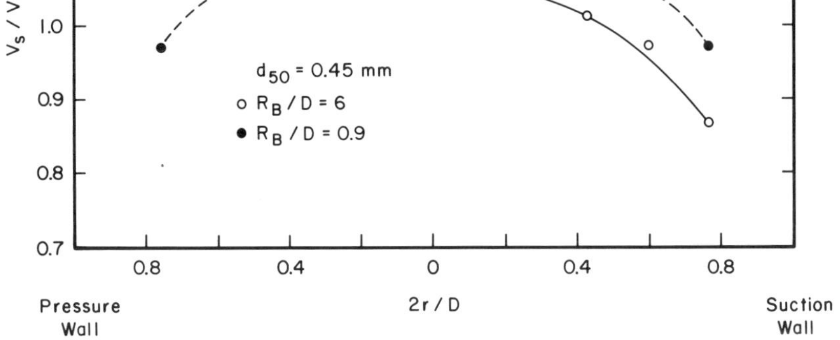

Figure 11–4. Axial velocity distributions in a slurry downstream of short radius and long radius H→V↑ 90 degree bends. (From Nasr-El-Din and Shook, 1987. *J. Pipelines* 6:250. Reprinted with permission.)

Because of these secondary flows, the effect of radius of curvature on pressure drop is complex. Kalyanaraman et al. (1973) suggested that the pressure drop passed through a minimum at $(R_B/D) \sim 2.5$ and a maximum near 5 for slurries of coarse particles.

These secondary flows suggest that wear may be highly localized within and downstream of bends. Attempts to predict particle trajectories and erosion rates in bends have been made (Benchaita, 1985), but until these have been perfected, selection of suitable new materials to resist wear must rely on simulation methods using standard testing techniques.

If a flow is to be divided with an asymmetrical separation, the solids concentrations in the resulting streams will generally differ because of the inertial effect described in Chapter 10. Figure 11–5 (Nasr-El-Din and Shook, 1986) illustrates the effect in the form of values of the ratio of the branch concentration to the upstream value, as functions of the flow rate ratio, for sands of diameter 0.17 mm and 0.33 mm. The approach velocity was 5.2 m/s. The effect of the ratio M = (branch pipe area/upstream pipe area) for the 0.33-mm particles is shown in Figure 11–6.

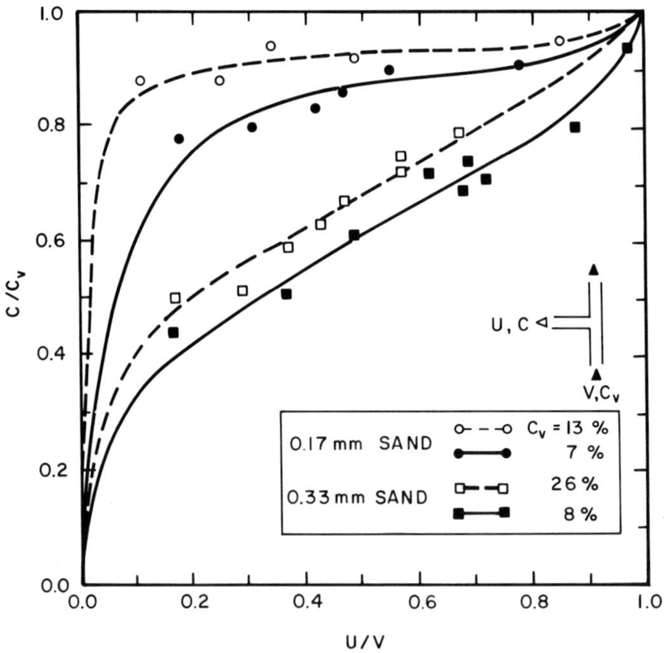

Figure 11–5. Effect of (branch/approach) velocity ratio, concentration, and sand particle diameter on (branch/approach) concentration ratio for slurry flow through a vertical tee. The approach velocity was 5.2 m/s. (Reprinted with permission from *Int. J. Multiphase Flow* 12:436, Nasr-El-Din, H., and Shook, C.A., Particle segregation in slurry flow through vertical tees, Copyright 1986, Pergamon Press PLC.)

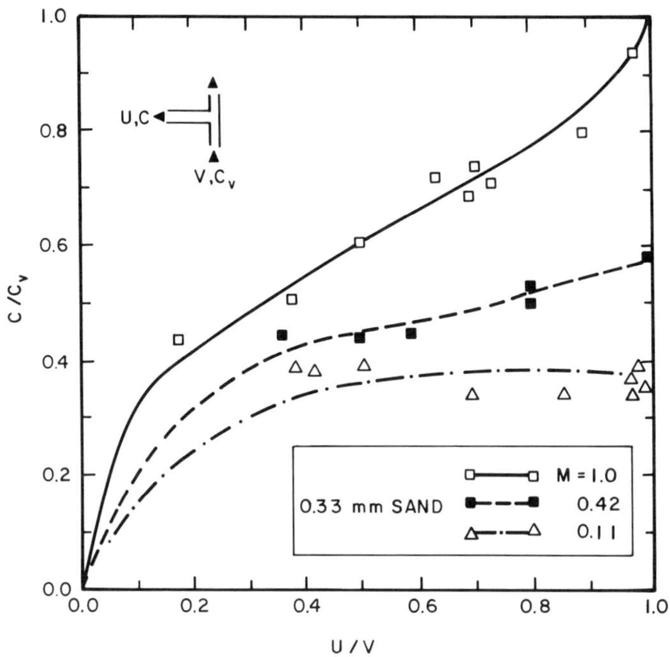

Figure 11–6. Effect of (branch/approach) velocity ratio and (branch/approach) area ratio on (branch/approach) concentration ratio for 0.33 mm sand. $C_v = 0.09$, $V = 5.2$ m/s. (Reprinted with permission from *Int. J. Multiphase Flow* 12:436, Nasr-El-Din, H., and Shook, C.A., Particle segregation in slurry flow through vertical tees, Copyright 1986, Pergamon Press PLC.)

These results were obtained with vertical tees and similar results were obtained with 45 degree branches. For horizontal flows, the concentration and velocity distribution in the approaching slurry complicates the separation. Experimental results for horizontal tees are given by Nasr-El-Din et al. (1989).

11.3 INCLINED PIPES

Headloss and deposit velocity prediction methods for these pipes were discussed in Chapters 5 and 6. However, there are additional factors to be considered if slurry pipelines must operate with significant elevation changes.

The first of these factors is related to the care which must be taken to ensure that the pressure in the pipeline does not fall to a value which allows vapor evolution to occur. Theoretically, this limits the pressure at any point in the pipeline to the vapor pressure of the fluid. In practice, liquids are usually saturated with air at the inlet to the piping system where the pressure is 1 atmosphere. Unless the temperature drops significantly, this dissolved gas will begin to be liberated when the pressure falls below the inlet pressure.

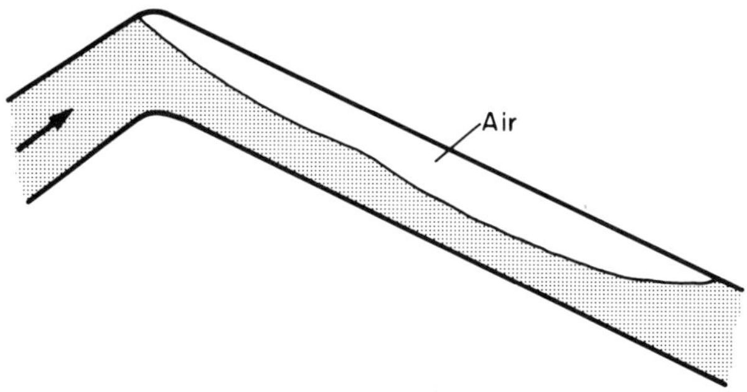

Figure 11–7. A gas bubble at a high spot in a pipeline.

This gas will rise in the pipe and collect at high spots. (Actually, it will collect on the downstream side of a high spot, as shown in Figure 11–7, where the downward drag exerted by the slurry is balanced by the buoyant effect of the low air density.) Slurries often contain small amounts of dispersed gas and this will separate rapidly at shutdown to collect at high spots. Bubbles of the type described previously will form and the drag force of the restarted flow may not be sufficient to remove them.

We note that the pressure will not vary significantly with axial position under the bubble so that the flow must be gravity driven in this region. These flows are sometimes called "slack" flows. If we consider steady flow, the volumetric flow rate Q of slurry is constant along the pipe so that under the gas bubble, where the cross-sectional area A is reduced, high velocities occur. The friction losses, written in terms of a friction factor f, are

$$\text{losses} = -\rho_m g \frac{\partial h}{\partial x} = \frac{2fQ^2\rho_m L}{A^2 D_{eq}}$$

where D_{eq} is the hydraulic equivalent diameter of the reduced area.

The flow disturbance at the bubble, and the high velocities associated with it, can increase erosion rates significantly. To ensure that bubble formation does not occur, the Bernouilli equation should be used to calculate the static pressure at all pipe diameter changes and at all high spots during steady flow.

To allow air to escape, so-called "vacuum-breaking" vents can be used at critical points. If the pressure approaches the vapor pressure of the fluid, it may even be necessary to insert a choking flow resistance at the pipe outlet to raise the pressure throughout the system.

Consider the schematic pipeline shown in Figure 11–8. The pressure at any point along the pipeline can be calculated from the extended Bernouilli equation. By rearranging this, we have an equation for the piezometric head

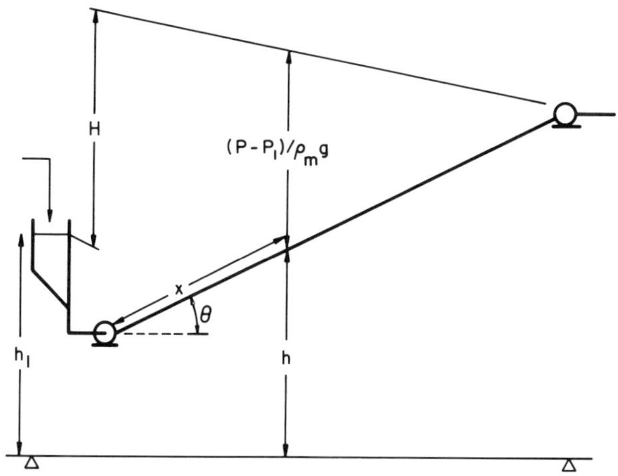

Figure 11–8. Schematic pipeline, illustrating piezometric head variation.

at a distance x from the pump outlet (x_s is the effective length of the suction piping)

$$\left(\frac{P}{\rho_m g}\right) + h = \left(\frac{P_1}{\rho_m g h_1}\right) + h_1 + \frac{V_1^2}{2g} - \frac{V^2}{2g} + H - \frac{i\rho_L(x_s + x)}{\rho_m} \tag{11–3}$$

If concentration (and therefore density) changes can occur during flow, the effect of these on the pressure at these critical points should be examined. Further discussion of the effect of such changes is given in Section 11–10.

Example 11.1

A tailings pipeline runs up a slope of 1.00 degrees to a second pump 1620 m away. The slurry density is 1540 kg/m^3 and the headloss is known to be 0.029 m water/m pipe at a velocity of 3.1 m/s. In terms of a datum located at the centerline of the inlet to pump 1, the elevation of the slurry surface in the sump tank (h_1) is 3.5 m in normal operation. Atmospheric pressure is 98.5 kPa at this location.

The pump discharges horizontally before entering the inclined pipe. The discharge port centerline is 0.5 m below the inlet. Find the pump head H required to ensure that the pressure at the inlet to the second pump is 200 kPa. Friction and entrance losses in the first pump suction pipe are estimated as 1.0 m slurry. The liquid density is 1000 kg/m^3.

Solution

In Equation 11–3

$$h_1 = 3.5 \text{ m}, \quad V_1 = 0, \quad V = 3.1 \text{ m/s}$$

$$\frac{V^2}{2g} = \frac{(3.1)^2}{2(9.8)} = 0.49 \text{ m}$$

$$\frac{i\rho_L x_s}{\rho_m} = 1.0 \text{ m}$$

$$\frac{i\rho_L x}{\rho_m} = \frac{0.029(1000)(1620)}{1540} = 30.51 \text{ m}$$

$$\frac{P_1}{\rho_m g} = \frac{98.5(1000)}{(1540)(9.8)} = 6.53 \text{ m}$$

At the inlet to pump 2, $h = -0.5 + 1620 \sin(1.00 \text{ degree}) = 27.77$ m

$$\frac{P}{\rho_m g} = \frac{200(1000)}{(1540)(9.8)} = 13.25 \text{ m}$$

Substituting in Equation 11–3 we have the required value of H: 62.99 m. The calculation illustrates the importance of a reliable prediction for line friction if satisfactory performance of the second pump is to be ensured.

A plot of $(P/\rho_m g + h)$ as a function of position is useful in defining where the pressure falls below atmospheric. If atmospheric pressure is defined as zero, and the piezometric head is plotted to the same scale as pipeline elevation, pressures below atmospheric are seen to occur where the piezometric head locus falls below the pipeline, i.e., $(P - P_1)$ becomes negative. A diagram of this type allows one to select H so that the NPSH criterion at the downstream pump is satisfied at all the concentrations and flow rates of interest.

The second factor to be considered for these pipelines is their response at shutdown. As axial motion dies out, settling of the slurry begins. If the sloped sections are inclined at angles lower than the static angle of repose of the settled slurry, the particles which settle to the bottom of the pipe remain stationary. Angles of repose vary between 22 and 30 degrees for most sediments.

Above the stationary sediment layer, there exists a mixture whose density decreases with height, reflecting the variation of solids concentration. This mixture will begin free convective flow, with the less dense mixture moving up the incline and the slurry (above the sediment layer) moving downslope. A layer model for this type of free convective flow was described in Section 6.7. The effect

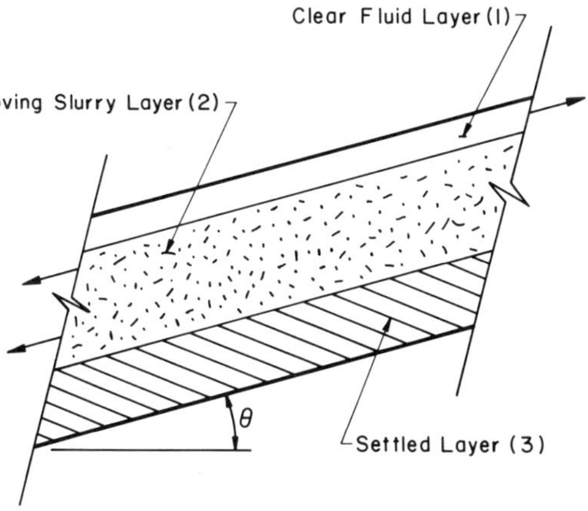

Figure 11–9. Settling and natural convective flow in an inclined pipeline. (From Shook and McLeod, 1975. *Can. J. Chem. Eng.* 53:594. Reprinted with permission.)

of the flow is to transport solids toward the bottom of the incline. Figure 11–9 illustrates the situation in the settling slurry.

Because of this free convective flow, the ultimate depth of the settled layer at the bottom of the incline will be increased, in comparison with the depth in a horizontal pipe. The magnitude of the increase depends on the settling tendency of the slurry.

Experimental studies of the region of high concentration which forms at the bottom of inclines suggest (Ercolani et al., 1987) that the total included angle of the incline should be less than about 30 degrees to prevent an obstruction from developing.

11.4 GAS-SLURRY FLOWS

Crude petroleum consists of a mixture of oil, gas, water, and sand or rock chips. Drilling fluids are non-Newtonian slurries or solutions which usually have yield stresses. Vertical flows occur in wells; horizontal and inclined flows can occur in wells and pipelines. To understand these very important and complex flows, it is desirable to begin with simpler situations, such as gas-slurry flow.

There are at least three particular applications where gas injection may have a practical advantage in slurry transport. The first of these occurs in vertical flows where it may be desirable to reduce the effective density of the particles to reduce the gravitational component of the total pressure drop. Faddick (1983) has reported tests in which gas bubbles were attached to particles with the aid of a surfactant to produce this effect.

A second situation occurs when a strongly shear-thinning mixture is to be transported in a laminar flow. In pipes of moderate or large diameter, slurries

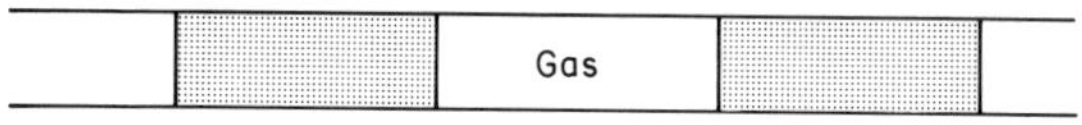

Figure 11–10. Idealized slug flow regime of gas–slurry flow.

of this type would display low values of n and high values of K in the power-law model described in Chapter 3. A simple model of the flow, in the form of alternating slugs of slurry and gas (after Carleton et al., 1973) is shown in Figure 11–10.

The wall shear stress for the flow of the slurry is proportional to the bulk velocity raised to the power q, where $q = n$ if the flow is laminar and $q \rightarrow 2$ if it is turbulent. Including all factors other than the pipe length in the dimensional coefficient K_1 we have, when slurry flows alone without gas injection, for horizontal flow

$$\Delta P_0 = K_1 V_0^q L \tag{11–4}$$

where L is the total length of the pipe.

With gas slugs present, the slurry wall stress is the same function of velocity but the velocity must be increased to provide the same throughput of slurry. If L_s is the total length of slurry plugs,

$$L V_0 = L_s V_s \tag{11–5}$$

The pressure drop within the gas slugs is negligible so that the pressure drop with gas injection is

$$\Delta P_s = K_1 V_s^q L_s \tag{11–6}$$

Using these relationships we find

$$\frac{\Delta P_s}{\Delta P_0} = \left(\frac{L_s}{L}\right)^{1-q} \tag{11–7}$$

The crude estimate of Equation 11–7 suggests that gas injection would not be advantageous for Newtonian liquids in laminar flow ($q = 1$) or for any liquid in turbulent flow, where q approaches 2. However, there is an advantage if $q < 1$. These flows have been studied by Farooqi et al. (1980), Heywood and Charles (1980), and Khatib and Richardson (1984).

With gas–liquid flows, a number of regime-specific methods for predicting the relationship between pressure drop and flow rate have been proposed (DeGance and Atherton, 1970). For horizontal liquid–gas flows, it is customary to relate the pressure drop ΔP to that which would be experienced by the quantity of gas flowing alone in the same pipeline, ΔP_G, and that which would be experienced by the liquid quantity flowing alone in the same pipeline, ΔP_L.

ΔP_G and ΔP_L are, of course, quantities that can be calculated. The mixture pressure drop due to friction can be nondimensionalized with either of these and is reported as

$$\Phi_L^2 = \frac{\Delta P}{\Delta P_L} \quad \text{or} \quad \Phi_G^2 = \frac{\Delta P}{\Delta P_G}$$

The Lockhart and Martinelli correlation described by DeGance and Atherton provides a method for estimating Φ_L or Φ_G from the value of X, defined by $X^2 = \Delta P_L/\Delta P_G$. For a slurry, experimental observations corroborating the effect estimated in Equation 11–7 would be evident as Φ_M^2 (M denotes slurry properties) values less than unity. Low friction results for shear thinning slurries with gas slugs have been reported by several workers.

In addition to the frictional contribution to pressure drop, an inertial or acceleration contribution will occur. Methods for estimating this effect for gas–liquid flows are given in standard references and, to a first approximation, these methods should be appropriate for slurries flowing with gases.

If the pipe is inclined, the gravitational effect will often be a very important component of the total pressure drop. The momentum equations of Chapter 2 show that it is the local in situ mean density of the mixture that determines the gravitational effect, for both dispersed and stratified flows:

$$\left(-\frac{\partial P}{\partial x}\right)_{\text{gravity}} = [\alpha_M \rho_M + (1 - \alpha_M)\rho_G] g \frac{\partial h}{\partial x} \tag{11–8}$$

where α_M is the volume fraction occupied by the slurry (the *holdup*). Time-averaged values of α_M have been measured and correlated with X for vertical flows (Khatib and Richardson, 1984) and values for horizontal flows have been reported. Equation 11–8 can be used to make estimates of gravitational effects more satisfactorily for upward-inclined flows than for those inclined downward. The reason for this difficulty is the problem described in conjunction with Figure 11–7, namely, that α_M may decrease substantially in a downward flow.

A third possible application of gas injection was described by Barnea et al. (1986). They considered a coarse particle slurry for which, as the correlations of Chapter 5 show (for example, Newitt's equation), the pressure gradient is insensitive to pipe diameter. If substantial quantities of gas are injected into a pipeline transporting a slurry, the simplest slug flow model (Figure 11–10) suggests that the total pressure drop depends on the relative amounts of gas and slurry. Thus, at a given total throughput, the *slurry* flow rate (but not the slurry velocity) and the total pressure drop are both reduced by the injected gas.

If we consider a situation in which the slurry flow rate (m^3/s) is fixed, Equation 11–5 shows that it is possible to operate a large pipeline at high slurry velocities (which prevent deposition) and low pressure drops by injecting substantial volumes of gas.

Although advantages can arise from the presence of gas in these special circumstances, air in slurry pipelines at low and moderate concentrations is usually undesirable. In horizontal pipelines, bubbles are formed which concentrate near the top of the pipe. These increase the flow resistance (the "losses," $i\rho_L(x_s + x)/\rho_m$, in Equation 11–3) for a given value of the total volumetric flow rate Q.

11.5 PULSATING FLOWS

Flow of a slurry with a pulsating component superimposed upon a mean velocity can produce lower energy consumption in certain circumstances. Round and El-Sayed (1985) found a reduction of up to 5% for Bentonite (approximately Bingham fluids) slurries using an oscillatory component roughly 20% of the average velocity. The minimum in the energy consumption was considered to be due to the competing effects of a structural breakdown in the high shear region at the pipe wall and of the additional energy required to maintain the fluctuation. Since the maximum reduction was observed with the most concentrated slurry, it is possible that greater reductions in energy consumption could have resulted from further increases in yield stress.

For "settling" slurries, Round (1981) showed that energy consumption could be reduced if the pulsation frequency was chosen correctly. The reduction was most pronounced in the low-velocity (near deposition) region at a solids concentration of 6% by volume. It was suggested that the technique could be useful with inclined pipes of the type used in dredging operations where the pulsating component of the axial velocity would oppose the settling tendency of the particles.

11.6 OPEN CHANNEL FLOWS

Most of the studies concerned with open channel flows have examined the case of flow with an erodible lower boundary formed by a deposit of particles. With its variety of roughness forms, this type of flow is interesting but the high and variable flow resistance is usually undesirable. In launder design, the problem is usually to select an appropriate combination of channel dimensions and slope which will transport a given slurry at a specified flow rate, without forming a permanent deposit of solids.

There have been a few experimental studies of bed-free flow open channel flow. The dependent variable of interest has usually been the solids throughput, expressed either as mass or volume per unit time. The independent variables include the mixture volumetric flow rate Q, the slope S, the hydraulic radius of the cross section R_h, and the particle and fluid properties.

We realize that the models discussed in Chapters 6 and 7 should be applicable to these flows. For example, it is not difficult to convert the computer program in Appendix 4 to a form which could be used for open channel flow. Tests of the model with some of the results in the literature (Lytle and Reed, 1984; Graf and

Acaroglu, 1968) give encouraging results but satisfactory comparisons require the coefficient η_s to be known.

Faddick (1986) considered the problem of launder design for a rectangular channel of width equal to twice the depth y. For this channel shape, the hydraulic radius R_h is $0.5y$. He suggested that supercritical flow is desirable because it is associated with lower lining costs and lower likelihood of particle deposition. The Froude number criterion is then used to select the flow depth from the specified volumetric flow rate Q:

$$\mathrm{Fr} = \frac{Q}{2y^2(gy)^{0.5}}$$

or

$$y^{5/2} = \frac{Q}{2g^{0.5}\mathrm{Fr}} \tag{11–9}$$

To determine the slope S, Manning's formula is used, with the conventional (pure fluid) value of n:

$$V = \frac{R_h^{2/3}S^{0.5}}{n} = \frac{(0.5y)^{2/3}S^{0.5}}{n} \tag{11–10}$$

The velocity V is selected from an equation derived from the Law of the Wall (Equation 1–8):

$$u = V + \left(\frac{u_*}{\kappa}\right)\left[1 + \ln\left(\frac{y'}{y}\right)\right] \tag{11–11}$$

The friction velocity u_* is $(gSR_h)^{0.5}$, u is the point velocity at a distance y' from the bottom of the channel, and κ is the von Karman coefficient. For satisfactory design, Faddick suggests u is related to the terminal falling velocity of the largest particle in the slurry, V_∞. He suggests as tentative values $u = 35\ V_\infty$, $y'/y = 0.1$, and $\kappa = 0.2$. Combining Equations 11–11, 11–10, and 11–9, we obtain with these simplifications:

$$S = \frac{(35V_\infty)^2}{[(0.5)^{2/3}(0.5Q/g^{0.5}\ \mathrm{Fr})^{4/15}/n - 6.513(0.5Q/g^{0.5}\mathrm{Fr})^{0.2}(g/2)^{0.5}]^2} \tag{11–12}$$

A value of 1.25 was suggested for Fr, so that S becomes a function of Q, n, and V_∞. Faddick gives an interesting example of a flow with $V_\infty = 0.06$ m/s, $n = 0.01$, and $Q = 1.39$ m^3/s. Using Equation 11–12, with Fr = 1.25, S is calculated to be 0.00506. This is very close to the actual value.

The initiation of particle motion and, subsequently, of solids transport by suspension in open channel flow, is usually expressed in terms of the Shields parameter $\tau/(\rho_s - \rho_L)gd$ (e.g., van Rijn, 1984a,b). For flow with a mobile bed of coarse solid particles, at values of the Shields parameter > 0.8, Wilson and

Nnadi (1990) suggest that Manning's formula can be replaced by

$$V = 11.55(S_s - 1)^{0.11}S^{0.39}(gR_h)^{0.5} \tag{11–13}$$

where S_s is the density ratio of solids to fluid. Equation 11–13 was derived using the concepts described in Chapter 6, i.e., proportionality between the contact load stress and the boundary shear stress. Equation 11–13 provides a second method for calculating the required slope for a given velocity in an open channel.

Wilson and Nnadi (1990) suggest that the thickness of the bed load layer or shear layer, δ_s, in a channel flow at high boundary stresses is $7.5Yd$, where Y is the Shields parameter. This layer is often thin but for the conditions of Faddick's example ($Y \sim 30.6$, $d = 0.5$ mm), it would be about 11.2 cm compared to the total depth of 47.4 cm. The particles above the bed load layer are considered to be suspended by fluid turbulence.

When a stationary bed of particles forms the lower boundary, the relationship between solids flow rate, channel slope, and the depth of the flow is of interest. There are numerous equations available, for which Yalin (1977) provides a review. Fan and Masliyah (1990) tested these equations when they examined solids deposition from flowing slurries, a situation which occurs when tailings pipelines are discharged on a beach above a pond. Theoretical predictions of the rate of deposition showed excellent agreement with experimental measurements.

11.7 PARTICLE DEGRADATION DURING FLOW

During transport, changes occur which arise from collision processes or from penetration of the liquid phase into the particles. The collisions include both particle–boundary and particle–particle interaction effects. These occur in the pipe itself, in pumps, fittings (valves, bends, tees), and during loading and discharge operations when the particles enter and leave the pipeline system. For both mechanisms, high relative velocities at collision can be assumed to increase rates of particle fracture. The friability of the particles will depend upon factors such as shape (angular corners will be eroded most easily), mechanical strength (as measured by toughness or energy absorption capacity), and the presence of flaws or cracks for fracture initiation.

Because coal is a friable material which has been considered frequently for pipeline transportation, a number of studies of the degradation process have used coal particles. Changes in particle size distribution during flow affect dewatering costs and could alter the hydraulic performance.

One of the earliest investigations (Lammers et al., 1958) examined the behavior of lignite. Its high friability was evident from size distribution changes which occurred during recirculation through a test loop 94 mm I.D. and from jar-tumbler tests. The latter were slightly modified from the ASTM test of coal friability (ASTM D441-45). In both types of test, abrasion seemed to be the dominant mechanism with particles "rounding" while generating fines. Rates appeared to increase with particle size in the tumbler tests (for which uniformity of composition between samples is achieved more easily). In the flow tests no attempt was made to distinguish degradation in the pipeline from that which occurred in the pump: changes in size distribution were expressed in terms of (velocity × time) or "distance transported." This potentially misleading independent variable implies that it was flow in the *pipeline,* rather than the rest of the system, which produced the degradation.

This deficiency was rectified in a later USBM investigation (Pipelen, 1966) using Pittsburgh Seam coal and a 150-mm I.D. pipeline system. The length of the pipeline could be varied, so that the relative importance of the pump and the pipeline could be established. The circuit included a pressurized receiver so that tests could be conducted at the same pump speed when the pipeline lengths and friction losses were substantially different. Using coal screened 50.8 × 25.4 mm, a significant amount of abrasive attrition could be attributed to flow through a pipe 266 m in length. In contrast with this tendency to generate fines in the pipeline, passage through the pump generated coarser fragments, suggesting fracture as the degradation mechanism.

In a later study using Canadian bituminous coal, Gillies et al. (1982) found that the loading process for a test pipeline also produced a significant amount of degradation. It seems likely that coal in the newly mined or crushed state is most susceptible to breakage and that this tendency decreases significantly with subsequent flow or handling. Evidence for this was also obtained from rates of degradation. These decreased with time in recirculation tests (Shook et al., 1979), with increasing equipment size or decreasing particle size. Figure 11–11 shows degradation in prolonged recirculation of a friable coal in a test loop.

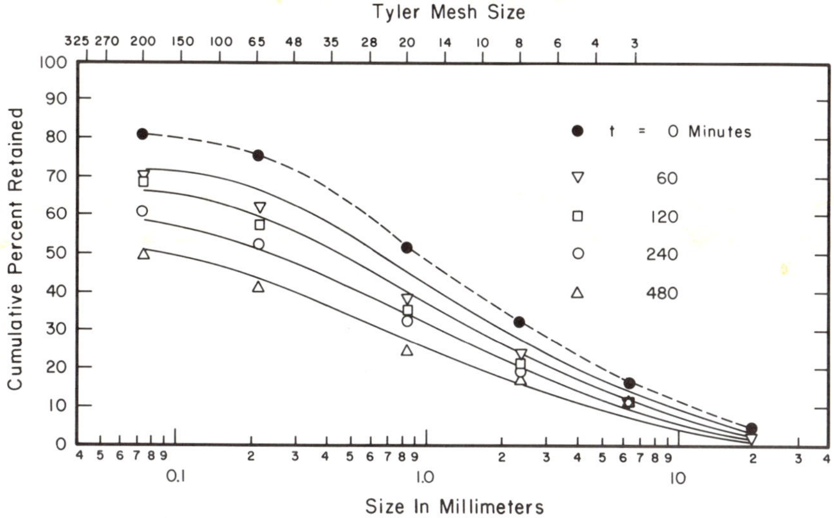

Figure 11–11. Change in particle size distribution resulting from continuous recirculation.

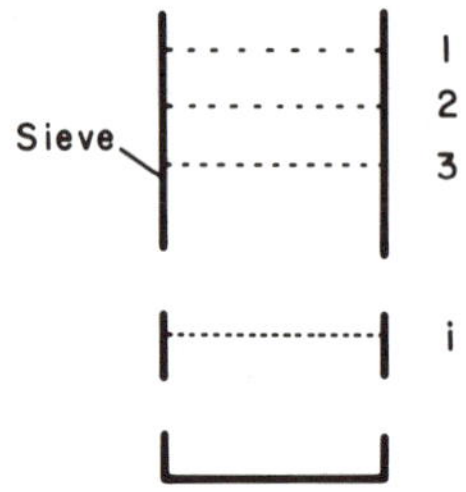

Figure 11–12. Sequence of test screens.

In order to discuss degradation quantitatively, we therefore require:

1. methods to establish the structure of the material and its size distribution;
2. a mathematical description of the degradation process.

A surface measurement technique (Mikhail et al., 1982) can provide an indication of structure and susceptibility to degradation. This was tested for friable coals of high Hardgrove grindability. It is an alternative to visual (microscopic) determination of fissuration.

The mathematical model used in grinding processes (Austin, 1971–72; Kelsall and Reid, 1965) can be employed in quantitative investigations. Assuming plug flow, the batch grinding rate equation describes the mass fraction of solids in the size intervals of a $\sqrt{2}$ series of test screens. If x_i is the fraction passing screen $i - 1$ and retained upon screen i (Figure 11–12)

$$\frac{dx_i}{dt} = \sum_{j=1}^{i-1} S_j b_{ij} x_j - S_i x_i \tag{11–14}$$

The first term on the right-hand side represents the contribution to size interval i from all the coarser fractions. S_j is the rate constant (or selection rate parameter) for the j fraction and b_{ij} is the fraction of the broken material from interval j which falls into interval i.

This model was used by Karabelas (1976) who investigated the degradation which resulted from shearing a suspension of fine coal particles in a Couette viscometer. He found the b values to be normalized, i.e., that the interval between i and j determines b_{ij}:

$$b_{i,j} = b_{i-j+1,1} \tag{11–15}$$

Cumulative breakage distributions (B_n values) are often reported. These are defined by

$$B_n = \sum_{j=i-1}^{i-n} b_{ij} \tag{11–16}$$

S values were found to be nearly proportional to particle size in these fine-particle studies. This contrasts with a stronger (approximately d^2) dependence observed in Gillies's tests (1982) of single-pass pump degradation of coarse coal.

The solid lines in Figure 11–11 were obtained by fitting the data to the model described above. Recirculation loop tests frequently show the S values of the largest particles to be insensitive to size. The fact that S values decrease exponentially with time suggests that significant "rounding" of the particles takes place with repeated recirculation. It thus appears that the degradation process shifts from an initial fracture process, determined by the immediate history of the particles, toward the "rounding" mechanism. In view of the fact that both pipeline and pump are important, loop recirculation flow tests require the pipe length to be varied to separate these effects. Evidence for the importance of the pump in producing changes in slurry viscosity, even when no change in particle size distribution occurred, was obtained by Brandis and Klose (1988) with coal–water fuels. S values for centrifugal pumps appear to increase with impeller tip speeds.

Simulations are useful in defining other phenomena of importance in degradation. For example, coal and coal waste material may contain swelling clays which are susceptible to fluid penetration. The presence of these would be reflected in the S values, but the effects upon rheology would be more profound at high slurry concentrations. Size distribution changes and their hydraulic effects were documented for colliery waste transport in a full-scale pipeline by Gies and Geller (1982). Coal–water fuels were found (Ercolani et al., 1988) to increase in viscosity with cumulative energy input per unit mass of slurry during recirculating flow tests. Qualitatively similar results were obtained with slurries stirred in a laminar flow mixer.

Slurries of soluble particles transported in saturated brines display an equilibrium size distribution in which the rate of generation of fines by abrasive degradation balances their rate of dissolution. The latter process (digestion) results from the increased solubility of very fine particles. Thus, fines dissolve and the larger particles grow through crystallization. If the solids contain a soluble impurity initially, this will disappear through leaching.

11.8 HEAT TRANSFER TO OR FROM SLURRIES

The ability of a fluid or slurry of bulk temperature T to exchange heat with a surface at T_s is expressed conveniently in terms of the heat transfer coefficient h defined in Equation 1–39:

$$\dot{q} = h(T_s - T) \tag{1–39}$$

where $\dot{q}$ is the heat flux (W/m^2) in the direction of the temperature decrease. h values for slurry-to-wall heat transfer are required for heat exchanger design.

A principal effect of the solid particles in slurries arises through the mixture heat capacity and density. Both are linear combinations of the properties of the individual phases, i.e.,

$$\rho = \rho_m = \rho_L(1 - C_r) + \rho_s C_r \tag{2-5}$$

and similarly

$$C_p = C_{pm} = C_{pL}(1 - C_r) + C_{ps} C_r \tag{11-17}$$

The density and heat capacity defined above should be used in heat balance calculations as long as the particles and fluid are at the same temperature. In calculating heat transfer coefficients, however, it is not so clear which properties should be used.

Viscosity, heat capacity, density, and thermal conductivity are known to affect the heat transfer coefficient for single-phase fluids and it seems that they should have similar effects in slurries. However, there is a difficulty because for most liquids the region at the pipe wall is even more important for heat transfer than it is in fluid flow. The particle packing considerations described in Section 1.4 and any tendency for particles to migrate away from the pipe wall would cause the appropriate properties of large-particle slurries to approach those in the fluid.

With coarse particles in turbulent flow, the relevant properties appear to be those of the carrier fluid. Figure 11–13 shows heat transfer coefficients [data of Vassie (1970)] for comparatively fine sand (S_s = 2.65) and polystyrene (S_s = 1.05) particles at concentrations to 30% by volume, correlated in terms of *fluid* properties. Both slurry flows were pseudo-homogeneous at high velocities (Reynolds numbers), but the contact load was significant for the sand at low

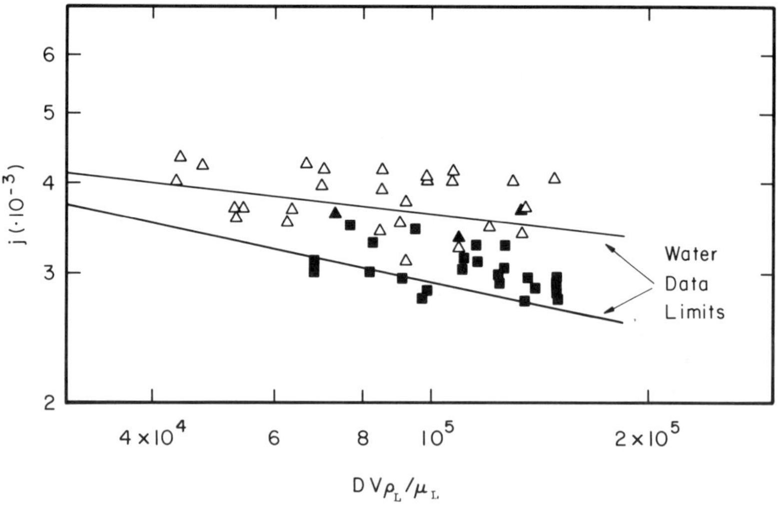

Figure 11–13. *j*-factor versus fluid Reynolds number for sand (open triangles) and 0.3-mm polystyrene beads (solid squares) slurries. (Adapted from Vassie, 1970. M.Sc. thesis.)

velocities. The heat transfer coefficients are expressed as j factors. In terms of liquid properties the j factor for heat transfer is

$$j = \left(\frac{h}{\rho_L C_{pL} V}\right)\left(\frac{\mu_L C_{pL}}{k_L}\right)^{2/3}\left(\frac{\mu_{Lw}}{\mu_L}\right)^{0.14} \tag{11–18}$$

where liquid viscosities μ_{Lw} and μ_L are evaluated at the wall and bulk slurry temperatures, respectively.

Figure 11–13 shows a small increase in the j factors, in comparison with the water values, for the sand slurries. This could reflect improved radial mixing and/or more rapid heat removal from the viscous sublayer region at the pipe wall. Similar results for a range of tube and particle diameters have been reported by Harada et al. (1985) and by Smith (1982) for vertical flows. However, temperature distributions in the core region measured by Liu et al. (1988) in turbulent slurries of nearly neutrally buoyant particles showed very poor radial mixing, although the heat transfer coefficients were within 15% of the values observed with particle-free flows.

For fine-particle slurries, Crewe and Simons (1973) found a distinct difference between the behavior of kaolin ($d < 2$ μm) and coal ($d < 30$ μm) slurries in turbulent flow. The kaolin was flocculated and the coal was unflocculated and both appeared to be Bingham fluids.

The kaolin slurry heat transfer coefficients could be correlated using the slurry density, heat capacity, thermal conductivity, and the plastic viscosity in the j-factor expression. Thermal conductivities were calculated from those of the liquid and solids using Maxwell's equation (1892):

$$\frac{k_m}{k_L} = \frac{2k_L + k_s - 2C_r(k_L - k_s)}{2k_L + k_s + C_r(k_L - k_s)} \tag{11–19}$$

The coal slurries gave j factors which were higher than the correlation which was satisfactory for the kaolin slurries. Furthermore, a plastic viscosity correction term $(\mu_{pw}/\mu_p)^{0.14}$ did not produce agreement between the heat transfer coefficients for heated and cooled slurries. At least part of the difference between the kaolin and coal slurry results can be explained in terms of the importance of the wall region and the difference between the particle diameters. With large particles and heat transfer resistances dominated by the wall region, the fluid properties are most important. As the particle size decreases the properties of the slurry, rather than those of the fluid, eventually dominate the heat transfer process.

Laminar slurry pipe flows are usually associated with high concentrations of fine particles. In these cases the slurry properties would be expected to determine the heat transfer coefficient. For laminar pipe flow of a power law fluid, the Graetz solution of the Newtonian flow problem can be reworked (Wilkinson, 1960). Defining δ as the ratio of the wall shear rate to that in a Newtonian fluid at the same mean velocity and tube diameter:

$$\delta = \frac{\dot{\gamma}_w}{(8V/D)}$$

Using the expression for $\dot{\gamma}_w$ from Equation 4–7

$$\delta = \frac{3n + 1}{4n} \tag{11–20}$$

and n' can be used for n in Equation 11–20. For laminar flow, the coefficient h_{av}, which uses the arithmetic mean value of $(T_s - T)$ in calculating q, is given by

$$\frac{h_{av}D}{k} = 1.75\delta^{1/3}\text{Gz}^{1/3}\left(\frac{m}{m_w}\right)^{0.14} \tag{11–21}$$

provided the Graetz number (Gz) is greater than 100. The Graetz number is defined as

$$\text{Gz} = \frac{(\pi D^2/4)\rho_m C_{pm} V}{k_m L}$$

The parameter m in Equation 11–21 is defined in terms of the Metzner–Reed K' value as a measure of the slurry effective viscosity

$$m = K' 8^{n'-1} \tag{11–22}$$

m_w is evaluated at the wall temperature so that the term $(m/m_w)^{0.14}$ in Equation 11–21 is an empirical temperature correction. There is some disagreement as to whether or not this correction should be used in slurry flows.

At lower Graetz numbers, heat transfer coefficients can be predicted by assuming a velocity distribution (flat or parabolic). A flat distribution would be likely to occur near the inlet to a pipe. Since the velocity distributions for shear thinning slurries are flatter than those of Newtonian fluids, this theoretical prediction is likely to apply to shear thinning slurries. The correlation can be found in standard heat transfer textbooks (Bennett and Myers, 1982).

At very low Graetz numbers (less than about 5), the discharge bulk temperature becomes equal to the wall temperature. In this limiting case, therefore

$$\frac{h_{av}D}{k_m} = \frac{2}{\pi}\text{Gz} \tag{11–23}$$

11.9 SYSTEM LAYOUT

The method used for introducing a slurry to the pipeline always deserves careful consideration. If dry solids are to be added to a fluid, this can be done in a sump tank open to the atmosphere. The sump provides a place for mixing and for air to escape from the slurry before it enters the pump. Stirring is often necessary to break up agglomerates if the solids are fine, have a broad size distribution, or are difficult to wet. With coarse, easily wetted particles, stirring probably cannot produce a uniform tank composition. Although a recirculated slurry stream can provide considerable mixing, baffles may be required to prevent vortexing in the sump.

Agitated flat bottom tanks can be used for slurries of very fine particles but if there is a high settling tendency, unagitated conical bottom sumps are preferable. To avoid tramp metal in metallurgical slurry pumping, Crisswell (1982) suggested that the bottom of the suction pipe should be at least one-quarter of the pipe diameter above the bottom of a flat bottom sump. Suction pipes should be as short as possible and preferably vertical or sloping downward to the pump inlet. An isolation valve and an easily disconnected coupling are desirable in the suction line.

To ensure sufficient suction head, deposition in the suction lines should be avoided at all costs. Deposition velocity equations or nomograms used for long pipes are probably conservative for selecting the diameter of short suction pipes because swirling flow produced by an elbow helps to prevent deposition from taking place in short horizontal lines.

If the slurry could possibly develop a yield stress (as a result of concentration fluctuations, variations in particle size distribution or settling during periods of shutdown), it may be desirable to provide a positive feeding device at the sump discharge to assist flow into the pump. Etchells (1986) suggests that, for slurries with high shear rate viscosities of 100 mPa s or less, and densities in the range from 1000 to 2000 kg/m^3, this difficulty becomes important at yield stresses greater than 30 to 40 Pa.

Lee (1977) suggested a sump volume equal to 1 minute of flow with a typical sump depth of 2.5 m for slurries which do not contain large quantities of gas. With particles to which gas bubbles may have become attached, such as flotation cell overflows, much larger sump volumes would be required. Crisswell suggested a volume between 1 and 2 minutes of flow and provided the table of sump dimensions shown in Figure 11–14. Particle deposition in the sump becomes more likely with large sumps and if substantial deposits of solids form in the sump, changes of inlet flow rate can cause changes in the concentration and density of the slurry entering the pump. This is, of course, highly undesirable.

Since wear is unlikely to be identical in pumps operating in parallel and feeding the same pipe, their characteristics are unlikely to remain the same with time. It would therefore be necessary to provide compensating speed control for both pumps if this method of operation were to be used. Crisswell suggested that use of separate parallel pipelines is probably preferable.

If a pipeline requires more than one pumping station, the minimum cost is associated with equally spaced stations (Yucel, 1984). Yucel provides a simplified design procedure to be used when the topography of the pipeline route is known. Sumps are not normally employed at the downstream pumping stations.

Valve selection for slurries is dominated by wear considerations and the need for reliable performance in emergencies. With very fine particles, and especially

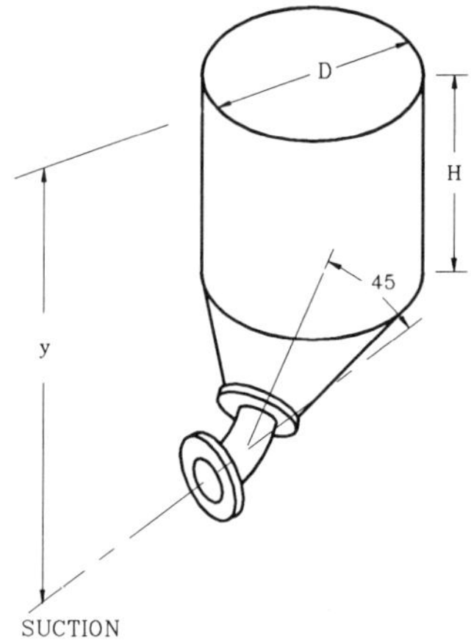

SUMP DIMENSIONS

Capacity, Q(l/s)	7	15	30	50	75	100	200
Diameter, D (mm)	765	915	1220	1525	1675	1980	2440
Height, H (mm)	1065	1220	1375	1830	1830	1980	2440
y (mm)	1830	2135	2540	3305	3430	3885	4725

Figure 11–14. Sump dimensions. (From Crisswell, 1982. *Proc. Hydrotransport 8 Conf.*, p. 334. Adapted with permission.)

with small pipes, diaphragm valves can be used. They have the advantage that the particles are separated from the movable stem so that binding after long periods of inaction does not occur. Pneumatically or hydraulically actuated ball valves, tapered plug valves, pinch valves, or knife gate valves are used for isolation purposes in larger pipes. Wherever possible, valves should open and close in clear fluid. The open area should be as close as possible to that of the pipe to avoid deposition and minimize erosion.

If a flow must be throttled for control purposes, pinch valves or symmetrically closed iris-diaphragm valves can be used for flows which contain coarse particles.

Pump speed control is the preferred method for flow rate adjustment. If centrifugal pumps are connected in series, speed control is not normally used with the first pump. This avoids NPSH difficulties with the second pump. However, Crisswell (1982) suggests that speed control on the first pump makes startup easier.

With sloped pipelines, Lee suggests that downward slopes for long lines should be limited to 0.5 degrees to minimize the problems arising from air in the slurry. With upward slopes, he suggests 4 degrees to reduce the likelihood of occasional coarse material collecting on upslopes. These criteria are conservative but they also reflect the way in which design is determined by possible variations in flow conditions.

Occasional oversize particles or objects such as hand tools can pass through pipelines but they are likely to lodge in valves, bends, or pumps. The additional cost of long radius bends is usually justified by the reduced energy consumption and wear as well as their ability to pass such objects. Centrifugal pumps discharging horizontally are preferable to ones discharging vertically in this respect.

Because they cannot provide the smooth walls and gradual transitions which are necessary to minimize impingement wear, screwed fittings are almost invariably undesirable. Flanged or quickly disconnected butt-joint couplings are required at high wear points.

Centrifugal pumps can usually be restarted with some settled solids in them but if the size distribution of the particles is broad and/or if the slurry concentration is high, the starting torque may become excessive. A flushing system is desirable so that the solids can be removed from the pump and replaced with clear fluid.

Horizontal or gently inclined pipelines can be restarted with settled deposits of solids in them without difficulty as long as there is a clear path for the fluid above the deposit. Restart takes place smoothly and quickly as long as the final velocity is above that which produces deposition.

Occasionally a complete blockage forms, such as upstream of an object wedged in a valve or bend. After the originating obstruction has been removed, a clear fluid path through the particle plug can be produced by a high-pressure fluid source connected upstream of the blockage. The flow rate required to form the path need not be high but it may take some time for the communicating path to form.

If the particles are coarse enough for the slurry to have a significant deposition velocity, the pipeline should be filled with the carrier fluid before slurry is admitted. A settling slurry is likely to form a blockage if it enters an empty pipe.

Flow rate and sump level are the variables which are usually sensed for control purposes; pump speed and diluent flow rate to the sump are the manipulated variables. Pump gland fluid flow rate, pressure rise across the pump, pressure difference between gland and pump outlet, power consumption, slurry flow rate, and slurry concentration should be measured to allow the performance of the system to be monitored.

Although pipelines are normally designed to avoid deposition, they can be operated with stationary deposits for considerable periods of time provided changes in flow rate and/or solids concentration occur slowly. Pipelines are much more difficult to control with a deposit present, however, Although the design equations quoted in Chapters 4 and 5 are useful for estimates and for extrapolation, Lee's comment on the value of test data is noteworthy. "The cost of building a line long enough to require two or more pumps in series can be considerable and the design should be based on reliable data. The expense of having a test done at a commercial laboratory, using a loop with a bore as close as feasible to the

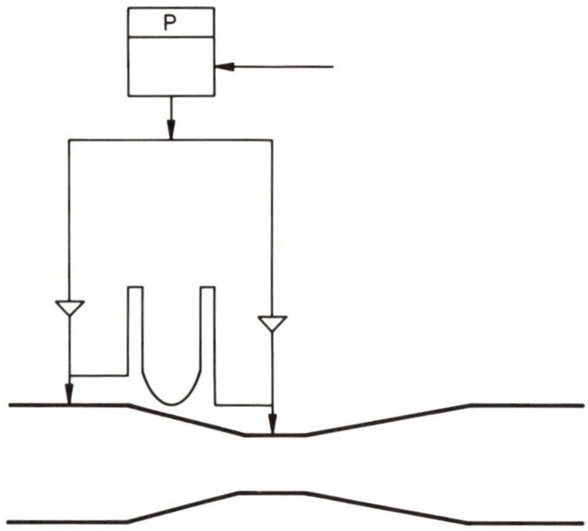

Figure 11–15. Continuous flushing system for pressure differential measurements.

anticipated full-scale bore, is money well spent." Sabbagha (1982) offers a similar opinion.

If a pipeline operates at constant solids concentration, pressure surge phenomena can be described by the homogeneous mixture equations given in Section 2.12, unless stratification is pronounced or the concentration approaches the maximum packing value. Surges generated by concentration variations are described in Section 11.10.

Instruments which rely on pressure difference measurements are subject to errors which result from blockage of pressure transmission lines. Flows with fluctuating concentrations are particularly susceptible to these difficulties. A backflush system that uses a reservoir of the clear carrier fluid at a high pressure P is shown in Figure 11–15. The pressure is reduced to produce a controlled continuous flow through both sensor tubes. Identical orifices of very high fluid resistance in these sensor lines provide identical backflush flows through them, despite their different downstream pressures. The difference between the pressures in these sensor lines is then the measured variable.

To minimize the possibility of air entering the pressure sensing lines, tapping holes through the pipe wall should always be located at the side (3 o'clock or 9 o'clock) of the pipe or venturi meter rather than at the top.

11.10 SCALE DEPOSITS IN SLURRY PIPES

The formation of deposits on the interior walls of slurry pipes may be caused by mechanical or chemical deposition mechanisms and can have significant effects on the operation of the pipe system. Particle deposition (at velocities below V_c) and wall adhesion (which may occur above V_c) are the simplest mechanisms, and consolidation of the deposits, by crystallization or chemical reactions, is particularly undesirable. In some situations the deposition rate may be related to the fluid temperature and temperature variation along the pipe, as well as the pipe wall temperature. Flotation plant installations at mining sites and ash hydrotransport pipelines from coal-burning electric power stations are typical slurry flow applications where deposits are encountered frequently.

We illustrate the scale deposition process with experimental data from an ash–water slurry pipeline of diameter 305 mm (Roco, 1982). The deposition rate was a function of ash and water chemistry as well as of the thermal regime during transport. Two kinds of deposits were observed: carbonates and silica. The rate of calcium carbonate $CaCO_3$ deposition was a function of the carbonate ion concentration in the water and of the contents of $CaCl_2$, CaO, and MgO_2 in the ash. The silica deposits resulted from mixing the high-temperature ash with water at about 15 °C. The resulting mixture had a temperature higher than the surroundings. During transport, the mixture cooled to become supersaturated with respect to silica, which deposited on all surfaces in contact with the mixture. The deposition process intensifies on rough surfaces where flow separation and reattachment occur. The deposits form quickly in winter. In the test case, which is representative for ash–water slurry pipelines, the deposit thickness measurements (see Figures 11–16 and 11–17) along the pipeline were correlated to an exponential law, except for the first 100 m of pipe:

$$g_d = g_{d0} \exp(-k\Delta L) \qquad (11\text{–}24)$$

where g_{d0} is the deposit thickness at $L_0 = 100$ m, and $k = 3.5\ \text{km}^{-1}$ is a constant specific to the system. The thickness was measured with a gamma ray apparatus and checked by direct cuts in the pipe. The time rate of scaling at a given location along the pipe decreased with time according to a quasi-exponential law. The inner surface exhibited a macroroughness Δ (mm) in Figure 11–16c which increased with the deposit thickness, g_d (mm)

$$\Delta = 0.3 \exp(0.4 g_d) \qquad (11\text{–}25)$$

Deposits are the result of the competing processes of deposition and removal. In some situations there are deposits at the top of the pipe while wear occurs at the bottom. Deposit adhesion to the pipe wall is a function of the pipe material and a typical prevention method is the continuous addition of inhibitor [as proposed, for instance, by ARMCO (1983)]. Pipe cleaning can be performed mechanically with a turbine traveling along the pipe or with high-pressure water jets. Alternatively, chemical treatment can be used with solutions recirculated in a closed circuit.

Figure 11–18 shows the effect of pipe scaling and pump wear on the mixture flow rate (Q) and pumping head (H). ΔH is the elevation increase along the pipeline. The operation point may take different positions within the domain defined by P_0, P_1, P_2, and P_3. All of these positions are characterized by flow rates smaller than the initial flow rate corresponding to P_0. At the same time, the

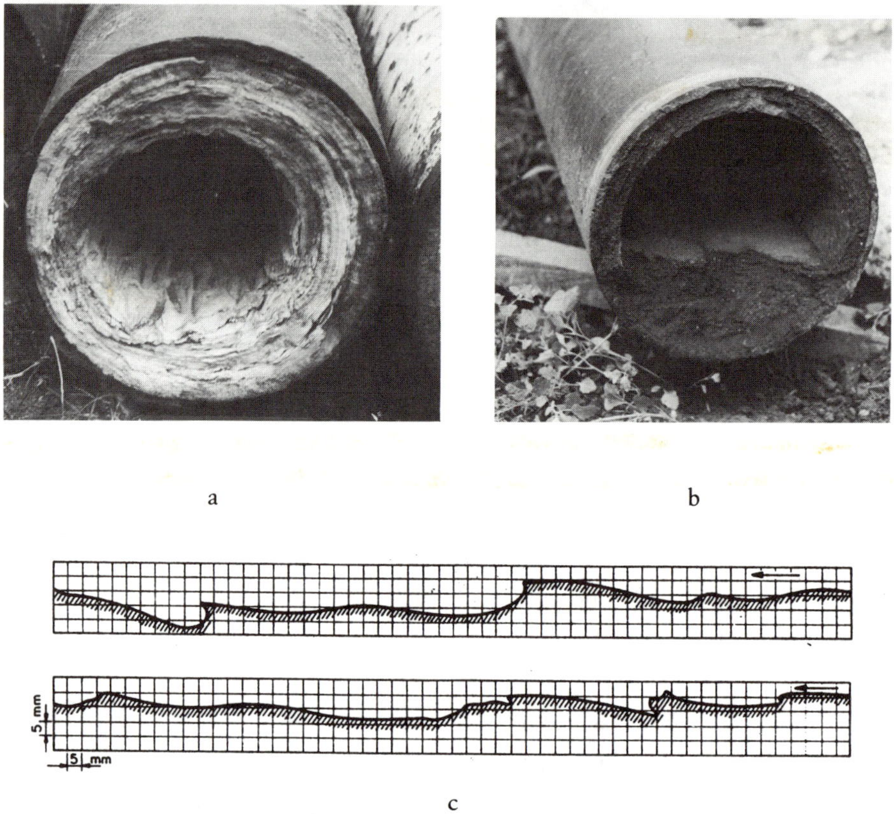

Figure 11–16. Deposits in a 305-mm pipe. (a) Deposits (about 90 mm in thickness), (b) deposits and sediments, (c) two longitudinal profiles along the deposits. (From Roco, 1982. *Proc. Hydrotransport 8 Conf.*, pp. 510, 511.)

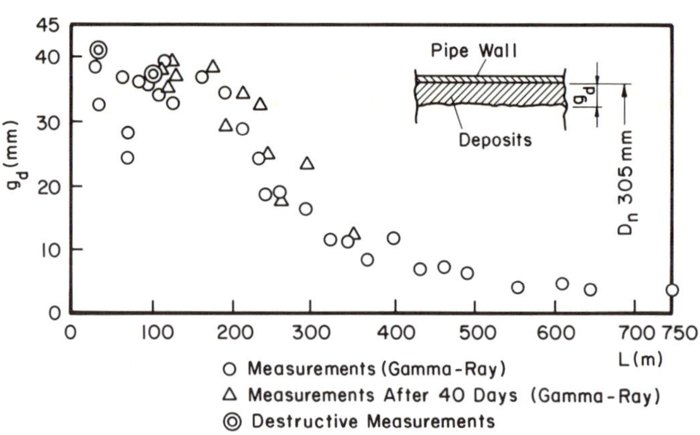

Figure 11–17. Variation of the deposit thickness along the pipe. (From Roco, 1982. *Proc. Hydrotransport 8 Conf.*, p. 510.)

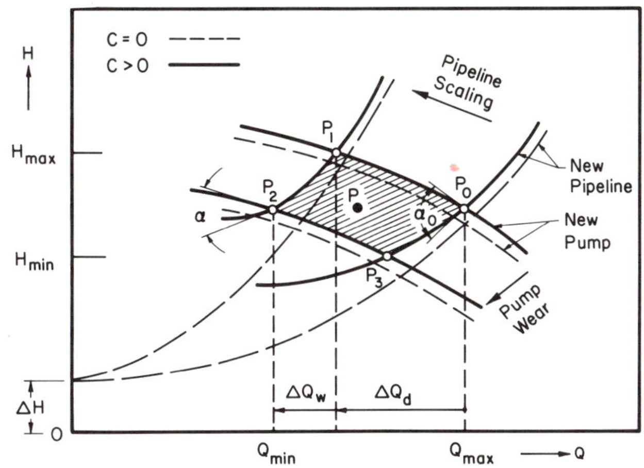

Figure 11–18. Effect of pipe deposits and pump wear on the operating point *P*. (From Roco, 1982. *Proc. Hydrotransport 8 Conf.*, p. 510.)

angle α, between the pump and pipeline curves, showing the stability of the operation, decreases when the point *P* moves to the left. It reaches a minimum at P_2 where a deposit is present and the pump impeller requires replacement. The dashed line in Figure 11–18 corresponds to operation with water ($C = 0$) and the solid line to operation with slurries (C).

11.11 EFFECT OF CONCENTRATION VARIATION OF PUMP–PIPE SYSTEMS

It is known that a variation in concentration has an effect on both pump and pipe characteristic curves. The pump characteristic curve changes immediately (in a fraction of a second) after a concentration variation at the pump suction. In contrast, the pipe characteristic curve responds slowly as a function of the average concentration in the pipe. The time scale of the concentration change in the pipe (Δt_{pipe}) is the ratio of the pipe length to the mean velocity. The difference between the time scales of response of the two curves may create dangerous transitory phenomena in the pump–pipe system. Figure 11–19 illustrates three typical situations measured on a 3-km-long ash slurry pipeline (Roco, 1978) when (a) a sudden variation of concentration occurs; (b) a progressive variation takes place, in a time about Δt_{pipe}; and (c) waves of variable amplitude and frequency occur. The last situation is often encountered in dredging and pit-operation systems. The solid lines correspond to a concentration C_0 and the dashed lines to $C_1 > C_0$. The pump and pipe curves are expressed in meters of water (m_w).

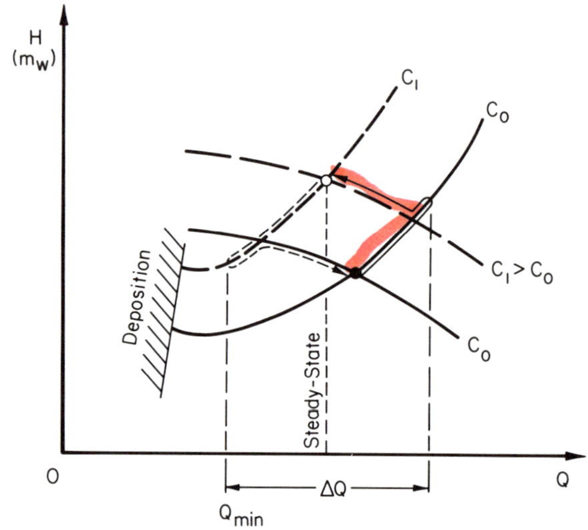

a

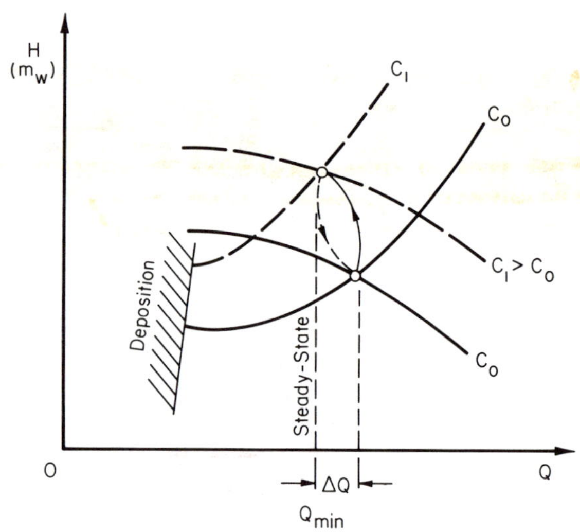

b

The effect of concentration variation can be evaluated with several criteria: the change of flow rate (ΔQ in Figure 11–19) caused by the concentration variation, the minimum flow rate during the transitory period (Q_{min}), and the pressure surge ($\Delta P_{max}/P_{steady}$). The change in mixture flow rate affects the solids transport rate and the pressure distribution in the pipe. Q_{min} is important for the

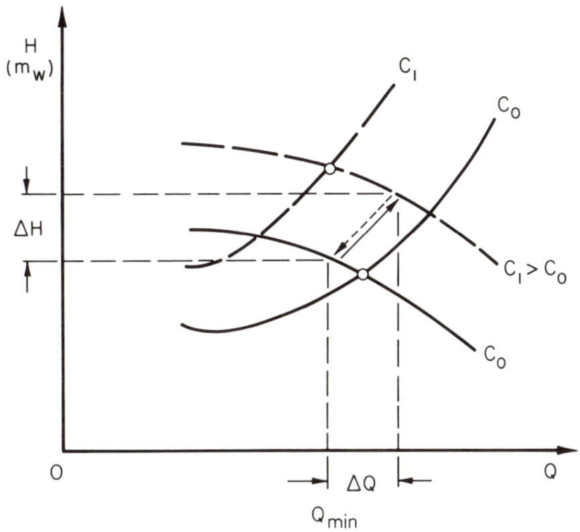

c

◄ **Figure 11–19.** Effect of concentration variation on the operating point (Q, H). (From Roco, 1982. *Proc. Hydrotransport 8 Conf.*, p. 760.)

operating stability of the system and should be greater than the deposition limit, as shown in Figure 11–19a. The sudden change of pump head produces a pressure surge along the pipeline in the same way as a sudden pipe blockage. Similar transitory effects may be produced by variations of particle size (d) or density (ρ_s) during operation.

The overshoots in Figure 11–19a are due to the inertia of the impeller, coupling system, and motor.

Appendix 1
Microscopic Equations of Motion for Homogeneous Media

CONSERVATION OF MASS

Assuming matter to be continuous, we can evaluate the terms in the mass balance for a control volume: Input = Output + Accumulation as

$$\frac{\partial \rho}{\partial t} + \nabla \cdot (\rho \mathbf{V}) = 0 \tag{A1–1}$$

where V is the velocity vector

$$\mathbf{V} = \mathbf{i}_1 v_1 + \mathbf{i}_2 v_2 + \mathbf{i}_3 v_3$$

In fact, the derivation is done most easily in Cartesian coordinates (shown above) and converted to other systems with a coordinate transformation. The operator $\nabla \cdot$ is expressed in terms of the metric coefficients, h_i, of the coordinate system. These relate elemental distances, dl_i, measured in the coordinate directions, to elemental increments in these coordinates:

$$dl_i = h_i \, dx_i$$

The generalized expression for the divergence is then

$$\nabla \cdot \mathbf{V} = \left(\frac{1}{h_1 h_2 h_3}\right)\left(\frac{\partial(v_1 h_2 h_3)}{\partial x_1} + \frac{\partial(v_2 h_1 h_3)}{\partial x_2} + \frac{\partial(v_3 h_1 h_2)}{\partial x_3}\right)$$

whereas the operator ∇ is

$$\left(\frac{\mathbf{i}_1}{h_1}\right)\frac{\partial}{\partial x_1} + \left(\frac{\mathbf{i}_2}{h_2}\right)\frac{\partial}{\partial x_2} + \left(\frac{\mathbf{i}_3}{h_3}\right)\frac{\partial}{\partial x_3}$$

In the Cartesian coordinate system ($x_1 = x$, $x_2 = y$, $x_3 = z$) the metric coefficients are all unity so that $\nabla \cdot \mathbf{V}$ can be written as $\Sigma\, \partial v_i/\partial x_i$ or, more simply, $\partial v_i/\partial x_i$ if the repeated subscript implies summation, as it does in tensorial notation.

The cylindrical coordinate (r,θ,x) system is useful for many situations of interest in slurry transport. In this case ($x_1 = r$, $x_2 = \theta$, $x_3 = x$), the expressions given above may be used with $h_1 = 1$, $h_2 = r$, $h_3 = 1$.

CONSERVATION OF MOMENTUM

Here we use the sign convention that a positive surface is one whose normal, in the direction of a coordinate axis, points *into* the control volume. A positive stress is one which acts in a positive coordinate direction on a positive surface or in a negative coordinate direction on a negative surface. For example, in Figure A1–1, τ_{xx}, τ_{yx}, and τ_{zx} act on positive surfaces.

A momentum balance for the control volume, assuming that there is no momentum generation, yields the vector equation

$$\frac{\partial(\rho \mathbf{V})}{\partial t} + \nabla \cdot \rho \mathbf{V}\mathbf{V} = -\nabla P + \rho \mathbf{b} - \nabla \cdot \mathbf{T} \tag{A1–2}$$

b is the body force (gravitational, electrostatic, or electromagnetic) per unit volume. If gravity is the only body force, $\mathbf{b} = -g\nabla h$ where h is the elevation above a datum and g is the (downward) acceleration due to gravity. **T** is the stress tensor, with nine components of the form τ_{ij}. The first subscript denotes the surface on which the stress component acts (a surface of constant i). The second subscript denotes the direction of the component. The stress components with i not equal to j are the shearing stresses.

The term on the left-hand side can be written as $\rho(\partial\mathbf{V}/\partial t + \mathbf{V} \cdot \nabla\mathbf{V})$ or $\rho\, D\mathbf{V}/Dt$, where D/Dt is the substantial derivative (following the fluid motion).

In Cartesian coordinates, the momentum equation can be written, using the convention that a repeated subscript implies summation,

$$\rho\left(\frac{\partial v_i}{\partial t} + v_j \frac{\partial v_i}{\partial x_j}\right) = -\frac{\partial P}{\partial x_i} - \rho g \frac{\partial h}{\partial x_i} - \frac{\partial \tau_{ji}}{\partial x_j} \tag{A1–3}$$

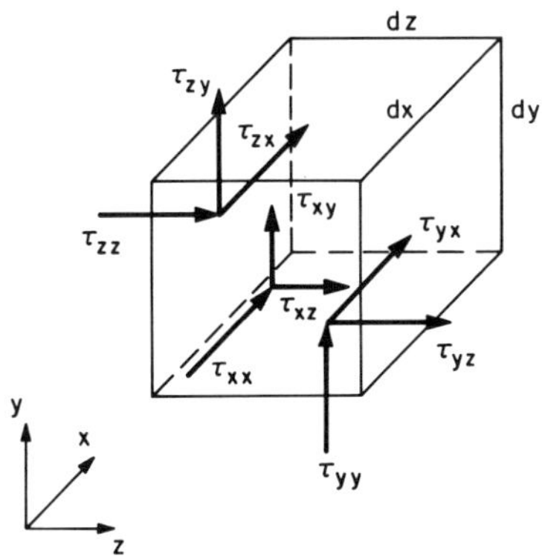

Figure A1–1. Differential control volume.

In cylindrical coordinates, the momentum equations are of interest. For the r-direction,

$$\rho\left[\frac{\partial v_r}{\partial t} + \frac{\partial v_r}{\partial r} + \frac{v_\theta}{r}\frac{\partial v_r}{\partial \theta} - \frac{v_\theta^2}{r} + v_x\frac{\partial v_r}{\partial x}\right] = -\frac{\partial P}{\partial r} - \rho g\frac{\partial h}{\partial r} - \left[\frac{1}{r}\frac{\partial(r\tau_{rr})}{\partial r} + \frac{1}{r}\frac{\partial \tau_{\theta r}}{\partial \theta} - \frac{\tau_{\theta\theta}}{r} + \frac{\partial \tau_{xr}}{\partial x}\right] \tag{A1–4}$$

For the θ-direction

$$\rho\left[\frac{\partial v_\theta}{\partial t} + v_r\frac{\partial v_\theta}{\partial r} + \frac{v_\theta}{r}\frac{\partial v_\theta}{\partial \theta} + \frac{v_\theta v_r}{r} + v_x\frac{\partial v_\theta}{\partial x}\right] = -\frac{1}{r}\frac{\partial P}{\partial \theta} - \frac{\rho g}{r}\frac{\partial h}{\partial \theta} - \left[\frac{1}{r^2}\frac{\partial(r^2\tau_{r\theta})}{\partial r} + \frac{1}{r}\frac{\partial \tau_{\theta\theta}}{\partial_\theta} + \frac{\partial \tau_{x\theta}}{\partial x}\right] \tag{A1–5}$$

For the x-direction

$$\rho\left[\frac{\partial v_x}{\partial t} + v_r\frac{\partial v_x}{\partial r} + \frac{v_\theta}{r}\frac{\partial v_x}{\partial \theta} + v_x\frac{\partial v_x}{\partial x}\right] = -\frac{\partial P}{\partial x} - \rho g\frac{\partial h}{\partial x} - \left[\frac{1}{r}\frac{\partial(r\tau_{rx})}{\partial r} + \frac{1}{r}\frac{\partial \tau_{\theta x}}{\partial \theta} + \frac{\partial \tau_{xx}}{\partial x}\right] \tag{A1–6}$$

These are the equations used to analyze tube and concentric cylinder flows.

CONSERVATION OF ENERGY

An energy balance, considering the specific internal energy (U), the kinetic energy of the flowing fluid, thermal conduction, and work done against gravity and surface forces yields the scalar equation

$$\rho\left(\frac{\partial U}{\partial t} + \mathbf{V} \cdot \nabla U\right) = \nabla \cdot (k\nabla T) - P\,\nabla \cdot \mathbf{V} - (\mathbf{T}:\nabla\mathbf{V}) \qquad \text{(A1–7)}$$

The left-hand term can be interpreted as the rate of accumulation of internal energy U for an element moving with the fluid. The next two terms represent energy input by thermal conduction and by compression. The final term is the rate of energy dissipation by viscous stresses per unit volume of fluid.

STRESSES IN FLUIDS

The deviator stress tensor **T** depends on the fluid deformation tensor **Δ**, defined as

$$\boldsymbol{\Delta} = (\nabla\mathbf{V} + \nabla\mathbf{V}^{\mathrm{T}}) \qquad \text{(A1–8)}$$

The components of **Δ** are sometimes denoted as $-\dot{\gamma}_{ij}$ and can be visualized as comprising the time rate of shear strain at a point.

For a Newtonian liquid, the relationship involves the viscosity

$$\mathbf{T} = -\mu\boldsymbol{\Delta} \qquad \text{(A1–9)}$$

as long as we can neglect density changes. In a non-Newtonian fluid the effective viscosity η can be used:

$$\mathbf{T} = -\eta\boldsymbol{\Delta} \qquad \text{(A1–10)}$$

where η is a function of the rate of energy dissipation $-(\mathbf{T}:\nabla\mathbf{V})$ as well as temperature, time, and possibly pressure. The quantity $-(\mathbf{T}:\nabla\mathbf{V})$ is rather inconvenient for use in non-Newtonian rheological equations of state and instead, we use $0.5(\boldsymbol{\Delta}:\boldsymbol{\Delta})$ or $0.5(\mathbf{T}:\mathbf{T})$.

For viscometric flows these quantities are not complex. The following table summarizes them for tube and concentric cylinder geometry.

Flow	$0.5(\boldsymbol{\Delta}:\boldsymbol{\Delta})$	$0.5(\mathbf{T}:\mathbf{T})$
Concentric cylinders	$\left(r\dfrac{d(v_\Theta/r)}{dr}\right)^2$	$\tau_{r\Theta}^2$
Tube	$\left(\dfrac{dv_x}{dr}\right)^2$	τ_{rx}^2

Comparing these expressions with the momentum equations in these geometries, we see that the effect of a variation of η is to create a more complex relationship between the nonzero components of **T** and **Δ**. The products $0.5(\mathbf{\Delta}:\mathbf{\Delta})$ are simply $\dot{\gamma}_{ij}^2$ in these flows.

A three-parameter expression for η is the Herschel–Bulkley (1926) yield-power law:

$$\eta = K|0.5(\mathbf{\Delta}:\mathbf{\Delta})|^{(n-1)/2} - \frac{\tau_y}{|0.5(\mathbf{\Delta}:\mathbf{\Delta})|^{1/2}} \qquad \text{(A1–11)}$$

provided $0.5(\mathbf{T}:\mathbf{T}) > \tau_y^2$. The parameter τ_y is the yield stress, which must be exceeded if shear is to occur. In Newtonian fluids, $K = \mu$, $n = 1$, and $\tau_y = 0$. In Bingham fluids, K is the plastic viscosity μ_p and $n = 1$. In simple power-law fluids, τ_y is zero.

Appendix 2

Microscopic Equations of Motion for Fluid-Particle Mixtures

AVERAGING

With single-particle motion, it is only necessary to consider the solid–liquid interaction but in slurries of finite concentration, particles can interact with each other or with fluid whose motion has been disturbed by the presence of other particles. The solid particles are distributed in space and time and the properties of the flow are expressed conveniently in terms of averages over time and mass or space. These averages are useful because they often can be assumed to be continuous functions for purposes of analysis. They can also be measured.

In terms of the particle diameter d and the mean flow length scale L, the *intrinsic mass average* of the flow property p for the flow component (solid or liquid) k is

$$\langle p_k \rangle = \frac{1}{\rho_k c_k \Delta v} \int_{\Delta v} p \rho_k \, K \, dv \tag{A2–1}$$

ρ_k and c_k are the density and the volume fraction of the phase and Δv is the averaging volume (Figure A2–1) for which $d \langle\langle \Delta v^{1/3} \langle\langle L$. K is the phase distribution function which is unity if the particular phase is present at the given location at the particular instant, and zero otherwise. If ρ_k is constant, mass and volume averaging are identical.

For single-phase turbulent flow, the mean flow equations were obtained by time averaging. Using the same definitions, but allowing for the fact that a

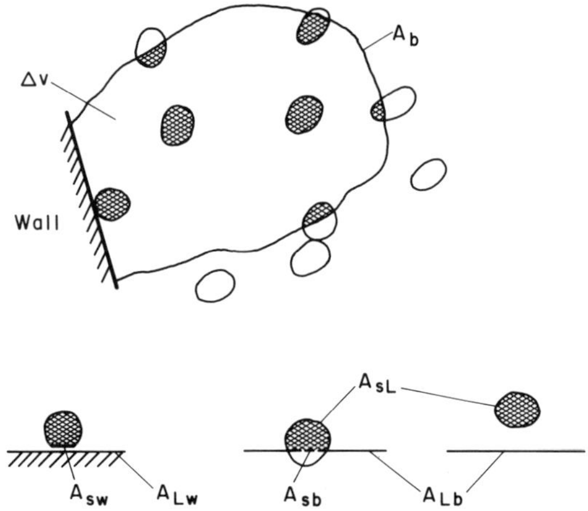

Figure A2–1. Averaging volume (Δv).

mixture exists, we have the *mass–time average* for component k over the time interval Δt:

$$\bar{p} = \frac{1}{\Delta t}\int_t^{t+\Delta t} \langle p_k \rangle \, dt \qquad \text{(A2–2)}$$

Other methods of averaging have been employed, notably "bulk averaging" or "ensemble averaging" (Batchelor, 1970; Buyevitch, 1971). Before considering the equations of motion, we introduce some definitions. The concentration by volume, c_k, and concentration by area c_k^a are defined in terms of the phase distribution function K:

$$c_k = \frac{1}{\Delta v}\int_{\Delta v} K \, dv = \frac{\Delta v_k}{\Delta v} \qquad \text{(A2–3)}$$

$$c_k^a = \frac{1}{A_b}\int_{A_b} K \, da = \frac{A_{kb}}{A_b} \qquad \text{(A2–4)}$$

where A_b is the area bounding Δv and A_{kb} is the portion of A_b which intersects phase k (Figure A2–1). The interfacial area per unit mixture volume a_{sL} is obtained from the total interfacial area A_{sL} within Δv as

$$a_{sL} = \frac{A_{sL}}{\Delta v} \qquad \text{(A2–5)}$$

The *intrinsic area average* of the flow property p is

$$\widetilde{p}_k = \frac{1}{A_b} \int_{A_b} K \mathbf{p}_k \cdot \mathbf{n}_b \, da \qquad \text{(A2–6)}$$

where $\mathbf{n}_b$ is the unit surface normal and $\widetilde{p}_k$ is an order lower than $\mathbf{p}_k$ on the tensorial scale.

By averaging the conservation equations, for flows with no mass transfer and moderate velocity variations with position, one obtains (Figure A2–1) for the solid phase ($k = s$, $c_s = c$):

$$\frac{\partial (c\rho_s)}{\partial t} + \nabla \cdot (c\rho_s \langle \mathbf{V}_s \rangle) = 0 \quad (\text{kg/m}^3 \text{ mixture}) \qquad \text{(A2–7)}$$

$$\frac{\partial (c\rho_s \langle \mathbf{V}_s \rangle)}{\partial t} + \nabla \cdot (c\rho_s \langle \mathbf{V}_s \rangle \langle \mathbf{V}_s \rangle)$$

$$= -\nabla (c^a P) + c\rho_s \mathbf{b} + \mathbf{i}_{sL} - \nabla \cdot \mathbf{T}_s + c^a \mathbf{f}_{sw} \quad (\text{N/m}^3 \text{ mixture}) \qquad \text{(A2–8a)}$$

P is the fluid pressure and $\mathbf{i}_{sL}$ is the diffusive flux of momentum on the solid–liquid interface within a unit control volume (Gray, 1975). It is convenient to define a liquid–solid interaction force $\mathbf{f}_{sL}$ as

$$c^a \mathbf{f}_{sL} = \mathbf{i}_{sL} \qquad \text{(A2–9)}$$

$c^a \mathbf{f}_{sw}$ is the diffusive flux of momentum resulting from interaction at the solid–wall interface A_{sw}. $\mathbf{T}_s$ is an *effective stress*, the solids stress tensor averaged over the whole of A_b. We define an interparticle force $\mathbf{f}_{ss}$, per unit volume of solids, as

$$c^a \mathbf{f}_{ss} = -\nabla \cdot \mathbf{T}_s \qquad \text{(A2–10)}$$

Simplifying the notation for intrinsic averaging: ($\langle \mathbf{V}_s \rangle \to \mathbf{V}_s$), with $c^a \to c$, the momentum equation for the solids becomes

$$\frac{\partial (c\rho_s \mathbf{V}_s)}{\partial t} + \nabla \cdot (c\rho_s \mathbf{V}_s \mathbf{V}_s) = -\nabla (cP) + c\rho_s \mathbf{b} + c(\mathbf{f}_{sL} + \mathbf{f}_{ss} + \mathbf{f}_{sw}) \qquad \text{(A2–8b)}$$

The corresponding liquid phase momentum equation is

$$\frac{\partial [(1 - c)\rho_L \mathbf{V}_L]}{\partial t} + \nabla \cdot [(1 - c)\rho_L \mathbf{V}_L \mathbf{V}_L]$$

$$= -\nabla [(1 - c)P] + (1 - c)\rho_L \mathbf{b} + (1 - c)(\mathbf{f}_{Ls} + \mathbf{f}_{LL} + \mathbf{f}_{Lw}) \qquad \text{(A2–11)}$$

Here, $\mathbf{f}_{Ls}$ is the interaction force acting on unit volume (N/m^3 liquid) caused by the solid phase. Since the net interfacial interaction force per unit volume of the mixture is zero, we have

$$c^a \mathbf{f}_{sL} + (1 - c^a)\mathbf{f}_{Ls} = 0 \tag{A2–12}$$

$\mathbf{f}_{LL}$ is defined from the fluid stress tensor acting at area A_{Lb}. For symmetry with the particle equations we define a fluid effective stress $\mathbf{T}_L$ as

$$(1 - c^a)\mathbf{f}_{LL} = -\nabla \cdot \mathbf{T}_L \tag{A2–13}$$

Appendix 3
Useful Data

1. Viscosity of Water*

$\mu = 0.1/\{2.1482\ \{(T - 8.435) + [8078.4 + (T - 8.435)^2]^{0.5}\} - 120\}$

where μ in Pa s, T in °C.

2. Density of water ($100 > T > 5$°C) in kg/m^3 ± 0.1.

$\rho = 999.7 - 0.10512(T - 10) - 0.005121(T - 10)^2 + 0.00001329(T - 10)^3$

3. Particle Size Table

U.S. Mesh	*Microns*	*Tyler Mesh*
4	4760	4
5	4000	5
6	3360	6
7	2830	7
8	2380	8
10	2000	9
12	1680	10
14	1410	12
16	1190	14
18	1000	16
20	841	20
25	707	24
30	595	28
35	500	32
40	420	35

*Bingham, 1922, from Perry, J.H. (Editor) 1963. *Chemical Engineers' Handbook*, 4th edition. New York: McGraw-Hill, 1963.

(*Particle Size Table* Continued)

U.S. Mesh	*Microns*	*Tyler Mesh*
45	354	42
50	297	48
60	250	60
70	210	65
80	177	80
100	149	100
120	125	115
140	105	150
170	88	170
200	74	200
230	62	250
270	53	270
325	44	325
400	37	400
500	31	

Appendix 4

BASIC Program for Two-Layer Model

The relationship between the symbols used in Chapter 6 and those employed in the program is shown below.

Symbol in Chapter 6	*Symbol in Program*
ρ_s	rs
ρ_L	rf
ρ_1	r1
ρ_2	r2
ρ_m	rm
D	dpipe
d	dp
g	9.8
$(-dP/dx)/\rho_L g$(upper layer)	p1
$(-dP/dx)/\rho_L g$(lower layer)	q1
μ_L	visl
$\tau_1/\rho_L g$	t1
$\tau_{12}/\rho_L g$	t12
$\tau_{2m}/\rho_L g$	t4
dh/dx	sin (theta)
η_s	etas
S_s	S

```
REM        *****        Two layer model         *****
                          v1.0, 1991
                          C.A. Shook

REM Begin, read in variables - data statements are at end of the program

READ dp,v,k,cr
READ dpipe,etas,visl,rs
READ clim,rf,e1,d3         :REM e1 used in convergence test
READ thetad                :REM thetad is angle of inclination,degrees
READ printflg              :REM If printing is desired, set printflg = 1

pi=4*ATN(1)

REM ----------------------- Begin Calculations -----------------------

IF printflg = 1 THEN
    LPRINT TAB(30);"Two Layer Model Solution"
    LPRINT:LPRINT
    LPRINT TAB(30);"Input Values:":LPRINT
    LPRINT "theta = ";thetad;" degrees"
    LPRINT "Particle diameter = ";dp;" m"
    LPRINT "In-situ conc = ";cr
    LPRINT "Velocity = ";v;" m/s"
    LPRINT
END IF

ar=4*rf*(rs-rf)*dp^3*9.8/visl^2/3
fr=v^2/9.8/dp
s=rs/rf

REM            + + + + + + Calculate contact load + + + + + +
    term=.124*ar^-.061*fr^.028
    term=term*(dp/dpipe)^-.431
    term=term*(s-1)^-.272
    yc=EXP(-term)
    cc=cr*yc
    c1=cr-cc
    c2=clim-c1

IF printflg = 1 THEN
    LPRINT TAB(30);" Calculated Results:":LPRINT
    LPRINT "Contact Load = ";cc,"ar = ";ar
    LPRINT "c1 = ";c1,c2 = ";c2
END IF
```

```
PRINT "in-situ conc", cr,"velocity";v;" m/s"
PRINT "Contact Load",cc,"ar", ar"
PRINT "c1",c1,"c2", c2

a = .25*pi*dpipe^2
r1 = rf*(1 + c1*(s - 1))
r2 = rf*(1 + clim*(s - 1))
rm = rf*(1 + cr*(s - 1))
theta = (thetad/180)*pi

REM find lower layer area
a2 = a*cc/c2
arat = a2/a

REM test for very small contact load
IF arat < .002 THEN
PRINT "Contact load very low, try homogeneous model":STOP

GOSUB sbet

betd = 180*beta/pi

IF printflg = 1 THEN:LPRINT "beta = " ;beta;" radians"

PRINT "beta,rad" beta

REM set limits for search
    l9 = 0
    t9 = 0

REM evaluate area fraction and friction factors
    a1 = a - a2
    rk1 = k/dpipe
    rey1 = dpipe*v*r1/visl
    term1 = ((7/rey1)^.9 + .27*rk1)
    term2 = LOG(1/term1)
    fa = (2.457*term2)^16
    b = (37530/rey1)^16

IF printflg = 1 THEN LPRINT "Reynolds number = ";rey1
PRINT "reynolds number = ", rey1

    IF rey1 > 2100 THEN
        fact = 0
    ELSE
        fact = (8/rey1)^12
    END IF
```

```
    f=(fact+1/(fa+b)^1.5)^.08333
    f1=2*f
    y=0

    IF (dp/dpipe)<0.0015 THEN
        y=0
    ELSE
        y=4+1.42*(LOG(dp/dpipe)/LOG(10))
    END IF

    f12 = 2 *(1+y) / (((4*LOG(dpipe/dp)/LOG(10))+3.36)^2)

    PRINT "f1",f1,"f12",f12

    IF printflg = 1 THEN
        LPRINT "f1 = ";f1,"f12 = ";f12
        LPRINT:LPRINT TAB(30);" Begin Velocity Iteration:": LPRINT
        LPRINT "p1","beta","v1","v2","q1"
    END IF

REM evaluate perimeters
    s2=beta*dpipe
    s1=(pi-beta)*dpipe
    s12=SIN(beta)*dpipe
    v1=v

REM iterate velocity
REM stresses are divided by fluid density and g

again:
    v2=(a*v-a1*v1)/a2
    t12=f12*(v1-v2)*ABS(v1-v2)*r1/rf
    t12=t12/2/9.8
    t1=f1*v1*ABS(v1)*r1/rf/2/9.8
    t4=f1*v2*ABS(v2)*r1/rf/2/9.8
    p1=(t1*s1+t12*s12)/a1
    p1=p1+r1*SIN(theta)/rf
    m=(rs-rf)*c2*(1-c1-c2)*dpipe^2
    m=m*etas*.5*COS(theta)/rf/(1-c2)
    q2=(SIN(beta)-beta*COS(beta))*m
    q1=(q2+t4*s2-t12*s12)/a2
    q1=q1+r2*SIN(theta)/rf

REM test p1 and q1
    u1=q1-p1
    u2=u1/p1

    IF ABS(u2)<e1 GOTO converged
```

```
REM if too many iterations, cc approaches zero

IF printflg = 1 THEN LPRINT p1,beta,v1,v2,q1
PRINT p1,beta,v1,v2,q1

    u3 = u1*u9
    IF u3 < -.0000005 THEN
        v1 = v1 + (t9 - v1)*11/(u1 - u9)
    ELSE
        u9 = u1
        t9 = v1
        v1 = v1 + d3*ABS(u1)/u1
    END IF
    GOTO again

REM print converged results

converged:
    pm = .5*(p1 + q1)
    h = pm - (rm/rf)*SIN(theta)
    cv = (a*c1*v + a2*c2*v2)/a/v

PRINT p1,beta,v1,v2,q1
PRINT"pressure gradient";pm;" m fluid/m pipe"
PRINT"frictional headloss";h;" m fluid/m pipe"
PRINT"delivered conc", cv

IF printflg = 1 THEN
    LPRINT
    LPRINT p1,beta,v1,v2,q1: LPRINT
    LPRINT:LPRINT TAB(30);" ---- Solution Has Converged ----"; LPRINT
    LPRINT "Pressure gradient = ";pm;" m fluid/m pipe"
    LPRINT "Frictional headloss = ";h;" m fluid/m pipe"
    LPRINT "Delivered conc = "; cv
END IF

END


REM       * * * * *     Subroutine Sbet     * * * * *
REM Interval-halving root finding technique

sbet:
tol = .0000001:REM Error Limit
beta0 = 0     :REM Search interval
beta1 = pi
```

```
REM Check interval for root
If a2<0 THEN PRINT"a2 negative":STOP
IF a2>a THEN PRINT"a2>a":STOP

search:
beta=(beta0+beta1)/2
aest=.25*dpipe*dpipe*(beta-COS(beta)*SIN(beta))

IF ABS(a2-aest)<tol GOTO worked:REM solution within tolerance
    IF aest<a2 THEN
        beta0=beta
    ELSE
        beta1=beta
    END IF
GOTO search

REM Solution has converged, return to calling routine
worked:
    PRINT "a2";a2
    RETURN


REM                 Data Statements

DATA .0005,4.0,.000045,.25   : REM dp,v,k,cr
DATA .25,.5,.001,2650        : REM dpipe,etas,visl,rs
DATA .6,999,.02,.5           : REM clim,rf,e1,d3
DATA -10                     : REM thetad
DATA 1                       : REM printflg
```

Appendix 5
Notation

A	cross-section area
A	constant, Eq. 1–6
A	Hamaker constant, Eq. 1–46
A_p	projected area of particle, Eq. 1–14
Ar	particle Archimedes number, $C_D \mathrm{Re}_p^2$
a	pressure wave velocity, Eq. 2–63
a	acceleration, Eq. 8–9
a_1	coefficient, Eq. 1–24
B	constant, Eq. 1–6
B	coefficient, Eq. 2–34
B	compressibility, Eq. 2–61
B	coefficient, Eq. 3–20
$\mathbf{B}$	magnetic induction, Section 10.11
Bi	Bingham number $\tau_y d / V_\infty \mu_p$
B_y	transverse magnetic induction component, Eq. 10–12
b_a	Tafel slope, Eq. 8–3
b_c	Tafel slope, Eq. 8–3
b_{ij}	particle breakage distribution parameter, Eq. 11–14
b_1	constant, Eq. 1–24
C	volume fraction of solids
$\bar{C}$	average concentration in test loop, Fig. 9–11
C_c	contact load concentration, Eq. 6–4
C_D	drag coefficient of an isolated particle, Eq. 1–15
C_{Ds}	drag coefficient of a particle in a mixture, Eq. 2–21
C_L	lift coefficient, Eq. 1–30
C_{Lb}	friction factor in Ergun equation, Eq. 2–25
C_p	heat capacity

C_r mean in situ or spatial volume fraction, Eq. 2–7
C_s shape factor, Eq. 1–34
C_t chord-average concentration, Eq. 10–4
C_v volume fraction solids in delivered mixture, Eq. 2–2
C_w mass fraction solids in delivered mixture, Eq. 2–6
c volume fraction of solids, at a point or in a homogeneous mixture
c tangential velocity, Eq. 9–1
c_* volume fraction of solids in a settled deposit or bed, Eq. 2–36
c_{max} maximum solids concentration
c_0 concentration upstream of sampler, Figs. 10–3 and 10–4

D pipe diameter
D diffusion coefficient, Eq. 1–36
D impeller diameter, Section 9.2
$\mathbf{D}$ electric displacement, Section 10.11
De Dean number, Eq. 11–2
D_{AB} molar diffusivity of binary pair A,B
D_{eq} hydraulic equivalent diameter of a flow region, Section 11.3
d particle diameter
d_{crit} critical drop diameter, Eq. 3–29
d_S particle diameter, from Eq. 1–13
d_{Stokes} (Stokes) particle diameter, from Eq. 1–22
d_t sampler tube diameter, Eq. 10–1
d_V particle diameter, from Eq. 1–12
d_{50} mass median particle diameter

E coefficient, Eq. 2–34
$\dot{E}$ time rate of energy dissipation, Eq. 8–6
E elastic modulus of pipe wall, Eq. 2–69
$\mathbf{E}$ electric field intensity, Eq. 1–38
e charge on an electron, Section 1.13

F dimensionless deposit velocity, Eq. 5–2
$\mathbf{F}_{am}$ "added mass" force, Eq. 1–34
$\mathbf{F}_{Bd}$ Basset drag force, Eq. 1–35
$\mathbf{F}_D$ drag force on a particle
$\mathbf{F}_L$ lift force on a particle
Fr Froude number, Eq. 11–9
Fr_* particle–fluid Froude number, Eq. 8–8
Fr_{*2} Froude number at impeller exit, Eq. 9–4
f Fanning friction factor, Eq. 1–5
f force per unit volume, Eq. 2–13
f slurry friction factor at the particular V and D values, Eq. 5–16
f frequency distribution function, Fig. 9–13b.
f_L liquid friction factor at the particular V and D values, Eq. 5–16
f_1 friction factor for upper layer flow, Eq. 6–9

Gz	Graetz number, Eq. 11–21
g	acceleration due to gravity
g_d	deposit thickness, Eq. 11–23
$\mathbf{g}_i$	gravitational acceleration in i direction, Eq. 10–1
H	pump head, Eq. 1–4
H	gap of Couette flow section, Eq. 1–48
H	correlating parameter, Eq. 3–25
$\mathbf{H}$	magnetic field intensity, Section 10.11
He	Hedstrom number, $\rho D^2 \tau_y / \mu_p^2$
h	elevation above a datum, Eq. 1–4
h	heat transfer coefficient, Eq. 1–39
h_i	metric coefficient, Appendix 1
h_0	distance between particle surface and a boundary, Eq. 1–33
i	slurry frictional headloss, m carrier fluid/m pipe, Eq. 5–6
i_L	liquid frictional headloss at the conditions pertaining to i (above)
J	particle flux, Eq. 1–36
$\mathbf{J}$	current density, Section 10.11
j_H	dimensionless pipe wall heat transfer parameter, Eq. 11–17
K	power law consistency parameter, Eq. A1–11
K	effective concentration parameter, Eq. 2–33
K	stress ratio, Eq. 2–48
K	compaction coefficient, Eq. 2–60
K	particle velocity at incipient erosion, Eq. 8–5
K	particle inertial parameter, Eq. 10–1
K	fitting loss coefficient, Eq. 11–1
K'	Metzner–Reed parameter, Eq. 4–6
k	equivalent sand roughness of a pipe wall, Eq. 1–6
k	Boltzmann's constant, Eq. 1–37
k	mass transfer coefficient, Eq. 1–40
k	permeability, Eq. 2–25
k	thermal conductivity, Eq. 11–17
l	mixing length for particle transfer, Eq. 7–2
M_A	molecular weight of species A
m	correlating exponent, Eq. 4–22
m	hindered settling exponent, Eq. 7–5, Eq. 10–1
$\dot{m}$	molar flux, Eq. 1–40
N	rotational speed (rotations per minute)
NPSH	Net Positive Suction Head
N_{BA}	(Brownian/attraction) force ratio

N_{BR}	(Brownian/repulsion) force ratio
N_{SA}	(shear/attraction) force ratio
N_{SB}	(shear/Brownian) force ratio
N_{SR}	(shear/repulsion) force ratio
N_s	specific speed, Section 9.2
N_s^*	dimensionless specific speed
n	power-law exponent, Eq. A1–11
n	exponent on $(1 - c)$, Eq. 2–21
n	Zandi–Govatos correlation exponent, Eq. 5–9
n	Manning formula coefficient, Eq. 11–10
n_i	ionic species concentration
P	pressure
P_s	cumulative fraction smaller
Q	volumetric flow rate
q	generalized dimensionless group governing C_D, Eq. 1–26
q	charge, Eq. 1–38
$\dot{q}$	heat flux, Eq. 1–39
R	radius of a particle or cylinder
R	correlation coefficient, Eq. 7–3
R	electrical resistance in slurry, Eq. 10–8
R_B	radius of pipe bend, Eq. 11–2
R_B	ballast resistance, Fig. 10–10
Re	Reynolds number for a pipe flow
Re_B	(Bingham) Reynolds number, Eq. 4–5
Re_n	(Torrance) Reynolds number, Eq. 4–17
Re_n'	(Metzner–Reed) Reynolds number, Eq. 4–14
Re_0	sampler Reynolds number, Eq. 10–1
Re_p	particle Reynolds number, Eqs. 1–15 and 9–6a
Re_p	dimensionless group defined in Eq. 5–1
$\mathrm{Re}_{\mathrm{pump}}$	pump Reynolds number, Eq. 9–7
Re_s	particle Reynolds number in a mixture, Eq. 2–22
Re_s	particle Reynolds number, Eq. 9–6a
R_h	hydraulic radius, Eq. 11–10
R_ρ	polarization resistance, Eq. 8–3
r	radial position coordinate
r	viscosity ratio, Eq. 3–28
r	radius of indentation, Section 8.8
r^*	dimensionless radial position in volute, Fig. 9–14
S	slope of channel, Eq. 11–10
S_{ij}	selection rate parameter, Eq. 11–14
S_m	relative density of mixture, ρ_m/ρ_L
S_s	density ratio ρ_s/ρ_L
s	distance between particle surfaces, Eq. 1–46

T	absolute temperature
T	viscometer torque per unit height, Eq. 3–17
T	pump torque, Eq. 9–3
T	$(2 \cdot \text{wall thickness}/d_t)$, Fig. 10–6
t	time
t	dummy variable, Eq. 3–16
t_r	particle relaxation time, Eq. 1–29
$\tan \beta$	coefficient of interparticle friction, Eq. 7–12
U	sampling velocity, Eq. 10–1
U	internal energy, Appendix 1
U_0	velocity upstream of sample, Fig. 10–3
U_s	particle impingement velocity
u	dimensionless velocity, Fig. 6–5
u	impeller blade velocity, Eq. 9–1
u_*	friction velocity, Eqs. 1–8 and 9–5b
V	mean velocity of pipe flow, Eq. 1–2
V_c	deposition velocity, Eq. 5–1
V_N	Newtonian fluid velocity at specified conditions, Eq. 4–11
V_P	particle volume, Eq. 1–12
V_s	slurry velocity with gas injection, Eq. 11–6
V_s	particle velocity, Fig. 11–4
V_{slip}	wall slip velocity, Section 4.3
V_w	particle velocity parallel to wall, Section 8.8
V_y	circumferential velocity, Fig. 9–13
V_∞	terminal settling velocity of a particle at infinite dilution, Eq. 1–17
v	local velocity
v'	velocity fluctuation, Eq. 1–7
v_r	relative velocity, $v_L - v_s$
w	mass flow rate
w	correlating exponent, Eq. 4–22
w	sample weight, Eq. 8–1
X	Lockhart-Martinelli parameter
x	axial position coordinate
x	mole fraction, Eq. 1–40
x	distance from particle surface, Eq. 1–43
x	distance measured in absorbing medium, Eq. 10–2
x_i	volume fraction particles in size range i
y	position coordinate in vertical direction, Figs. 2–2 and 7–10
y	distance from wall, Eq. 7–13
y	distance measured downward, Eqs. 7–1 and 7–7
y	particle penetration depth, Section 8.8
y	channel depth, Eq. 11–9

y'	dimensionless distance measured downward from pipe axis, Section 7.4
y'	distance from bottom of channel, Eq. 11–11
Z	stability parameter, Eq. 3–14
Z_i	valence of ionic species
z	position coordinate
z^*	dimensionless axial position, Fig. 9–14

Greek Symbols

α	kinetic energy correction factor, Eq. 1–4
α	angle of internal friction for sheared particles, Eq. 2–55
α	parameter in Eq. 2–58
α	correlating parameter (exponent) in Eq. 3–28
α	impingement angle, Eq. 8–4
α^*	dimensionless position between blades
β	blade angle, Eq. 9–5b
β	correlating parameter (exponent) in Eq. 3–28
β	angle defining hypothetical interface, Eq. 6–14
β	ratio (D_2/D_1) for Venturi meter, Eq. 10–9
β_1	velocity profile correction factor, Eq. 5–12
β_2	concentration profile correction factor, Eq. 5–12
γ	correlating parameter (exponent) in Eq. 3–28
$\dot{\gamma}$	shear rate, component of Δ, Appendix 1
$\boldsymbol{\Delta}$	deformation rate tensor, Appendix 1
Δ	deposit roughness, Eq. 11–24
Δ_D	F correction for inclined pipes, Eq. 5–13
ΔP	pressure surge amplitude, Eq. 2–66
ΔP	pressure drop due to friction, Eq. 3–7
Δs	decrease in wall thickness, Eq. 8–5
Δs^*	$\Delta s/\Delta s_{max}$, Fig. 8–18
ϵ	permittivity
ϵ_s	particle diffusion coefficient, Eq. 7–1
ϵ_L	fluid turbulent diffusion coefficient, Eq. 7–3
ζ	zeta potential
η	pump efficiency, Section 9.2
η	volumetric efficiency, Eq. 9–8
η_s	coefficient of particle–wall friction, Eq. 2–47
Θ_p	particle pseudo-temperature, Eq. 2–57
θ	angular position in cross section, Eq. 6–14

θ	inclination of pipe to horizontal, Eq. 6–22
θ_i	impeller angle, Fig. 9–13
κ	von Karman constant, Eq. 1–8
κ	reciprocal double layer thickness, Eq. 1–45
κ	correlating parameter, Eq. 3–27
κ_p	pseudo-thermal conductivity, Eq. 2–58
λ	constant, Eq. 2–34
λ	linear concentration, Eq. 2–52
λ	structural parameter, Eq. 3–26
λ	delay time, Eq. 10–15
μ	viscosity
$[\mu]$	intrinsic viscosity, Section 3.5
μ_c	continuous phase viscosity, Eq. 3–28
μ_d	dispersed phase viscosity, Eq. 3–28
μ_f	viscosity of mixture of liquid and fines, determining C_D in Eq. 5–3
μ_0	limiting viscosity at low shear rates, Eq. 3–25
μ_p	plastic viscosity, Eq. A1–11
μ_r	relative viscosity, Eq. 3–18
μ_∞	Casson model parameter, Eq. 3–11
μ_∞	limiting viscosity at high shear rates, Eq. 3–25
ν	kinematic viscosity
ν_t	turbulent kinematic viscosity, Eq. 1–10
ξ	stress ratio, Eq. 3–15
ρ	mass density
ρ	space charge density, Eq. 1–43
ρ_f	density of mixture of fluid and fines, Eq. 5–2
σ	standard deviation
σ	interfacial tension, Eq. 3–29
σ	particle–wall stress, Eq. 8–7
σ_y	solid wall yield stress, Section 8.8
$\mathbf{T}$	stress tensor, Appendix 1
τ	component of stress tensor $\mathbf{T}$
τ	dimensionless time, Eq. 10–1
τ	time, Eq. 10–15
τ	boundary stress, Section 11.7
τ_0	Casson model parameter, Eq. 3–11
τ_w	shear stress at pipe wall
τ_y	yield stress, Eq. A1–11

ϕ'	coefficient, Eq. 8–5
ϕ	angle of internal friction, Eq. 2–50
ϕ	specific energy for material removal, Eq. 8–4
Φ	dimensionless excess headloss, Eq. 5–6
Φ	electrical potential, Eq. 10–11
Ψ	sphericity, Eq. 1–20
ψ	potential, Eq. 1–43
Ψ	modified Froude number, Eq. 5–7
ψ_0	surface potential, Eq. 1–44
ω	angular velocity

Subscripts

B	denotes Bagnold stress
BEP	denotes best efficiency point
C	denotes Coulombic (strain rate independent) stress
G	denotes gas
L	denotes the liquid phase
Li	denotes liquid approach velocity
Ls	exerted by solids on liquid
LL	transmitted within the liquid
Lw	exerted on the liquid by the wall
M	denotes mixture containing gas
m	denotes a mixture
s	denotes solid particles
si	denotes solids approach velocity
sL	exerted by liquid on solids
ss	transmitted by solid-solid contact
std	standard deviation
sw	exerted on the solids by the wall
th	theoretical
w	pipe wall
w	weight basis, concentrations
1	denotes inner cylinder radius, Eq. 3–16
1	denotes entrance to impeller, Chapter 9
1	denotes upper layer, Section 6.2
2	denotes outlet from impeller, Chapter 2
2	denotes outer cylinder radius, Eq. 3–16
2	denotes lower layer, Section 6.2 or middle layer, Section 6.8
3	denotes bottom layer, Section 6.3
12	denotes interface separating layers 1 and 2
23	denotes interface separating layers 2 and 3

Bibliography

Bird, R.B., Stewart, W.E., and Lightfoot, E.N. 1960. *Transport Phenomena*. New York: Wiley.

Govier, G.W. and Aziz, K. 1972. *The Flow of Complex Mixtures in Pipes*. New York: Van Nostrand Reinhold.

Happel, J. and Brenner, H. 1986. *Low Reynolds Number Hydrodynamics*. Dordrecht, The Netherlands: Martinus Nijhoff Publ.

Hetsroni, G. 1982. *Handbook of Multiphase Flow*. New York: McGraw-Hill.

Karassik, I.J., Krutzsch, W.C., Fraser, W.H., and Massina, J.P. 1976. *Pump Handbook*. New York: McGraw-Hill.

Kim, S. and Karilla, S.J. 1991. *Microhydrodynamics*. Boston: Butterworth.

Kruyt, It.R. 1952. *Colloid Science*. Amsterdam: Elsevier.

Launder, B.E. and Spalding, D.B. 1972. *Lectures in Mathematical Models of Turbulence*. London: Academic Press.

Russel, W.B., Saville, D.A., and Schowalter, W.R. 1989. *Colloidal Dispersions*. Portchester, NY: Cambridge Univ. Press.

Skelland, A.H.P. 1967. *Non Newtonian Flow and Heat Transfer*. New York: Wiley.

Smoldyrev, A. Ye. 1982. *Pipeline Transport Principles of Design*. W.C. Cooley (ed.). Rockville, MD: Terraspace Inc. [Translated from Russian].

Soo, S.L. 1989. *Particulates and Continuum, Multiphase Fluid Dynamics*. New York: Hemisphere.

Svarovsky, L. 1987. *Powder Testing Guide: Methods of Measuring the Physical Properties of Bulk Powders*. New York: Elsevier.

Wasp, E.J., Kenny, J.P., and Gandhi, R.L. 1977. *Solid-Liquid Flow: Slurry Pipeline Transportation,* Trans Tech Publ. Series on Bulk Materials Handling, Vol. 1, Clausthal, Germany: Trans Tech Publications.

Wylie, E.B. and Streeter, V.L. 1977. *Fluid Transients*. New York: McGraw-Hill.

References

Abbott, J.R., Brenner, H., Graham, A.L., and Mondy, L.A. 1991. Techniques for analyzing the behavior of concentrated suspensions, Proc. Fourth Int. Symp. on Liquid-Solid Flows, Eds. M.C. Roco and T. Masuyama, ASME, FED Vol. 118, New York: pp. 1–11.

Abou Arab, T.W. and Roco, M.C. 1989. Turbulent kinetic energy equation for dilute suspensions. In *Constitutive Relationships and Models in Continuum Theories of Multiphase Flows,* NASA Conf. Publ. 3047, Huntsville, AL, pp. 147–162; (also *J. Fluids Eng.,* **112**(3): 351–361, 1990).

Adrian, R.J. 1986. Multi-point optical measurements of simultaneous vectors in unsteady flow—A review. *Int. J. Heat Fluid Flow* **7:** 127–145.

Adrian, R.J. 1991. Particle-imaging techniques for experimental fluid mechanics. *Ann. Review of Fluid Mech.* **23:** 261–304.

Allen, T. 1981. *Particle Size Measurement.* 3rd ed. London: Chapman and Hall.

Ansley, R.W. and Smith, T.N. 1967. Motion of spherical particles in a Bingham plastic. *AIChE J.* **13:** 1193–1196.

ARMCO. 1983. *Scale Inhibitor,* Product Bulletin, Houston, TX.

Asszonyi, C., Kapolyi, L., Kantas, C., and Meggyes, T. 1972. An Experimental Method to Produce a Size Distribution Ensuring Optimum Pipeline Capacity. Proc. Hydrotransport 2, BHRA, Cranfield, UK, Paper D3, pp. 23–31.

ASTM Standard D2776–79. 1989. Standard Test Methods for Corrosivity of Water in the Absence of Heat Transfer (Electrical Methods), 03.02: 30–34.

ASTM Standard G75–89. 1989. Test Method for Slurry Abrasivity by Miller Number, 03.02: 286–306.

ASTM Standard G59–78. 1989. Standard Practice for Conducting Potentiodynamic Polarization Resistance Measurements, 03.02: 221–233.

ASTM Standard G5–87. 1989. Standard Reference Test Method for Making Potentiostatic and Potentiodynamic Anodic Polarization Measurements, 03.02: 79–85.

ASTM Standard G102–89. 1989. Standard Practice for Calculation of Corrosion Rates and Related Information from Electrochemical Measurements, 03.02: 394–400.

Atapattu, D.D., Chhabra, R.P., and Uhlherr, P.H.T. 1988. Particle Drag and Equilibrium in Viscoplastic Fluids. Proc. Int. Symp. on Transportation of Coal and Other Minerals, Indian Institute of Metals, Bhubaneswar, India, pp. 253–260.

Athinson, C.M. and Kytomaa, H.K. 1991. Acoustic Properties of Solid-Liquid Mixtures and the Limits of Ultrasound Diagnostics: Experiments Liquid-Solid Flows, ASME, FED Vol. 118, New York, 145–158.

Auburn International Inc., Danvers, MA. 1982. *Instruction Manual for the Auburn Model 1090 Two-Phase Flow Analyzer.*

Austin, L.G. 1971–72. A review-introduction to the mathematical description of grinding as a rate process. *Powder Technol.* **5:** 1–17.

Ayukawa, K. 1973. Pressure drop in the hydraulic conveyance of solid materials through a bend in a vertical plane. *Bull. JSME* **12**(54): 1388–1396.

Bachalo, W.D. 1980. Method for measuring the size and velocity for dual-beam light scattering interferometry. *Applied Optics* **19**(3): 363–370.

Bagnold, R.A. 1954. Experiments on a gravity-free dispersion of large solid spheres in a Newtonian fluid under shear. *Proc. Roy. Soc., Ser. A* **225:** 40–63.

Bain, A.G. and Bonnington, S.I. 1970. *The Hydraulic Transport of Solids by Pipelines.* New York: Pergamon Press.

Barker, D.B. and Fourney, M.E. 1977. Measuring fluid velocities with speckle patterns. *Optic Letters* **1:** 135–137.

Barnea, D., Granica, D., Doron, P., and Taitel, Y. 1986. Hydraulic Transport of Coarse Particles with Gas Injection. Proc. Hydrotransport 10 Conf., BHRA, Cranfield, UK pp. 303–310.

Batchelor, G.K. 1970. The stress system in a suspension of force-free particles. *J. Fluid Mech.* **41:** 545–570.

Beck, M.S., Gough, J.R., and Mendies, P.J. 1970. Flow Velocity Measurement in a Hydraulic Conveyor. Proc. Hydrotransport 1 Conf., BHRA Fluid Eng., Cranfield, U.K., Paper B3, pp. 25–52.

Beddow, J.K. 1984. *Particle Characterization in Technology.* Boca Raton, FL: CRC Press.

Benchaita, M.T. 1985. Erosion of a two-dimensional channel bend by a solid-liquid stream. *Trans. CSME* **9:** 98–104.

Bennett, C.O. and Myers, J.E. 1982. *Momentum, Heat and Mass Transfer.* 3rd ed. New York: McGraw-Hill.

Bergeron, P. 1952. Consideration of the factors influencing wear due to hydraulic transport of solid materials, Societe Hydrotechnique de France.

Beris, A.N., Tsampoulous, J.A., Armstrong, R.C., and Brown, R.A. 1985. Creeping motion of a sphere through a Bingham plastic. *J. Fluid Mech.* **188:** 199–218.

Berthe, D., Flamand, L., Foucher, D., and Godet, M. 1980. Micropitting in Hertzian contacts. *J. Lub. Tech., Trans. ASME* **102:** 478–489.

Bhattacharayya, A. and Imrie, I. 1986. Development of the ASEA Mineral Slurry Transport System for Coarse Coal. Proc. Hydrotransport 10 Conf., BHRA, Cranfield, UK, pp. 63–67.

Bird, R.B., Stewart, W.E., and Lightfoot, E.N. 1960. *Transport Phenomena,* New York: Wiley.

Bitter, J.G.A. 1963a. A study of erosion phenomena, part I. *Wear* **6**(1): 5–21.

Bitter, J.G.A. 1963b. A study of erosion phenomena, part II. *Wear* **6**(3): 61–190.

Bobkowicz, A.J. and Gauvin, W.H. 1965. The turbulent flow characteristics of model fiber suspensions. *Can. J. Chem. Eng.* **43:** 87–91.

Boger, D.V., Leong, Y.K., Mainwaring, D.E., and Christie, G.B. 1987. Victorian Brown Coal Suspensions as Liquid Fuels. Proc. Third European Conf. on Coal Liquid Mixtures, *I. Chem. E. (London) Symposium Series,* No. 107, pp. 1–13.

Booth, F. 1950. The electroviscous effect in suspensions of solid spheres. *Proc. Roy. Soc.* **203:** 533–551.

Bowen, R. 1961. Designing turbulent flow systems. *Chem. Eng.* **68**(15): 143–150.

Boyd, W., Bell, S., and Thompson, D. 1985. A Study of Design and Development of a Coal Injector for Coarse Slurry Transport. Report of U.S. Dept. of Interior, Pittsburgh, PA.

Brandis, V. and Klose, R. 1988. Scale-up of Pressure Drop Curves, Influence of Temperature and Long-Distance Transport of Highly Concentrated Coal-Water Slurries in Pipelines. Proc. Hydrotransport 11 Conf., BHRA, Cranfield, UK, pp. 237–250.

Brenner, H. 1966. Hydrodynamic resistance of particles at small Reynolds number. *Advances in Chem. Eng.* **6:** 287–438.

Bristol, J.M. 1981. Diaphragm metering pumps. *Chem. Eng.* **88**(19): 124–128.

Brown, N.P. 1980. Friction Mechanisms in Hydraulic Conveying at High Solids Concentrations. Ph.D. Thesis. University of London.

Brown, N.P. 1988. Three Scale-up Techniques for Stabilized Coal Water Slurries. Proc. Hydrotransport 11 Conf., BHRA, Cranfield, UK, pp. 267–283.

Brown, N.P., Streat, M., and Wilson, K.C. 1979. Hydraulic Hoisting at High Concentrations: A New Study of Friction Mechanisms. Proc. Hydrotransport 6 Conf., BHRA Fluid Eng., Cranfield, UK, Paper F2, pp. 269–282.

Brown, N.P., Shook, C.A., Peters, J., and Eyre, D. 1983. A probe for point velocities in slurry flows. *Can. J. Chem. Eng.* **61:** 597–602.

Buchlin, J.M. (ed.). 1984. *Digital Image Processing in Fluid Mechanics.* Rhode Saint Genese, Belgium: Von Karman Institute of Fluid Mechanics.

Buhrow, R.P. 1985. Computer Forecasting Inspection Dates from Metal Corrosion Data. ASME Paper No. 85–Pet–3.

Burwell, J.T. and Strang, C.D. 1952. On the empirical law of adhesive wear. *J. Applied Phys.* **23:** 18–28.

Buyevitch, Yu. A. 1971. Statistical hydrodynamics of disperse systems. Part 4. Physical background and general equations. *J. Fluid Mech.* **40**(3): 340–507.

Cader, T. 1989. Predicting Wear Distribution in Slurry Pipelines. M.Sc. Thesis. University of Kentucky.

Cader, T., Masbernat, O., and Roco, M.C. 1991. LDV Measurements in a Centrifugal Pump: Water and Dilute Slurry Flow. In *Fourth Int. Conf. on Laser Anemometry* (Cleveland) Ed. A. Dybbs and B. Ghorashi, ASME Vol. FED, New York, pp. 763–773.

Cameron, A. 1982. *Basic Lubrication Theory.* 3rd ed.. Chichester, UK: Ellis Horwood Ltd.

Campbell, C.S. and Brennen, C.E. 1985. Computer simulation of granular shear flows. *J. Fluid Mech.* **151:** 167–188.

Caprihan, A. and Fukushima, E., 1990. Flow measurements by NMR, *Physics Reports,* **198**(4): 195–200.

Carleton, A.J. and Cheng, D.C. 1974. Design Velocities for Hydraulic Conveying of Settling Suspensions. Proc. Hydrotransport 3 Conf., BHRA Fluid Eng., Cranfield, UK, Paper E5, pp. 57–74.

Carleton, A.J., Cheng, D.C.-H., and French, R.J. 1973. Pneumatic Transport of Thick Pastes. Proc. Pneumotransport 2 Conf., BHRA, Cranfield, UK, pp. 11–19.

Carleton, A.J., Cheng, D.C.-H., and Whittaker, W. 1974. Determination of the Rheological Properties and Start-up Pipeline Flow Characteristics of Waxy Crude and Fuel Oils. Technical Paper No. 1P74–009, Institute of Petroleum, London.

Carleton, A.J., French, R.J., James, J.G., Broad, B.A., and Streat, M. 1978. Hydraulic Transport of Large Particles Using Conventional and High Concentration Conveying. Proc. Hydrotransport 5 Conf., BHRA Fluid Eng., Cranfield, UK, Paper D2, pp. 15–28.

Casson, N. 1959. A flow equation for pigment-oil suspensions of the printing ink type. In *Rheology of Dispersed Systems.* London: Pergamon, pp. 84–104.

Castillo, C. and Williams, M.C. 1979. Rheology of very concentrated coal suspensions. *Chem. Eng. Commun.* **3:** 529–547.

Cave, I. 1976. Effects of Suspended Solids on the Performance of Centrifugal Pumps. Proc. Hydrotransport 4 Conf., Cranfield, UK, pp. 35–52.

Cawkwell, M.G. and Charles, M.E. 1989. An improved model for start-up of pipelines containing gelled crude oil. *J. Pipelines* **7:** 41–52.

Cawkwell, M.J. and Charles, M.E. 1989. Characterization of Canadian arctic gelled crude oils utilizing an eight-parameter model. *J. Pipelines* **7:** 251–264.

Chapman, S. and Cowling, T.G. 1952. *The Mathematical Theory of Non-uniform Gases.* 2nd ed., Cambridge: Cambridge University Press.

Charles, M.E. 1970. Transport of Solids by Pipelines. Proc. Hydrotransport 1 Conf., BHRA Fluid Eng., Cranfield, UK, Paper A3, pp. 25–36.

Chen, S.S., Fan, L.T., and Hwang, C.L. 1970. Entrance region flow of the Bingham fluid in a circular pipe. *AIChE J.* **16:** 293–299.

Cheng, D.C.-H. 1984. Flow Properties of Suspensions of Inert Spheres. Report LR 548 (MH) M, Warren Spring Laboratory, Stevenage, UK.

Cheng, D.C.-H. and Evans, F. 1965. Phenomenological characterization of the rheological behaviour of inelastic reversible thixotropic and antithixotropic fluids. *Br. J. Applied Phys.* **16:** 1599–1617.

Cheng, D.C.-H. and Richmond, R.A. 1978. Some observations on the rheological behavior of dense suspensions. *Rheologica Acta* **17:** 446–453.

Chhabra, R.P., Farooqi, S.J., Richardson, J.F., and Wardle, A.P. 1983. Co-current flow of air and shear thinning suspensions in pipes of large diameter. *Chem. Eng. Res. Des.* **61:** 56–61.

Chhabra, R.P. and Richardson, J.F. 1985. Hydraulic transport of coarse particles in viscous Newtonian and non-Newtonian media in a horizontal pipe. *Chem. Eng. Res. Des.* **63:** 390–397.

Cho, Z.H., Kim, H.S., Song, H.B., and Cumming, J. 1982. Fourier transform nuclear magnetic resonance tomographic imaging. *Proc. IEEE* **70:** 1152–1173.

Churchill, S.W. 1977. Friction factor equation spans all fluids regimes. *Chem. Eng.* **84**(24): 91–92.

Clapp, R.M. 1961. Turbulent heat transfer in pseudoplastic non-Newtonian fluids. *International Developments in Heat Transfer, ASME,* Part III, 652–661.

Clift, R., Grace, J.R., and Weber, M. 1978. *Bubbles, Drops and Particles.* New York: Academic Press.

Collins, M. and Schowalter, W.P. 1963. Behavior of non-Newtonian fluids in the entry region of a pipe. *AIChE J.* **9:** 804–809.

Colwell, J.M. 1987. Slurry Flowmeter and Entry Length Tests. M.Sc. Thesis. University of Saskatchewan.

Colwell, J.M., Shook, C.A., Gillies, R.G., and Small, M. 1989. Ultrasonic flowmeter tests with slurries. *J. Pipelines* **7:** 127–140.

Cooke, R. and Lazarus, J.H. 1988. Isokinetic Sampling Probe for Slurry Flows. Proc. Hydrotransport 11 Conf., BHRA, Cranfield, UK, pp. 117–130.

Crewe, J.F. and Simons, H.P. 1973. Heating and cooling of flocculated and non-flocculated slurries in tubular heat exhangers. *Can. J. Chem. Eng.* **51:** 556–561.

Crisswell, J.W. 1982. Practical Problems Associated with Selection and Operation of Slurry Pumps. Proc. Hydrotransport 8 Conf., BHRA, Cranfield, UK, pp. 317–338.

Crowe, C.T. 1982. Review-numerical models for dilute gas-particle flow. *J. Fluids Eng.* **104:** 297–303.

Daniel, S.M. 1965. Flow of Suspensions in a Rectangular Channel. Ph.D. Thesis. University of Saskatchewan.

DeBree, S.E.M., Rosenbrand, W.F., and DeGee, A.W.J. 1982. On the Erosion Resistance of Water-sand Mixtures of Steels for Application in Slurry Pipelines. Proc. Hydrotransport 8 Conf., BHRA, Cranfield, UK, pp. 161–180.

Dedegil, M.Y. 1986. Drag coefficient and settling velocity of particles in non-Newtonian suspensions. In *International Symposium on Slurry Flows,* M.C. Roco and W. Wiedenroth, eds. New York: ASME, Vol. FED-38, pp. 9–15.

DeGance, A.E. and Atherton, R.W. 1970. Chemical engineering aspects of two-phase flow. *Chem. Eng.* **77** ((6): 135–139; (8): 151–158; (10): 113–120; (15): 95–103; (17): 119–126; (21): 87–94; and (24): 101–108).

Devanathan, N., Moslemian, D., and Dudukovic, M.P. 1990. Flow mapping in bubble columns using CARPT, *Chem. Eng. Sci.,* **45:** 2285–2291.

Ding, J. and Gidaspow, D. 1990. A bubbling fluidization model using kinetic theory of granular flow, *AIChE J.* **36:** 523–538.

Dodge, D.W. and Metzner, A.B. 1959. Turbulent flow of non-Newtonian systems. *AIChE J.* **5:** 189–204.

Doron, P., Granica, D., and Barnea, D. 1987. Slurry flow in horizontal pipes—experimental and modeling, *Int. J. Multiphase Flow,* **13:** 535–547.

Duffy, G.G. and Titchener, A.L. 1974. Design procedures for obtaining pipe friction loss for chemical pulps. *Tappi* **57**(5): 162–166.

Duffy, G.G., Moller, K., and Titchener, A.L. 1972. The determination of pipe friction loss. *Appita* **26:** 191–195.

Durand, R. and Condolios, E. 1952. Experimental Study of the Hydraulic Transport of Coal and Solid Materials in Pipes. Proc. Colloq. on the Hydraulic Transport of Coal, Nat. Coal Board (UK), Paper IV, pp. 39–55.

Durst, F. 1973. Scattering phenomena and their application in laser Doppler anemometry. *ZAMP* **24**(4): 619–643.

Durst, F. 1982. Review-Combined measurements of particle velocities, size distributions, and concentrations. *J. Fluids Eng.* **104:** 284–296.

Edwards, R.V. and Dybbs, A. 1984. Refractive index matching for velocity measurements in complex geometries. *TSI Quarterly* **10**(4): 1–4.

Einstein, H.A. and Chien, N. 1955. Effects of Heavy Sediment Concentration near the Bed on Velocity and Sediment Distribution. University of California Institute of Engineering Research, M.S.D. Sediment Series, Report No. 8.

Ekmann, J.M., Wildman, D.J., and Klinzing, G.E. 1985. Study of Unsteady Flow Conditions for Slurry Fuels. Fine Particle Society, 16th Annual Meeting, Miami Beach, pp. 1–15.

Eng. and Mining J., by editor, 1987. Slurry transport pumps: separation of moving parts and slurry is the key to economical operation, **188**(11): 74–75.

Engerlin, C., Gore, R., Legrande, R., Steinbach, D., Crowe, C.T., Lee, J., Ushimaru, K., and Bernstein, S. 1988. An experimental study of performance and erosion of an annular slurry jet pump. In *Third International Symposium on Liquid-Solid Flows*. New York: ASME, FED Vol. 75, pp. 195–202.

Ensminger, D. 1973. *Ultrasonics: The Low and High Intensity Applications*. New York: M. Dekker.

Ercolani, D., Ferrini, F., and Arrigoni, V. 1979. Electric and Thermic Probes for Measuring the Limit Velocity. Proc. Hydrotransport 6 Conf., BHRA Fluid Eng., Cranfield, UK, Paper A3, pp. 27–42.

Ercolani, D., Giachette, G., and Pareschi, A. 1987. Experimental analysis of solid sediment behavior in sloping pipes for solid liquid mixtures. *J. Pipelines* **6**: 205–216.

Ercolani, D., Carniani, E., Meli, S., Pellegrini, L., Primicerio, M., and Gabrialli, G. 1988. Effect of Mechanical and Thermal Energy on CWS Rheology. Proc. Hydrotransport 11 Conf., BHRA, Cranfield, UK, pp. 215–228.

Ergun, S. 1952. Fluid flow through packed columns. *Chem. Eng. Progress* **48**: 89–94.

Eroglu, H. and Tabakoff, W. 1991. 3-D LDV measurements of particle rebound characteristics, Proc. 29th Aerospace Sciences Meeting, Paper AIAA-91-0011, Washington, DC, pp. 1–8.

Etchells, A.W. 1986. Some Slurry Handling Problems in the Chemical Industry. Proc. Hydrotransport 10 Conf., BHRA, Cranfield, UK, pp. 229–231.

Faddick, R.R., 1975. Pipeline wear from abrasive slurries, Proc. 1st Int. Conf. on the Internal and External Protection of Pipes, BHRA, Cranfield, UK, Paper G3, pp. 29–37.

Faddick, R.R. 1982. Lecture Notes for Short Course. Proc. Hydrotransport 8 Conf., BHRA, Cranfield, UK.

Faddick, R.R. 1983. Enlightened Coal Slurries. Proc. 8th Int. Tech. Conf. on Slurry Transportation, Slurry Transport Association, pp. 165–174.

Faddick, R.R. 1986. Slurry Flume Design. Proc. Hydrotransport 10 Conf., BHRA, Cranfield, UK, pp. 143–147.

Faddick, R.R., Pouska, G., Connery, J., DiNapoli, L., and Punis, G. 1979. Ultrasonic Velocity Meter. Proc. Hydrotransport 6 Conf., BHRA, Cranfield, UK, pp. 113–124.

Fan, X. and Masliyah, J.H. 1990. Laboratory investigation of beach profiles in tailing disposal, *J. Hyd. Eng.*, **116**: 1357–1373.

Fanger, H.-U., Michaelis, W., Pipelnik, R., and The, H.L. 1978. Applications of Gamma-ray Absorptiometry in the Hydraulic Transport of Solids. Proc. Hydrotransport 5 Conf., BHRA, Cranfield, UK, Paper G4, pp. 43–58.

Fanger, H.-U., Michaelis, W., Pipelnik, R., and The, H.L. 1986. A two-channel gamma ray method applied to three component analyses in hydraulic transport systems. *Atomkernenergie-Kerntechnik* **48**: 67–74.

Farooqi, S.I., Heywood, N.I., and Richardson, J.F. 1980. Drag reduction by air injection for suspension flow in a horizontal pipeline. *Trans. Inst. Chem. Engrs.* **58:** 16–27.

Ferrini, F., Giommi, C., and Ercolani, D. 1982. Corrosion and Wear Measurements in Slurry Pipelines. Proc. Hydrotransport 8 Conf., BHRA, Cranfield, UK, Paper C1, pp. 133–144.

Finnie, I. 1960. Erosion of surfaces by solid particles. *Wear* 3(3–4): 87–103.

Finnie, I. 1972. Some observations on the erosion of ductile metals. *Wear* **19:** 81–90.

Follansbee, P.S., Sinclair, G.B., and Williams, J.C. 1980. Fundamental Studies of Erosion. DOE-ER10468 Report.

Frankel, N.A. and Acrivos, A. 1967. On the viscosity of a concentrated suspension of solid spheres. *Chem. Eng. Science* **22:** 847–853.

Frankiewicz, T., Perry, G., Gillies, R.G., Small, M., and Shook, C.A. 1990. Studies on the Rheology of Coal-Oil (transCOM) Slurries. Proc. 15th Int. Conf. on Coal and Slurry Technologies, Clearwater, FL, pp. 551–564.

Frankiewicz, T., Gillies, R.G., Shook, C.A., and Small, M., 1990. Pipeloop and pipeline flow tests on coal-oil and coal-condensate slurries, Proc. 16th Int. Conf. on Coal and Slurry Technologies, Coal and Slurry Technology Assoc., Clearwater, FL, pp. 205–216.

Friedlander, S.K. 1977. *Smoke, Dust and Haze.* New York: Wiley.

Funk, J. 1981. Coal-water Slurry and Method for its Preparation. US Patent 4,282,006.

Gahlot, V.K. 1987. Studies on the Effect of Particle Size Distribution and Material Properties on the Rheological Parameters of Slurries and the Performance of Slurry Pumps. Ph.D. Thesis. Indian Institute of Technology, Delhi.

Gaigolas, A.K. and Whetstone, J.R. 1986. Measurement of the dielectric constant of slurries in pipes. *Chem. Eng. Commun.* **40:** 85–95.

Gay, E.C., Nelson, P.A., and Armstrong, W.P. 1969. Flow properties of suspensions at high concentrations. *AIChE J.* **15:** 815–822.

Gandhi, R.L., Snoek, P.E., and Carney, J.C. 1980. An Evaluation of Slurry Pumps. Proc. 5th Conf. on Slurry Transportation, Slurry Transport Assoc., Washington, DC, pp. 267–275.

Gies, R. and Geller, F.J. 1982. Pressure Gradient and Degradation at the Hydrotransport of Coarse Washery Shales. Proc. Hydrotransport 8 Conf., BHRA, Cranfield, UK, pp. 227–240.

Gillies, R.G. 1990. Private communication.

Gillies, R.G. and Shook, C.A. 1991. A deposition velocity correlation for water slurries. *Can. J. Chem. Eng.* In press.

Gillies, R.G., Small, M., and Husband, W.H.W. 1984. 300 mm Gravity-based Slurry Densimeter Tests. Saskatchewan Research Council Rep. R833-3-C84.

Gillies, R.G., Small, M., Shook, C.A., and Eyre, D. 1988. Determining Delivered Slurry Concentration with a Venturi Meter. Proc. Hydrotransport 11 Conf., BHRA, Cranfield, UK, pp. 157–172.

Gillies, R.G., Haas, D.B., Husband, W.H.W., Small, M., and Shook, C.A. 1982. A System to Determine Single Pass Particle Degradation by Pumps, Proc. Hydrotransport 8 Conf., BHRA, Cranfield, UK, pp. 415–432.

Goldsmith, W. 1960. *Impact: the Theory and Physical Behaviour of Colliding Solids,* London: Edward Arnold.

Goldsmith, H.L., Karnis, A. and Mason, S.G. 1963. The flow of suspensions through tubes, *Can. J. Chem. Eng.* **44:** 181–193.

Goldstein, S. 1938. *Modern Developments in Fluid Dynamics.* Vol. 2. Oxford: Clarendon Press.

Goodwin, J.W. 1982. Some uses of rheology in colloid science. *Publ. Roy. Soc. Chem. (Great Britain)* **43:** 165–195.

Govier, G.W. and Aziz, K. 1972. *The Flow of Complex Mixtures in Pipes,* New York: Van Nostrand Reinhold.

Graf, W.H. and Acaroglu, E.R. 1968. Sediment transport in conveyance systems. *Bull. Int. Assoc. Sci. Hydrol.* **13**(2):20–39.

Graham, A.L. and Bird, R.B. 1984. Particle Clusters in Concentrated Suspensions. 1. Experimental Observations of Particle Clusters, *Ind. Eng. Chem. Fundamentals,* **23:** 406–410.

Gray, W.G. 1975. A derivation of the equations for multiphase transport. *Chem. Eng. Sci.* **30:** 229–233.

Hanes, D.M. and Inman, D.L. 1985. Observations of rapidly flowing granular-fluid materials. *J. Fluid Mech.* **150:** 357–380.

Hanks, R.W. 1963. The laminar-turbulent transition for fluids with a yield stress. *AIChE J.* **9:** 306–309.

Hanks, R.W. and Dadia, B.H. 1982. Theoretical analysis of turbulent flow of non-Newtonian slurries in pipes. *AIChE J.* **28:** 547–565.

Hanks, R.W. and Hanks, K.W. 1982. A New Viscometer for Determining the Effect of Particle Size Distributions and Concentrations on Slurry Rheology. Proc. 7th Int. Tech. Conf., Slurry Transport Assoc., Washington, DC, pp. 151–161.

Hanks, R.W. and Hanks, K.W. 1986. Accurate Pipeflow Pressure Drop Predictions from Bench Scale Tests Require Correct Rheological Characterization. Proc. 11th Int. Conf. on Slurry Technology, Slurry Transport Assoc., Washington DC, pp. 115–122.

Hanks, R.W. and Sen, S. 1983. The Influence of Yield Stress and Fluid Rheology on Particle Drag Coefficient. Proc. 8th Int. Tech. Conf., Slurry Transport Assoc., Washington DC, pp. 71–79.

Hanks, R.W. and Sloane, D.G. 1981. A Rheology-based Correlation for Minimum Deposition Velocity. Proc. 6th Int. Conf. on Slurry Transportation, Slurry Transport Assoc., Washington DC, pp. 107–112.

Happel, J. and Brenner, H. 1965. *Low Reynolds Number Hydrodynamics.* Englewood Cliffs, NJ: Prentice-Hall.

Harada, E., Toda, M., Kuriyama, M., and Konno, H. 1985. Heat transfer between wall and solid-water suspension flow. *J. Chem. Eng. Japan* **18:** 33–38.

Harvey, A. 1982. Injector for a Material Transport System. US Patent 4,363,572.

Hashimoto, H., Noda, L., Masuyama, T., and Kawashima, T. 1980. Influence of Pipe Inclination on Deposit Velocity. Proc. Hydrotransport 7 Conf., BHRA, Cranfield, UK, pp. 231–244.

Herschel, W.H. and Bulkley, R. 1926. Measurement of consistency as applied to rubber-benzene solutions. *Proc. ASTM,* **26,** Part 2, 621–633.

Hetsroni, G. 1988. Particle-turbulence interaction. In *Third International Symposium on Liquid-Solid Flows.* New York: ASME, FED Vol. 75, pp. 1–12.

Heywood, N.I. and Charles, M.E. 1980. Effects of Gas Injection on the Vertical Pipe Flow of Fine Slurries. Proc. Hydrotransport 7 Conf., BHRA, Cranfield, UK, pp. 173–194.

Heywood, N.I. and Mehta, K. 1988. A Survey of Non-invasive Flowmeters for Pipeline Flow of High Concentration Settling Slurries. Proc. Hydrotransport 11 Conf., BHRA, Cranfield, UK, pp. 131–156.

Hinze, J.O. 1975. *Turbulence.* 2nd ed. New York: McGraw-Hill.

Hirleman, E.D., Ed. 1990. *Proc. Second Int. Congress on Optical Particle Sizing,* Arizona St. Univ., Tempe, AZ.

Hirsimaki, O. 1978. Determination of radial velocity profile and flow disturbance of pulp suspension by ultrasonic echo correlation, Paperi ja Puu-Papper och Tra, Finland, (2), 95–97.

Hoey, G.R. and Bednar, J.S. 1980. Erosion-corrosion of Carbon Steel and Stainless Steel by Coal-Water Slurries. Report EPR/PMRL 80–74, CANMET, Energy Mines and Resources Canada.

Hogmark, S. and Soderberg, S. 1983. On the Combined Effects of Erosion and Corrosion. Proc. 6th Int. Symp. on Erosion by Liquid-Solid Impact, Cambridge, England, pp. 37–1 to 37–8.

Holthuis, C.H. and Simons, P.W.H. 1981. The Economics of Positive Displacement Slurry Pumps. Proc. 6th Int. Tech. Conf. on Slurry Transportation, Slurry Transport Assoc., Washington, DC, pp. 156–161.

Horsley, R.R. 1982. Viscometer and Pipe Loop Tests on Gold Slime Slurry at Very High Concentrations by Weight, With and Without Additives. Proc. Hydrotransport 8 Conf., BHRA, Cranfield, UK, pp. 367–382.

Hubbard, L.T. 1972. Transient Flow of a Slurry. M.Sc. Thesis. University of Saskatchewan.

Hutchings, I.M. 1981. A model for the erosion of metals by spherical particles at normal incidence. *Wear* 70(3): 269–281.

Ito, H. 1959. Friction factors for turbulent flow in curved pipes. *Trans. ASME, J. Basic Eng.* **81D:** 123–132.

Ito, H. 1960. Pressure losses in smooth pipe bends. *Trans. ASME, J. Basic Eng.* **82D:** 131–139.

Ito, H. 1987. Flow in Curved Pipes. *JSME Int. J.* **30**(262): 543–552.

Iverson, R.M. and LaHusen, R.G. 1989. Dynamic pore-pressure fluctuations in rapidly shearing granular materials. *Science* **246:** 796–799.

Jacques, R.B. and Neil, W.R. 1977. Internal Corrosion of Slurry Pipelines, Cause-Control-Economics. Proc. 2nd Int. Tech. Conf. on Slurry Transportation. Slurry Transport Assoc., Washington, DC, pp. 1–32.

Jastrzebski, Z.P. 1967. Entrance effects and wall effects in an extrusion rheometer during the flow of concentrated suspensions. *Ind. Eng. Chem. Fundamentals* **6:** 445–454.

Jeffrey, G.B. 1922. On the motion of ellipsoidal particles immersed in a viscous fluid, *Proc. Roy. Soc.,* Ser. A, **102:** 161–179.

Jenkins, J.T. and Savage, S.B. 1983. A theory for the rapid flow of identical, smooth, nearly elastic spherical particles. *J. Fluid Mech.* **140:** 197–202.

Johnson, P.C. and Jackson, R. 1987. Frictional Constitutive Equations for Granular Materials, with Applications to Plane Shearing. *J. Fluid Mech.* **176:** 67–93.

JSME, Ed. 1988. *Visualized Flow,* Oxford: Pergamon Press.

Kachel, G.C. 1986. The slurry monitor—A new development in the precision determination of the density of slurries flowing in pipelines. In *International Symposium on*

Slurry Flow, Eds. M.C. Roco and W. Wiedenroth. New York: ASME, FED. Vol. 38, pp. 167–173.

Kale, D.D. and Metzner, A.B. 1974. Drag reduction in the turbulent flow fiber suspensions. *AIChE J.* **20:** 301–306.

Kalyanaraman, K., Ghosh, D.P., and Rao, A.J. 1973. Characteristics of sand-water slurry in 90 degree horizontal sections. *IE (India) J.* **54:** 73–79.

Kanamori, S., Ito, K., Woda, Y., and Shibata, K. 1990. Study on the Relationship Between Coal Properties and Characteristics of Coal Water Mixtures. Proc. 15th Int. Conf. on Coal and Slurry Technology, Coal and Slurry Technology Assoc., Clearwater, FL, pp. 573–584.

Kao, D.T. and Wood, D.J. 1978. Pressure Surge Generation Due to Rapid Shutdown of Pipelines Conveying Slurries, Proc. Hydrotransport 5 Conf., BHRA Cranfield, UK, Paper E1, pp. 1–14.

Kao, S.V., Neilsen, L.E., and Hill, C.T. 1975. Rheology of concentrated suspensions of spheres. II. Suspensions agglomerated by an immiscible second liquid. *J. Colloid Interface Sci.* **55:** 367–373.

Karabelas, A.J. 1976. Particle attrition in shear slurry flow of concentrated slurries. *AIChE J.* **22:** 765–771.

Kazanski, I. and Bruhl, H. 1972. The Influence of Highly Concentrated Rigid Particles on the Macroturbulent Characteristics in Pipe Flow. Proc. Hydrotransport 2 Conf., BHRA Fluid Eng., Cranfield, UK, Paper A2, pp. 15–30.

Kelsall, D.F. and Reid, K.J. 1965. The Derivation of a Mathematical Model for Breakage in a Small Continuous Wet Ball Mill. *Joint AIChE/IChE Meeting,* London, (4) 4:14–4:20.

Kemblowski, Z. and Kolodziejski, J. 1973. Flow resistances of non-Newtonian fluids in transitional and turbulent flow. *Int. Chem. Eng.* **13:** 265–279.

Kenchington, J.M. 1972. The Design of Large Pipelines for Cement Slurries. Proc. Hydrotransport 2 Conf., BHRA, Cranfield, UK, Paper C4, pp. 41–60.

Keska, J. 1978. Measurement of the Distribution and Overall Spatial Solid Body Content in the Process of Hydrotransport of Mixtures in Pipelines. Proc. Hydrotransport 5 Conf., BHRA, Cranfield, UK, Paper G4, pp. 59–72.

Khan, S.A. and Armstrong, R.C. 1989. Rheology of foams: effect of gas volume fraction, *J. Rheology,* **33:** 881–911.

Khatib, Z. and Richardson, J.F. 1984. Vertical co-current flow of air and shear thinning suspension of kaolin. *Chem. Eng. Res. Des.* **62:** 139–154.

Khuntia, S. and Murty, J.S. 1988. Integrated slurry mixing and feeding system using jet pump. In *Third International Symposium on Liquid-Solid Flows.* New York: ASME, Vol. FED-75, pp. 211–216.

Kocsanyi, L. and Maurer, H. 1972. Rohraufgeber für die Hydraulische Forderung. *Gluckauf* **108**(24).

Kraynik, A.M. 1988. Foam Flows, *Ann. Rev. Fluid Mech.,* **20:** 325–357.

Kraynik, A.M. and Hansen, M.G. 1987. Foam rheology: a model of viscous phenomena, *J. Rheology* **31:** 175–205.

Krieger, I.M. 1968. Shear rate in the Couette viscometer. *Trans. Soc. Rheology* **12:** 5–11.

Krieger, I.M. 1972. Rheology of monodisperse latices. *Advances Colloid Interface Sci.* **3:** 111–136.

Krieger, I.M. and Dougherty, T.J. 1959. Mechanism for non-Newtonian flow in suspensions of rigid spheres. *Trans. Soc. Rheology* **3:** 137–152.

Kruyt, It. R., 1952. "Colloid Science," Amsterdam: Elsevier.

Laapas, H.R. and Aarnio, A.A. 1988. A study on the preparation of highly loaded coal-water slurry. Proc. Hydrotransport 11 Conf., BHRA, Cranfield, UK, pp. 229–238.

Lambe, T.W. and Whitman, R.W. 1969. *Soil Mechanics*. New York: Wiley.

Lammers, G.C., Allen, R.R., Donaven, D.J., Wagner, E.O., and Parry, V.F. 1958. A Study of the Feasibility of Hydraulic Transport of a Texas Lignite. U.S. Bureau of the Mines RI No. 5404, U.S. Dept. of the Interior, 35 pp.

Law, H.-S., Masliyah, J.H., MacTaggart, R.S., and Nandakumar, K. 1987. Gravity separation of bidisperse suspensions: Light and heavy particles. *Chem. Eng. Sci.* **42:** 1527–1538.

Lazarus, M.D. and Lazarus, J.H. 1988. Development of an Ultrasonic Doppler Bed Load Velocimeter. Proc. Hydrotransport 11 Conf., BHRA, Cranfield, UK, pp. 173–185.

Lazarus, J.H. and Sive, A.W. 1984. A Novel Balanced Beam Viscometer and the Rheological Characterization of High Concentration Fly Ash Slurries. Proc. Hydrotransport 9 Conf., BHRA, Cranfield, UK, pp. 207–226.

Lazarus, J.H. and Slatter, P.T. 1986. Comparative Rheological Characterization Using a Balance Beam Tube Viscosimeter and Rotary Viscosimeter. Proc. Hydrotransport 10 Conf., BHRA, Cranfield, UK, pp. 291–302.

Lee, J.E. 1978. Slurry Pumping at INCO Metals Company. 10th Ann. Mtg. Can. Mineral Producers, Ottawa, pp. 241–263.

Lee, P.F.W. and Duffy, G.G. 1976. An analysis of the drag reducing regime of pulp suspension flow. *TAPPI* **59**(8): 119–122.

Lee, P.F.W. and Duffy, G.G. 1976. Relationship between velocity profiles and drag reduction in turbulent fiber suspension flow. *AIChE J.* **22:** 750–752.

Lee, P.F.W. and Duffy, G.G. 1977. An analysis of the transition regime of fiber suspension flow in pipes. *Can. J. Chem. Eng.* **55:** 361–363.

Lee, S.L. 1987. Particle drag in a dilute turbulent two-phase suspension flow. *Int. J. Multiphase Flow* **13:** 247–256.

Leighton, D. and Acrivos, A. 1987. The shear-induced migration of particles in concentrated suspensions. *J. Fluid Mech.* **181:** 415–439.

Leong, Y.K., Creasy, D.E., Boger, D.V., and Nguyen, Q.D. 1987. Rheology of brown coal. *Rheologica Acta* **26:** 291–300.

Levy, A.V. and Hickey, G. 1987. Liquid-solid particle slurry erosion of steel. *Wear* **117:** 146–199.

Liou, C.P. 1984. Acoustic wave speeds for slurries in pipelines. *J. Hydr. Eng.* **110:** 945–957.

Liu, K.V., Choi, U.S., and Kasza, K.E. 1988. Pressure Drop and Heat Transfer Characteristics of Nearly Neutrally Buoyant Slurry for Advanced Energy Transmission Fluids. In *Third International Symposium on Liquid-Solid Flows*. New York: ASME, FED Vol. 75, pp. 107–113.

Loadstar Publ., Ed. 1988. *Abrasion-Resistant and Low-Friction Lining,* London.

Longwell, P.A. 1966. *Mechanics of Fluid Flow*. New York: McGraw-Hill.

Lumley, J.L. (ed.) 1990. *Whither Turbulence? Turbulence at the Crossroads,* Lecture Notes in Physics Series No. 357. New York: Springer-Verlag.

Lytle, M.B. and Reed, A.J. 1984. Conveyance of Run-of-mine Coal by Open Channel Flow. Proc. Hydrotransport 9 Conf., BHRA Fluid Eng., Cranfield, UK, Paper C4, pp. 129–148.

Ma, D.N. and Roco, M.C. 1990. Three dimensional model for slow particulate flow with interstitial fluid. In *Particle Technology Review—1990,* Ed. J.K. Beddow. New York: Hemisphere.

Ma, T.-W. 1987. Stability, Rheology and Flow in Pipes, Fittings, Valves and Venturi Meters of Concentrated Non-Newtonian Suspensions. Ph.D. Thesis. University of Illinois, Chicago.

Madsen, B.W. 1988. A portable slurry wear test for the field. *Third International Symposium on Liquid-Solid Flows.* New York: ASME, FED Vol. 75, pp. 247–254.

Majors, P.D., Givler, R.C., and Fukushima, E. 1989. Velocity and concentration measurements in multiphase flows by NMR. *J. Magnetic Resonance* **85:** 235–243.

Majumdar, A., Graham, A.L., Roco, M.C., and Stroeve, P. 1987. Experimental study on solid particle dynamics in shear flow. *Powder Technol.* **49:** 217–226.

Manley, R.St.J. and Mason, S.G. 1955. Particle motions in sheared suspensions III: Further observations on collisions of spheres. *Can. J. Chem.* **33:** 763–773.

Mannheimer, R.J. 1985. Flow Characteristics of Slurries at Shear Stress Near the Yield Value. Proc. 10th Int. Conf. on Slurry Technology, Slurry Transport Assoc., Washington, DC, pp. 123–133.

Masliyah, J.H. 1979. Hindered settling in a multi-species particle system. *Chem. Eng. Sci.* **34:** 1166–1168.

Masliyah, J.H. and Shook, C.A. Two phase laminar zero net flow in a circular inclined pipe. *Can. J. Chem. Eng.* **56:** 165–175.

Mathur, M.P. and Ekmann, J.M. 1982. Preliminary Evaluation of Instruments for On-line Characterization of Coal-oil Mixture Fuel. DOE/PETC/TR-82/2, (DE82004486), U.S. Department of Energy, Pittsburgh Energy Technology Center, Pittsburgh, PA.

Matsumoto, S. and Saito, S., 1970. On the mechanism of suspension of particles in horizontal pneumatic conveying: Monte Carlo simulation based on the irregular bouncing model, *J. Chem. Eng. Japan,* **3:** 83–92.

Maude, A.D. and Whitmore, R.L. 1958. A Generalized Theory of Sedimentation, *Br. J. Applied Phys.* **12**(9): 477–482.

Maxwell, J.C. 1892. A Treatise on Electricity and Magnetism. 3rd ed. Oxford: Clarendon Press.

McComb, W.D. and Rabie, L.H. 1982. Local drag reduction due to injection of polymer solutions into turbulent flow in a pipe. *AIChE J.* **28:** 547–565.

Metzner, A.B. and Reed, J.C. 1955. Flow of non-Newtonian fluids—correlation of the laminar, transition and turbulent flow regions. *AIChE J.* **1:** 434–440.

Michalik, A., Praxmeyer, T., and Prezewlocki, K. 1968. A radiometric method of determination of the spatial distribution of density in closed media. *Nukleonika* **XII:** 107–114.

Migniot, C. 1968. Etude des properties physiques de differents sediments tres fins et de leur comportement sous des actions hydrodynamiques. *La Houille Blanche* (7), 591–620.

Mikhail, S., Marcos, S.M., and Shaheen, Y.A. 1988. Experimental investigations of a slurry jet pump. In *Third International Symposium on Liquid-Solid Flows.* New York: ASME, FED Vol. 75, pp. 203–210.

Mikhail, M.W., Mikula, R.J., and Husband, W.H.W. 1982. Prediction of Size Degradation During Slurry Pipeline Transportation. Proc. 6th Int. Tech. Conf. on Slurry Transportation, Slurry Transport Assoc., Washington, DC, pp. 399–404.

Miller, J.D., Lin, C.L., and Cortes, A.B. 1990. Review of x-ray computed tomography and its application. *Mineral Processing and Extractive Metallurgy Review* **7:** 1–18.

Miller, J.E. 1986. The mechanism of wear in slurry pumping. In *International Symposium on Slurry Flow.* New York: ASME, FED Vol. 38, pp. 181–187.

Miller, J.E. 1987. *The Reciprocating Pump.* New York: Wiley.

Milliken, W., Mondy, L.A., Gottlieb, M., Graham, A.L., and Powell, R.L. 1989. The effect of the diameter of falling balls on apparent viscosity of suspensions of spheres and rods, *Physico-Chemical Hydrodynamics,* **11**(3): 341–355.

Minani, L.K. 1989. Wall Phenomena in Liquid-Solid Flow. PhD Thesis. University of Kentucky.

Moore, F. 1959. Rheology of ceramic slips and bodies. *Trans. Br. Ceram. Soc.* **58:** 470–492.

Moore, D.F. 1975. *Principles and Applications of Tribology.* New York: Pergamon Press.

Mueller, N.H.G. 1964. Water jet pump. *Proc. ASCE, J. Hydr. Div.,* **90(HY3):** 83–114.

Nakabayashi, K., Yamada, Y., and Kishimoto, T. 1982. Viscous frictional torque in the flow between two concentric rough cylinders. *J. Fluid Mech.* **119:** 409–422.

Nasr-El-Din, H. 1984. Sampling from Slurry Pipelines, PhD Thesis, Chem. Eng., Univ. of Saskatchewan.

Nasr-El-Din, H. and Shook, C.A. 1984. Sampling from slurry pipelines: thick-wall and straight probes. *J. Pipelines* **5:** 113–124.

Nasr-El-Din, H. and Shook, C.A. 1986. Particle segregation in slurry flow through vertical tees. *Int. J. Multiphase Flow* **12:** 427–443.

Nasr-El-Din, H. and Shook, C.A. 1987. Effect of a 90 degree bend on slurry velocity and concentration distributions. *J. Pipelines* **6:** 239–252.

Nasr-El-Din, H., Afacan, A., and Masliyah, J.H. 1988. Particle Distribution in Slurry Flow Through a Tee with a Horizontal Approach. *Third International Symposium on Liquid-Solid Flows.* New York: ASME, Vol. FED-75, pp. 129–138 [also *Int. J. Multiphase Flow* **15:** 659–671, 1989].

Nasr-El-Din, H., Afacan, A., and Masliyah, J.H. 1989. Wall sampling from horizontal slurry pipelines. *Chem. Eng. Commun.* **82:** 203–221.

Nasr-El-Din, H., Shook, C.A., and Colwell, J.A. 1987. A conductivity probe for local concentration measurement in slurry flows. *Int. J. Multiphase Flow* **13:** 365–378.

Nasr-El-Din, H., Shook, C.A., and Colwell, J. 1987. The lateral variation of solids concentration in horizontal slurry pipeline flow. *Int. J. Multiphase Flow* **13:** 661–670.

Nasr-El-Din, H., Shook, C.A., and Esmail, N. 1984. Isokinetic sampling from slurry pipelines. *Can. J. Chem. Eng.* **62:** 179–185.

Nasr-El-Din, H., Shook, C.A., and Esmail, N. 1985. Wall sampling in slurry systems. *Can. J. Chem. Eng.* **63:** 746–753.

Nasr-El-Din, H., Masliyah, J.H., Nandakumar, K., and Law, D.H.-S. 1988. Continuous gravity separation of a bidisperse suspension in a vertical column. *Chem. Eng. Sci.* **43:** 3225–3234.

Neilson, J.H. and Gilchrist, A. 1968. Erosion by a stream of solid particles. *Wear* **11:** 111–222.

Newitt, D.M., Richardson, J.F., and Gliddon, B.J. 1961. Hydraulic conveying of solids in vertical pipes. *Trans. Inst. Chem. Engrs.* **39:** 93–100.

Newitt, D.M., Richardson, J.F., Abbott, M., and Turtle, R.B. 1955. Hydraulic conveying of solids in horizontal pipes. *Trans. Inst. Chem. Engrs.* **33**(2): 93–113.

Nir, A. and Acrivos, A. 1989. Sedimentation and sediment flow on inclined surfaces. *J. Fluid Mech.* **212:** 139–153.

Nouri, J.M., Whitelaw, J.H., and Yianneskis, M. 1987. Particle motion and turbulence in dense two-phase flows. *Int. J. Multiphase Flow* **13:** 729–739.

Odrowaz-Pieniazek, S. 1982. Pumps for abrasive media. *The Chartered Mech. Engineer.* **29**(4): 21–27.

Okayama, Y. 1986. A study of the wear of a steel dredge pipeline and wear resistance of a polyurethane lined pipe. *Dredging and Port Construction,* 49–52.

Oroskar, A.R. and Turian, R.M. 1980. The critical velocity in pipeline flow of slurries. *AIChE J.* **26:** 550–558.

Pal, R. and Masliyah, J.H., 1990. Rheology of oil in water emulsions with added solids. I, *Can. J. Chem. Eng.,* **68:** 24–28.

Park, J.T., Mannheimer, R.J., Grimley, T.A., and Morrow, T.B. 1988. Pipe flow measurements in a transparent non-Newtonian slurry. In *Third International Symposium on Liquid-Solid Flows.* New York: ASME, FED Vol. 75, pp. 83–90.

Patwardhan, V.S. and Tien, C. 1985. Sedimentation and fluidization in solid-liquid systems: A simple approach. *AIChE J.* **31:** 146–149.

Peters, J. and Shook, C.A. 1981. Electromagnetic sensing of slurry concentration. *Can. J. Chem. Eng.* **59:** 430–437.

Pettyjohn, E.S. and Christiansen, E.B. 1948. Effect of particle shape on free settling rates of isometric particles. *Chem. Eng. Progress* **44:** 157–172.

Pipelen, A.P., Weintraub, M., and Orning, A.A. 1966. Hydraulic transport of coal. U.S. Bureau of Mines TN No. 6743, U.S. Dept. of the Interior.

Porges, K.G.A., Cox, S.A., Herzenberg, C., and Kampschoer, C. 1988. Flow Speed Measurement and Rheometry by Pulsed Neutron Activation. In *Third International Symposium on Liquid-Solid Flows.* New York: ASME, FED Vol. 75, pp. 175–179.

Postlethwaite, J. 1979. Electrochemical studies of corrosion inhibitors in aqueous slurries of sand, iron ore and coal. *Corrosion* **35:** 475–480.

Postlethwaite, J. 1981. Effect of chromate inhibition on the mechanical and electrochemical components of erosion-corrosion in aqueous slurries of sand. *Corrosion-NACE* **37:** 1–5.

Postlethwaite, J., Dobbin, M.H., and Bergevin, K. 1986. The role of oxygen mass transfer in the erosion-corrosion of slurry pipelines. *Corrosion-NACE* **42:** 514–521.

Pourahmadi, F. and Humphrey, A.C. 1983. Modeling solid-fluid turbulent flows with application to predicting erosive wear. *Physico-Chemical Hydrodynamics* **4:** 191–219.

Powell, R.L., Mondy, L.A., Stoker, G.G., Milliken, W., and Graham, A.L. 1989. Development of a falling-ball rheometry with application to opaque systems, *J. Rheology,* **33:** 1173–1188.

Powell, R.L. and Walla, L.A. 1991. Viscosity measurement of suspensions of rod mixtures, *Proc. Fourth Int. Symp. on Liquid-Solid Flows.* New York: ASME, FED Vol. 118, pp. 108–113.

Prettin, W. and Gaessler, H. 1976. Bases of Calculation and Planning for the Hydraulic Transport of Run-of-mine Coal in Pipelines. Proc. Hydrotransport 4 Conf., BHRA, Cranfield, UK, Paper E2, pp. 13–26.

Price, R.C. 1971. Flow improvers for waxy crudes. *J. Inst. Petroleum* **57**(554): 106–109.

Probstein, R.F. and Sengun, M.Z. 1987. Dense slurry rheology with application to coal slurries. *Physico-Chemical Hydrodynamics* **8:** 299–313.

Rabinovich, E. 1965. *Friction and Wear of Materials.* New York: Wiley.

Radin, I., Zakin, J.L., and Patterson, G.K. 1975. Drag reduction in solid-fluid systems. *AIChE J.* **21:** 358–371.

Raudkivi, A.J. 1976. *Loose Boundary Hydraulics.* New York: Pergamon Press.

Richardson, R.C.D. 1967. The wear of metals by hard abrasives. *Wear* **10:** 291–309.

Richardson, J.F. and Meikle, R. 1961. Sedimentation and fluidization: Part III. *Trans. Inst. Chem. Engrs.* **39:** 348–356.

Richardson, J.F. and Zaki, W.N. 1954. Sedimentation and fluidization. *Trans. Inst. Chem. Engrs.* **32:** 35–52.

Richmann, M.W. 1988. The source and second moment in dilute granular flow of highly inelastic spheres. *J. Rheology* **33**(8): 1293–1306.

Riddick, T.M. 1968. *Control of Colloid Stability Through Zeta Potential.* New York: Zeta-Meter Inc.

Ridgeway, K. and Tarbuck, K.J. 1968. Voidage fluctuations in randomly packed spheres. *Chem. Eng. Sci.* **23:** 1147–1155.

Roco, M.C. 1977. Experimental investigation of critical velocity in pipelines for hydrotransport. *St. Cerc. Mec. Appl.* **36:** 111–112.

Roco, M.C. 1978. Time evolution of the operation characteristics of ash transportation pipelines. *St. Cerc. Mec. Appl.,* Part I, **37:** 749–766, Part II, **38:** 567–594.

Roco, M.C. 1980. Ein neues Rechenmodell zur Untersuchung der Turbulenzstromung, 3-Rohre-International J., F.R. of Germany, **19**(6): 347–350.

Roco, M.C. 1982. Scale Deposits and Their Effects in Ash and Slag Slurry Pipelines. Proc. Hydrotransport 8 Conf., BHRA, Cranfield, UK, pp. 503–513.

Roco, M.C. 1990. One-equation turbulence modeling of incompressible mixtures. In *Encyclopedia of Fluid Mechanics.* Houston, TX: Gulf Publishing Co., Vol. 10, Chapter 1, pp. 1–68.

Roco, M.C. and Balakrishnan, N. 1985. Multi-dimensional flow analysis of solid-liquid mixtures. *J. Rheology* **29**(4): 431–456.

Roco, M.C. and Cader, T. 1988. Numerical method to predict wear distribution in slurry pipelines. In *Advances in Pipe Protection.* BHRA Fluid Eng., Cranfield, UK, pp. 53–85.

Roco, M.C. and Cader, T. 1990. Energy Approach for Wear Distribution in Slurry Pipelines. *Jap. J. Multiphase Flow* **4**(1): 2–20.

Roco, M.C. and Dehaven, S. 1989. Marching approach for channel flow with secondary currents and lateral inflow. *Appl. Math. Modelling* **13:** 468–478.

Roco, M.C. and Frasineanu, G. 1977. Computational method for two-phase liquid-solid flow in pipes and channels. *St. Cerc. Mec. Appl.* **36:** 311–324.

Roco, M.C. and Mahadevan, S. 1986. Scale-up technique of slurry pipelines. Part I. Turbulence modeling, Part II: Numerical simulation. *J. Energy Resources Technol.* **108:** 269–283.

Roco, M.C. and Minani, L.K. 1989. Effect of particle size distribution and gravitation on wear in centrifugal pump casings. New York: ASME Paper 89-FE8.

Roco, M.C. and Reinhart, E. 1980. Calculation of Solid Particles Concentration in Centrifugal Slurry Pump Impellers Using Finite Element Technique. Proc. Hydrotransport 7 Conf., BHRA, Cranfield, UK, pp. 359–376.

Roco, M.C. and Shook, C.A. 1981. New Approach to Predict Concentration Distribution in Fine Particle Slurry Flow. Proc. AIChE, Detroit; also *Physico Chemical Hydrodynamics.* **8:** 43–60, 1987.

Roco, M.C. and Shook, C.A. 1983. Modeling of slurry flow: the effect of particle size. *Can. J. Chem. Eng.* **61:** 494–503.

Roco, M.C. and Shook, C.A. 1985a. Turbulent flow of incompressible mixtures. *J. Fluids Eng.* **107:** 224–231.

Roco, M.C. and Shook, C.A. 1985b. Critical deposit velocity in slurry flow. *AIChE J.* **31:** 1401–1404.

Roco, M.C., Hamelin, P., and Davidson, G. 1990. Experimental Investigation of a Centrifugal Slurry Pump. Eds. In *Freight Pipelines,* H. Liu and G.F. Round. New York: Hemisphere, pp. 235–244.

Roco, M.C., Maffet, J.R., and Addie, G.R. 1986b. Experimental study on centrifugal slurry pump performance. In *International Symposium on Slurry Flows.* New York: ASME, Vol. FED-38, pp. 157–160.

Roco, M.C., Makhijani, V., and Ma, D.N. 1989. Probabilistic micromechanical model for two-dimensional slow particulate flow: Dry friction. *Powder Technol.* **60:** 223–243.

Roco, M.C., Nair, P., and Addie, G.R. 1987. Test approach for dense slurry erosion. In *Slurry Erosion: Uses, Application and Test Methods.* Philadelphia: Am. Soc. Testing Materials, pp. 185–210.

Roco, M.C., Addie, G.R., Visintainer, R., and Ray, E.L. 1986. Optimum Wearing High Efficiency Design of Phosphate Slurry Pumps. Proc. 11th Int. Conf. on Slurry Technol., Slurry Transport Assoc., Washington, DC, pp. 277–285.

Roco, M.C., Marsh, M., Addie, G.R., and Maffet, J.R. 1986a. Dredge pump performance prediction. *J. Pipelines* **5:**171–190.

Roco, M.C., Nair, P., Addie, G.R., and Dennis, J. 1984a. Erosion of concentrated slurries in turbulent flow. *J. Pipelines* **4:** 213–221.

Roco, M.C., Nair, P., Addie, G.R., and Dennis, J. 1984b. Modeling Erosion Wear in Centrifugal Slurry Pumps. Proc. Hydrotransport 9 Conf., BHRA, Cranfield, UK, pp. 291–316.

Round, G.F. 1981. Some factors affecting the decomposition of three commercial zinc organic phosphates. *ASLE Trans.* **18:** 79–89.

Round, G.F. 1981. Pulsed slurry flow in pipelines. *J. Pipelines* **1:** 307–318.

Round, G.F. and El-Sayed, E. 1985. Pulsating flows of solid-liquid suspensions: Bentonite-clay/water suspensions. *J. Pipelines* **15:** 95–106.

Rouse, H. 1937. Modern conceptions of the mechanics of fluid turbulence. *Trans. ASCE* **102:** 463–505.

Ryan, N.W. and Johnson, M.M. 1959. Transition from laminar to turbulent flow in pipes. *AIChE J.* **5:** 433–435.

Sabbagha, C.M. 1982. Practical Experiences in Pumping Slurries at ERGO. Proc. Hydrotransport 8 Conf., BHRA, Cranfield, UK, pp. 1–16.

Sadler, L.Y. III and Bethany, B.A. 1987. Effect of soluble ash components on the viscosity of lignite-water mixtures. *Can. J. Chem. Eng.* **65:** 142–147.

Saffman, P.G. 1965. The lift on a small sphere in shear flow. *J. Fluid Mech.* **22:** 385–400.

Sakamoto, M., Inoue, H., Uchida, K., Kamino, Y., and Saito, M. 1983. Progress of Hydrohoist for Coarse Coal Slurry Transportation. Proc. 8th Int. Tech. Conf. on Slurry Transportation, Slurry Transport Assoc., Washington, DC, pp. 297–303.

Sanderson, M.L. and Hemp, J. 1981. Ultrasonic Flowmeters—A Review of the State of the Art. Int. Conf. on Advances in Flow Measurement Techniques, BHRA, Cranfield, UK, pp. 157–178.

Sargent, G.A., Spencer, D.K., and Sagues, A.A. 1982. Slurry Erosion of Materials. Proc. Corrosion-Erosion Wear of Materials, NACE, Berkeley (CA).

Savage, S.B. 1979. Gravity flow of cohesionless granular materials in chutes and channels. *J. Fluid Mech.* **92:** 53–96.

Savage, S.B. and McKeown, S. 1983. Shear stresses developed during rapid shear of concentrated suspensions of large spherical particles between concentric cylinders. *J. Fluid Mech.* **127:** 453–472.

Schmidt, W. 1925. Der Massenaustausch in Freien Luft und Verwandte Erscheinungen. *Probleme der Kosmischen Physik* **7.**

Scott, K.J. 1984. Sedimentation. In *Handbook of Powder Science and Technology.* M.E. Fayed and L. Otten (eds.). New York: Van Nostrand Reinhold, Chap. 14, pp. 607–675.

Scrivener, O., Reitzer, H., Hazzab, A., and Idrissi, A. 1986. Optical and Ultrasonic Instruments for Mixture Flow Investigation. Proc. Hydrotransport 10 Conf., BHRA, Cranfield, UK, pp. 275–282.

Sellgren, A. 1982. The Choice of Operating Velocity in Vertical Solid-Water Pipeline Systems. Proc. Hydrotransport 8 Conf., BHRA, Cranfield, UK, pp. 211–226.

Sellgren, A. and Vappling, L. 1986. Effects of highly concentrated slurries on the performance of centrifugal pumps. In *International Symposium on Liquid-Solid Flows.* Eds. M.C. Roco and W. Wiedenroth. New York: ASME, FED Vol. 38, pp. 143–148.

Sengun, M.Z. and Probstein, R.F. 1989. High-shear-limit viscosity and the maximum packing fraction in concentrated monomodal suspensions. *Physico Chemical Hydrodynamics J.* **11:** 229–241.

Sharma, M.P. and Crowe, C.T. 1989. Application of computer modelling in the design of a multiphase flow metering system. *J. Fluids Eng.* **111:** 184–190.

Shercliff, J.A. 1962. *The Theory of Electromagnetic Flow Measurement.* Cambridge: Cambridge University Press.

Shook, C.A. 1982. Flow of stratified slurries in a Venturi meter, *Can. J. Chem. Eng.* **60:** 342–345.

Shook, C.A. 1985. Experiments with concentrated suspensions of particles with densities slightly greater than the carrier fluid. *Can. J. Chem. Eng.* **63:** 861–869.

Shook, C.A. 1988. Segregation and Plugging in Hydraulic Hoisting of Solids. Proc. Hydrotransport 11 Conf., BHRA, Cranfield, UK, pp. 359–380.

Shook, C.A. and Liebe, J.O. 1976. Determination of dispersed gases in slurries. *Can. J. Chem. Eng.* **54:** 118–120.

Shook, C.A. and Masliyah, S.H. 1974. Flow of a slurry through a Venturi meter. *Can. J. Chem. Eng.* **52:** 228–233.

Shook, C.A. and McLeod, D.J. 1975. The effect of line length for inclined slurry pipelines at shutdown. *Can. J. Chem. Eng.* **53:** 594–598.

Shook, C.A., McKibben, M., and Small, M. 1990. Experimental investigation of some hydrodynamical factors affecting slurry pipeline wall erosion. *Can. J. Chem. Eng.* **68:** 17–23.

Shook, C.A., Gillies, R., Husband, W.H.W., and Small, M. 1982b. Pipe Slope and Roughness Effects for Fine Sand Slurries. Proc. Symp. on Slurry Transport and Pneumatic Handling, ASME, Phoenix AR, pp. 1–8.

Shook, C.A., Haas, D.B., Husband, W.H.W., and Small, M. 1979. Degradation of Coarse Coal Particles during Hydraulic Transport. Proc. Hydrotransport 6 Conf., BHRA, Cranfield, UK, pp. 125–136.

Shook, C.A., Geller, L., Gillies, R.G., Husband, W.H.W., and Small, M. 1986. Experiments with Coarse Particles in 250 mm Pipeline. Proc. Hydrotransport 10 Conf., BHRA Fluid Eng., Cranfield, UK, pp. 219–227.

Shook, C.A., Gillies, R.G., Haas, D.B., Husband, W.H.W., and Small, M. 1982a. Flow of coarse and fine sand slurries in pipelines. *J. Pipelines* **3:** 13–21.

Skelland, A.H.P. 1967. *Non-Newtonian Flow and Heat Transfer*. New York: Wiley.

Smith, J.W. and Paradi, J.C. 1982. Heat transfer to settling slurries in vertical transport. *J. Pipelines* **3:** 43–52.

Spedding, H. and Watkins, A. 1983. The antiwear mechanism of ZDDP's, Part I. *Tribology Int.* **15:** 9–12.

Streat, M. and Bantin, R.A. 1972. Mechanism of Hydraulic Conveying at High Concentrations in Vertical and Horizontal Pipes. Proc. Hydrotransport 2 Conf., BHRA, Cranfield, UK, Paper B2, pp. 11–24.

Sumner, R.J., McKibben, M., and Shook, C.A. 1991. Concentration and Velocity Distributions in Vertical Slurry Flows. Solid-Liquid Flow (Ecoulements Solide-Liquide) in press.

Sumner, R.J., Shook, C.A., and Roco, M.C. 1989. Concentration fluctuation spectra in turbulent slurry pipeline flow. *J. Fluids Eng.* **107:** 243–250.

Suh, N.P. 1973. The delamination theory of wear. *Wear* **25:** 111–124.

Szilas, A.P., Bobok, E., and Navratil, L. 1981. Determination of turbulent pressure loss of non-Newtonian oil in rough pipes. *Rheologica Acta* **20:** 487–496.

Tadros, T.F. 1985. Use of Surfactants and Polymers for Preparation and Stabilization of Coal Suspensions. Proc. 2nd European Conf. on Coal-Liquid Mixtures, London, pp. 1–66.

Takeda, T. 1986. Velocity profile measurement by ultrasonic Doppler shift method. *Int. J. Heat Fluid Flow* **7:** 313–318.

Tangsathitkulchai, C. and Austin, L.G. 1988. Rheology of concentrated slurries of natural size distribution produced by grinding. *J. Powder Technol.* **56:** 293–299.

Tatsis, A. 1990. An Investigation of the Pipe Flow Regime for High Concentration Slurries with Coarse Particles. Proc. 15th Int. Conf. on Coal and Slurry Technology, Coal and Slurry Technology Assoc., Clearwater, FL, pp. 573–584.

Taylor, G.I. 1932. The viscosity of a fluid containing small drops of another fluid. *Proc. Roy. Soc., Ser. A,* **138:** 41–48.

Taylor, G.I. 1936. Fluid friction between rotating cylinders. *Proc. Roy. Soc., Ser. A* **157:** 546–578.

Thomas, A.D. 1979. Predicting the deposit velocity for horizontal turbulent pipe flow of slurries. *Int. J. Multiphase Flow* **5:** 113–129.

Thomas, A.D. 1979. The Role of Laminar-turbulent Transition in Determining the Critical Deposit Velocity for Long Distance Slurry Pipelines. Proc. Hydrotransport 6 Conf., BHRA Fluid Eng., Cranfield, UK, Paper A2, pp. 13–26.

Thomas, A.D. and Wilson, K.C. 1987. New analysis of non-Newtonian turbulent flow–yield power law fluids. *Can. J. Chem. Eng.* **65:** 335–338.

Thomas, D.G. 1963. Transport characteristics of suspensions VII. Relation of hindered-settling floc characteristics to rheological parameters. *AIChE J.* **9:** 310–316.

Thomas, D.G. 1965. Transport characteristics of suspensions: VIII. A note on the viscosity of Newtonian suspensions of uniform spherical particles. *J. Colloid Sci.* **20:** 267–277.

Thompson, B.J., Ward, J.H., and Zinky, W.R. 1967. Application of hologram techniques for particle size analysis. *App Optics* **6:** 519–526.

Tilly, G.P. 1979. Erosion caused by impact of solid particles. In *Treatise on Materials Science and Technology*. New York: Academic Press, Vol. 13, pp. 287–319.

Tilly, G.P. and Sage, W. 1970. The interaction of particle and material behavior in erosion processes. *Wear* **16:** 447–465.

Titus, P.E. and DiGregorio, J.S. 1972. Non-Corrosive Coal Slurries. Patent applic.

724617, Republic of South Africa (June 1972), also US Patent 3,865,547 (Feb. 11, 1975).

Toda, M., Komori, N., Saito, S. and Maeda, S. 1972. Hydraulic conveying of solids through pipe bends. *J. Chem. Eng. Japan* **5:** 4–13.

Toda, M., Komori, N., Saito, S., and Maeda, S. 1973. Hydraulic conveying of solids through a bend in a vertical plane. *Bulletin of JSME* **12**(54): 1388–1396.

Torobin, L.B. and Gauvin, W.H. 1959. Fundamental aspects of solids-gas flow, Part III: Accelerated motion of a particle in a fluid. *Can. J. Chem. Eng.* **37:** 229–236.

Torrance, B. McK. 1963. Friction factors for turbulent non-Newtonian fluid flow in circular pipes. *The South African Mech. Engineer.* **13:** 89–91.

Tsai, S.C. and Zammouri, K. 1988. Roles of interparticle van der Waals force in rheology of concentrated suspensions. *J. Rheology,* **32:** 737–750.

Tsai, W., Humphrey, J.A.C., Cornet, I., and Levy, A. 1980. Experimental Measurement of Accelerated Erosion in a Slurry Pot Tester. Proc. Symp. on Polyphase Flow and Transport, Emerging Energy Technology Conf, ASME, pp. 9–16.

Turian, R.M. and Yuan, T.-F. 1977. Flow of slurries in pipelines. *AIChE J.* **23:** 232–243.

Tuzson, J.J. 1982. Laboratory slurry erosion tests and pump wear rate calculations. In *Particulate Laden Flow in Turbomachinery.* Ed. W. Tabakoff. New York: ASME, pp. 49–52.

Ultrasonics, Ed. 1970. *Blood-flow Velocity Profiles,* p. 203.

Van Rijn, L.C. 1984a. Sediment transport, Part 1: Bed load transport. *J. Hydraulic Eng.* **110:** 1431–1455.

Van Rijn, L.C. 1984b. Sediment transport, Part 2: Suspended load transport. *J. Hydraulic Eng.* **110:** 1613–1641.

Vaseleski, R.C. and Metzner, A.B. 1974. Drag reduction in the turbulent flow of fiber suspensions. *AIChE J.* **20:** 1218–1219.

Vassie, G.S. 1970. Heat Transfer to Flowing Suspensions. M.Sc. Thesis. University of Saskatchewan.

Verbinski, V.V., Cassapakis, C.G., and de Lesdernier, D.L. 1979. Three Component Coal Slurry Sensor for Coal-rock-water Concentrations in Underground Mining Operations. Proc. Hydrotransport 6 Conf., BHRA, Cranfield, UK, pp. 161–168.

Vetter, G. and Hering, L. 1980. Leakfree pumps for the chemical process industries. *Chem. Eng.* **22:** 149–154.

Vetter, G. and Schweinfurter, F. 1987. Pressure Pulsations in the Piping of Reciprocating Pumps. *Chem. Eng. Technol.* **10:** 262–271.

Vocadlo, J.J. 1976. Role of Some Parameters and Effective Variables in Turbulent Slurry Flow. Proc. Hydrotransport 4 Conf., BHRA, Cranfield, UK, Paper D4, pp. 49–62.

Vocadlo, J.J., Koo, J.K., and Prang, A.J. 1974. Performance of Centrifugal Pumps in Slurry Service. Proc. Hydrotransport 3 Conf., BHRA, Cranfield, UK, Paper J2, pp. 17–32.

Walker, C.I. and Goulas, A. 1984. Performance characteristics of centrifugal pumps when handling non-Newtonian homogeneous slurries. *Proc. Inst. Mech. Engrs.* **198A:** 41–49.

Wallis, G.B. 1969. *One Dimensional Two Phase Flow.* New York: McGraw-Hill.

Walton, O.R. and Braun, R.L. 1986. Viscosity, granular-temperature and stress calculations for shearing assemblies of inelastic, frictional disks. *J. Rheology* **30:** 949–980.

Wasp, E.J., et al. 1970. Deposition Velocities, Transition Velocities and Spatial Distributions of Solids in Slurry Pipelines. Proc. Hydrotransport 1 Conf., BHRA, Cranfield, UK, Paper H4, pp. 53–76.

Weber, M. 1972. Feststoffinjektoren Genauer Berechnet. *MM-Industriejournal* **78**(13).

Weidner, G. 1955. Grundsatzliche Untersuchung uber die Verhaltnisse bei Beschleunigung und Umlenkung. *Forschung und Ingenieurwesen,* No. 21.

Weiland, R.H., Fessas, Y.P., and Ramarao, B.V. 1984. On instabilities arising during sedimentation of two component mixtures of solids. *J. Fluid Mech.* **142:** 383–389.

Weiland, R.H. and McPherson, R.R. 1979. Accelerated settling by adding of buoyant particles. *Ind. Eng. Chem. Fund.* **18:** 45–49.

Wellinger, K. and Uetz, H. 1955. Gleitverschleiβ, Spulverschleiβ, Strahverschleiβ unter der Wirkung von Kornigen Stoffen, VDI-Forschungsheft, **21**(449): 40.

Wiedenroth, W. 1970. The Influence of Sand and Gravel on the Characteristics of Centrifugal Pumps, and Some Aspects of Wear in Hydrotransport. Proc. Hydrotransport 1 Conf., BHRA, Cranfield, UK, Paper E1, pp. 1–28.

Wildemuth, C.R. and Williams, M.C. 1985. A new interpretation of viscosity and yield stress in dense slurries: coal and other irregular particles. *Rheologica Acta* **24:** 75–91.

Wildman, D.J., Ekmann, J.M., and Klinzing, G.F. 1984. Flow Studies of Highly Loaded Coal-water Mixtures Around Bends and Across Horizontal Sections. Proc. 9th Int. Tech. Conf. on Slurry Techn., Slurry Transport Assoc., Washington, DC, pp. 209–216.

Wilkinson, W.L. 1960. *Non-Newtonian Fluids: Fluid Mechanics, Mixing and Heat Transfer.* New York: Pergamon Press.

Willetts, B.B. and Naddel, K.F. 1976. Measurements of lift on spheres fixed in low Reynolds number flows. *J. Hydr. Res.* **24:** 425–435.

Wilson, G. 1972. The Design Aspects of Centrifugal Pumps for Abrasive Slurries. Proc. Hydrotransport 2 Conf., BHRA, Cranfield, UK, Paper H2, pp. 25–52.

Wilson, K.C. 1970. Slip point of beds in solid-liquid pipeline flow, *Proc. ASCE, J. Hyd. Div.* **96:** 1–12.

Wilson, K.C. 1971. Parameters for bed slip point in two-phase flow, *Proc. ASCE, J. Hyd. Div.,* **97:** 1665–1679.

Wilson, K.C. 1976. A unified physically based analysis of solid-liquid pipeline flow. Proc. Hydrotransport 4 Conf., BHRA, Cranfield, UK, Paper A1, pp. 1–16.

Wilson, K.C. 1979. Deposition Limit Nomograms for Particles of Various Densities in Pipeline Flow. Proc. Hydrotransport 6 Conf., BHRA, Cranfield, UK, pp. 1–12.

Wilson, K.C. 1980. Analysis of Slurry Flows with a Free Surface. Proc. Hydrotransport 7 Conf., BHRA, Cranfield, UK, pp. 123–132.

Wilson, K.C. 1988. Evaluation of interfacial friction for pipeline transport models, Proc. Hydrotransport 11 Conf., BHRA Fluid Eng., Cranfield, UK, pp. 107–116.

Wilson, K.C. 1989. Two mechanisms for drag reduction. In *Drag Reduction in Fluid Flows.* R.H.J. Sellin and P.T. Moses (eds.). Chichester, UK: Ellis Horwood Ltd.

Wilson, K.C. and Nnadi, F.N. 1990. Behaviour of Mobile Beds at High Shear Stress. Proc. 22nd Int. Conf. on Coastal Eng., Delft.

Wilson, K.C. and Thomas, A.D. 1985. A new analysis of the turbulent flow of non-Newtonian fluids. *Can. J. Chem. Eng.* **63:** 539–546.

Wilson, K.C. and Tse, J.K.P. 1984. Deposition Limit for Coarse Particle Transport in Inclined Pipes. Proc. Hydrotransport 9 Conf., BHRA, Cranfield, UK, pp. 149–162.

Wilson, K.C., Brown, N.P., and Streat, M. 1979. Hydraulic Hoisting at High Concentration: A Study of Friction Mechanisms. Proc. Hydrotransport 6 Conf., BHRA, Cranfield, UK, pp. 269–282.

Wilson, K.C., Streat, M., and Bantin, R.A. 1972. Slip-model Correlation of Dense Two-phase Flow. Proc. Hydrotransport 2 Conf., BHRA, Cranfield, UK, Paper B1, pp. 1–10.

Wright, G.J. and Garrett, G.G. 1986. Wear and Performance of Materials in Pipelines for the Hydraulic Conveyance of Fused Ash. Council for Mineral Technology, Randburg, South Africa, Report MINTEK, M242.

Yalin, M.S. 1971. *Mechanics of Sediment Transport,* 2nd Ed., Pergamon Press.

Yan, Y., Pal, R., and Masliyah, J. 1991. Rheology of oil in water emulsions with added solids. II, *Chem. Eng. Sci.* **46:** 985–994.

Yang, W.J., Ed. 1989. *Handbook of Flow Visualization,* New York: Hemisphere.

Ye, J., Roco, M.C., and Hamelin, P. 1989. Experimental investigation of dilute suspension flow in a rectangular venturi section, New York ASME, FED Vol. 79 *Cavitation and Multiphase Flow* Ed. O. Furuya, 93–100.

Ye, J., Roco, M.C., and Stafford, D. 1990. Particle rotation in Couette slurry flow, New York, ASME, FED Vol. 98 *Cavitation and Multiphase Flow,* Ed. O. Furuya, 19–27.

Yucel, O. 1984. Simplified optimum cost design of a slurry pipeline along a given route, *J. Pipelines* **4:** 241–247.

Zandi, I and Govatos, G. 1967. Heterogeneous flow of solids in pipelines, Proc. ASCE, *J. Hyd. Div.* **93:** 145–159.

Zisselmar, R. and Molerus, O. 1979. Investigation of Solid-Liquid Pipe Flow with Regard to Turbulence Modification, *Chem. Eng. J.* **8:** 233–239.

Zuber, N. 1964. On the dispersed two-phase flow in the laminar regime, *Chem. Eng. Sci.* **19:** 897–917.

Index